Troubleshooting and Servicing Modern Air Conditioning and Refrigeration Systems

John Tomczyk

ESCO Press
A division of
Educational Standards Corporation
http: //www.escoinst.com

Library of Congress Cataloging in Publication Data

Tomczyk, John.
    Troubleshooting and servicing modern air conditioning
and refrigeration systems / John Tomczyk.
      p. cm.
    Includes index.
    ISBN 1-930044-06-2
    1. Air conditioning--Equipment and supplies--Maintenance
and repair.   2. Refrigeration and refrigeration machinery--
Maintenance and repair.   I. Title.
TH7687.7.T66   1995   94-26238
697.9´3´0288--dc20   CIP

Editor: Joanna Turpin
Art Director: Mark Leibold
Copy Editor: Carolyn Thompson

Printed in the United States of America
7 6 5 4 3 2 1

Dedication

To my beautiful wife, Laurie: Thanks for all of your patience, support, kindness, encouragement, and love during this endeavor.

To my children, Brian and Stacy: Thanks for giving Dad the time he needed to write this book. Hopefully the quality time sacrificed from the family will help make this world a better place to live.

Acknowledgments

Special thanks to the following companies and associations for providing artwork and reference materials for use in this book:

AC&R Components, Inc.; Addison Products Company; Air Conditioning and Refrigeration Institute (ARI); Air Conditioning, Heating, and Refrigeration News; AlliedSignal Genetron® Refrigerants; Alnor Instrument Company; Alternative Fluorocarbons Environmental Acceptability Study (AFEAS); American Society of Heating, Refrigeration and Air-Conditioning Engineers (ASHRAE); Blissfield Manufacturing Company; BVA Oils; Carrier Corporation; Castrol North America; Copeland Corporation; Danfoss Automatic Controls, Division of Danfoss, Inc.; The Delfield Company; DuPont Canada; DuPont Company; Engineered Systems Magazine; Environmental Protection Agency (EPA); ESCO Institute; Ferris State University; Frigidaire Company; Hy-Save, Inc.; Pitzographics; Programme for Alternative Fluorocarbon Toxicity Testing (PAFT); Refrigeration Research; Refrigeration Service and Contracting (RSC) Journal; Refrigeration Service Engineers Society (RSES); Refrigeration Technologies; Rheem Air Conditioning Division; Robinair Division, SPX Corporation; Spectronics Corporation; Sporlan Valve Company; Tecumseh Products Company; Thermal Engineering Company; TIF Instruments, Inc.; Trane Company; Tyler Refrigeration, Inc.

About the Author

John A. Tomczyk received his Associate's degree in refrigeration, heating, and air conditioning from Ferris State University in Big Rapids, Michigan; his Bachelor's degree in mechanical engineering from Michigan State University in East Lansing, Michigan; and his Master's degree in education from Ferris State University. He currently teaches refrigeration, heating, and air conditioning technology at Ferris State University.

Mr. Tomczyk has worked in refrigeration, heating, and air conditioning service, project engineering, and technical writing consultation for both education and industry. He is the author of many technical journal articles and is a member of the American Society of Heating, Refrigerating and Air Conditioning Engineers (ASHRAE) and the Refrigeration Service Engineers Society (RSES).

In his spare time, Mr. Tomczyk enjoys boating, scuba diving, and other outdoor activities in northern Michigan.

Foreword

There is an old tale about a young sparrow who, upon being told that the sky was falling, lay down on his back with his legs sticking up into the air. This caused a lot of laughter among the animals watching, and one of them asked, "Do you think you are going to keep the sky from falling with your skinny legs?" The sparrow replied, "One does what one can."

This is the spirit with which this book on understanding and troubleshooting modern refrigeration and air conditioning systems was written. The hvac/r industry is experiencing a massive transition with new refrigerants and oils, and the author hopes that this book will help educate hvac/r students and service technicians to find solutions to the ozone depletion and global warming problems the world now faces.

Table of Contents

CHAPTER ONE

Refrigerant Pressures, States, and Conditions

The typical vapor compression refrigeration system shown in Figure 1-1 can be divided into two pressures: condensing (high side) and evaporating (low side). These pressures are divided or separated in the system by the compressor discharge valve and the metering device. Listed below are field service terms often used to describe these pressures:

Condensing Pressure	Evaporating Pressure
High side pressure	Low side pressure
Head pressure	Suction pressure
Discharge pressure	Back pressure

CONDENSING PRESSURE

The condensing pressure is the pressure at which the refrigerant changes state from a vapor to a liquid. This phase change is referred to as *condensation*. This pressure can be read directly from a pressure gauge connected anywhere between the compressor discharge valve and the entrance to the metering device, assuming there is negligible pressure drop. In reality, line and valve friction and the weight of the liquid itself cause pressure drops from the compressor discharge to the metering device. If a true condensing pressure is needed, the technician must measure the pressure as

close to the condenser as possible to avoid these pressure drops. This pressure is usually measured on smaller systems near the compressor valves, Figure 1-2. On small systems, it is not critical where a technician places the pressure gauge (as long as it is on the high side of the system), because pressure drops are negligible. The pressure gauge reads the same no matter where it is on the high side of the system if line and valve losses are negligible.

EVAPORATING PRESSURE

The evaporating pressure is the pressure at which the refrigerant changes state from a liquid to a vapor. This phase change is referred to as *evaporation* or *vaporizing*. A pressure gauge placed anywhere between the metering device outlet and the compressor (including compressor crankcase) will read the evaporating pressure. Again, negligible pressure drops are assumed. In reality, there will be line and valve pressure drops as the refrigerant travels through the evaporator and suction line. The technician must measure the pressure as close to the evaporator as possible to get a true evaporating pressure. On small systems where pressure drops are negligible, this pressure is usually measured near the compressor (see Figure 1-2). Gauge placement on small systems is usually not critical as long as it is placed on the low side of the refrigeration system, because the refrigerant

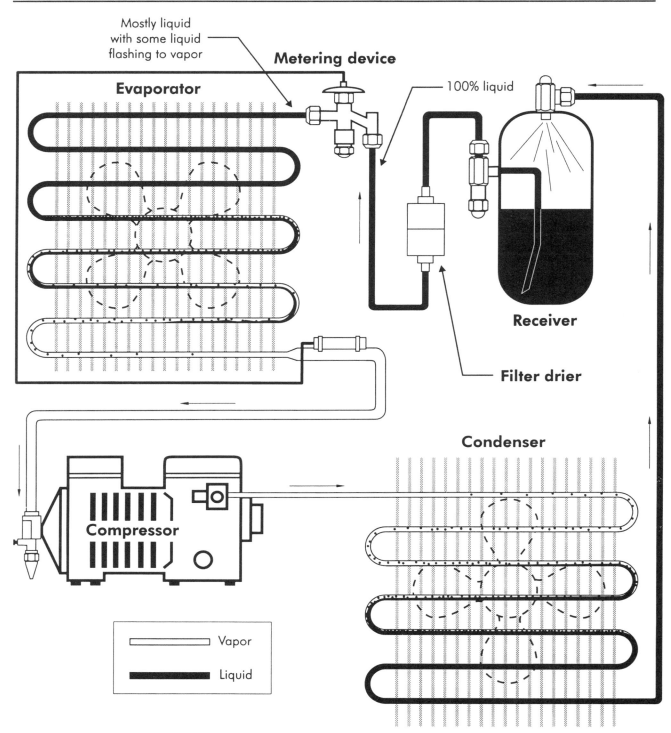

Figure 1-1. Typical compression refrigeration system

vapor pressure acts equally in all directions. If line and valve pressure drops become substantial, gauge placement becomes critical. In larger more sophisticated systems, gauge placement is more critical because of associated line and valve pressure losses. If the system has significant line and valve pressure losses, the technician must place the gauge as close as possible to the component that requires a pressure reading.

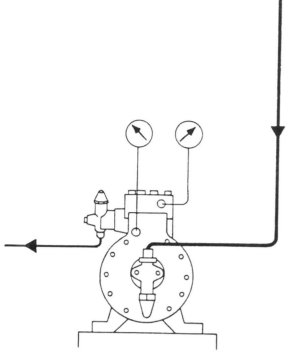

Figure 1-2. Semi-hermetic compressor showing pressure access valves (Courtesy, Danfoss Automatic Controls, Division of Danfoss, Inc.)

REFRIGERANT STATES AND CONDITIONS

Modern refrigerants exist either in the vapor or liquid state. Refrigerants have such low freezing points that they are rarely in the frozen or solid state. Refrigerants can co-exist as vapor and liquid as long as conditions are right. Both the evaporator and condenser house liquid and vapor refrigerant simultaneously if the system is operating properly. Refrigerant liquid and vapor can exist in both the high or low pressure sides of the refrigeration system.

Along with refrigerant pressures and states are refrigerant conditions. Refrigerant conditions can be *saturated*, *superheated*, or *subcooled*.

Saturation

Saturation is usually defined as a temperature. The saturation temperature is the temperature at which a fluid changes from liquid to vapor or vapor to liquid. At saturation temperature, liquid and vapor are called saturated liquid and

saturated vapor, respectively. Saturation occurs in both the evaporator and condenser. At saturation, the liquid experiences its maximum temperature for that pressure, and the vapor experiences its minimum temperature. However, both liquid and vapor are at the same temperature for a given pressure when saturation occurs. Saturation temperatures vary with different refrigerants and pressures. All refrigerants have different vapor pressures. It is vapor pressure that is measured with a gauge.

Vapor Pressure

Vapor pressure is the pressure exerted on a saturated liquid. Any time saturated liquid and vapor are together (as in the condenser and evaporator), vapor pressure is generated. Vapor pressure acts equally in all directions and affects the entire low or high side of a refrigeration system.

As pressure increases, saturation temperature increases; as pressure decreases, saturation temperature decreases. Only at saturation are there pressure/temperature relationships for refrigerants. Table 1-1 shows the pressure/temperature relationship at saturation for refrigerant 134a (R-134a). If one attempts to raise the temperature of a saturated liquid above its saturation temperature, vaporization of the liquid will occur. If one attempts to lower the temperature of a saturated vapor below its saturation temperature, condensation will occur. Both vaporization and condensation occur in the evaporator and condenser, respectively.

The heat energy that causes a liquid refrigerant to change to a vapor at a constant saturation temperature for a given pressure is referred to as *latent heat*. Latent heat is the heat energy that causes a substance to change state without changing the temperature of the substance. Vaporization and condensation are examples of a latent heat process.

Temperature (°F)	Pressure (psig)	Temperature (°F)	Pressure (psig)
-10	1.8		
-9	2.2		
-8	2.6	30	25.6
-7	3.0	31	26.4
-6	3.5	32	27.3
-5	3.9	33	28.1
-4	4.4	34	29.0
-3	4.8	35	29.9
-2	5.3	40	34.5
-1	5.8	45	39.5
0	6.2	50	44.9
1	6.7	55	50.7
2	7.2	60	56.9
3	7.8	65	63.5
4	8.3	70	70.7
5	8.8	75	78.3
6	9.3	80	86.4
7	9.9	85	95.0
8	10.5	90	104.2
9	11.0	95	113.9
10	11.6	100	124.3
11	12.2	105	135.2
12	12.8	110	146.8
13	13.4	115	159.0
14	14.0	120	171.9
15	14.7	125	185.5
16	15.3	130	199.8
17	16.0	135	214.8
18	16.7		
19	17.3		
20	18.0		
21	18.7		
22	19.4		
23	20.2		
24	20.9		
25	21.7		
26	22.4		
27	23.2		
28	24.0		
29	24.8		

Table 1-1. R-134a saturated vapor/liquid pressure/ temperature chart

Superheat

Superheat always refers to a vapor. A super-heated vapor is any vapor that is above its saturation temperature for a given pressure. In order for vapor to be superheated, it must have reached its 100% saturated vapor point. In other words, all of the liquid must be vaporized for superheating to occur; the vapor must be removed from contact with the vaporizing liquid. Once all the liquid has been vaporized at its saturation temperature, any addition of heat causes the 100% saturated vapor to start super-heating. This addition of heat causes the vapor to increase in temperature and gain *sensible heat*. Sensible heat is the heat energy that causes a change in the temperature of a substance. The heat energy that superheats vapor and increases its temperature is sensible heat energy. Super-heating is a sensible heat process. Superheated vapor occurs in the evaporator, suction line, and compressor.

Subcooling

Subcooling always refers to a liquid at a temperature below its saturation temperature for a given pressure. Once all of the vapor changes state to 100% saturated liquid, further removal of heat will cause the 100% liquid to drop in temperature or lose sensible heat. Subcooled liquid results. Subcooling can occur in both the condenser and liquid line and is a sensible heat process. Another method of subcooling liquid, called liquid pressure amplification™, is covered in Chapter Two. This method increases the pressure on subcooled liquid, causing it to be subcooled even more. This creates a liquid with a temperature below its new saturation temperature for the new higher pressure.

A thorough understanding of pressures, states, and conditions of the basic refrigeration system enables the service technician to be a good systematic troubleshooter. It is not until then that a service technician should even attempt systematic troubleshooting.

BASIC REFRIGERATION SYSTEM

Figure 1-3 illustrates a basic refrigeration system. The basic components of this system are the compressor, discharge line, condenser, receiver, liquid line, metering device, evaporator, and suction line. Mastering the function of each individual component can assist the refrigera-

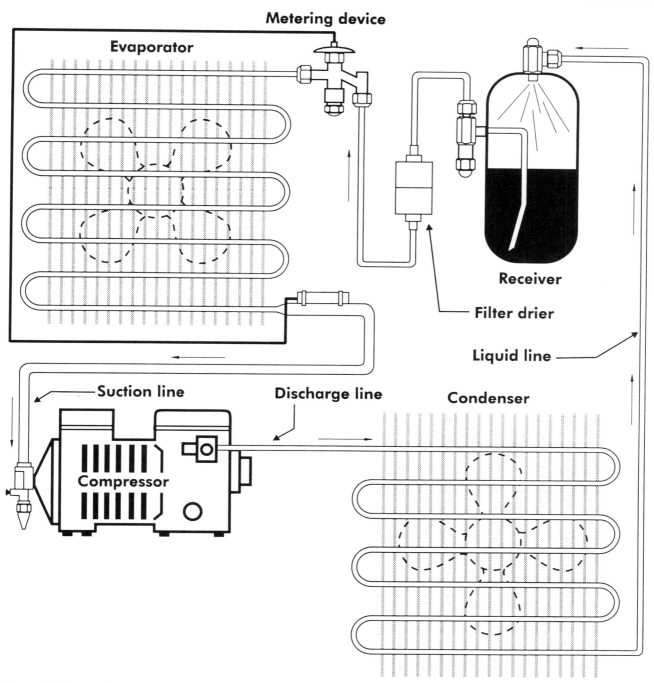

Figure 1-3. Basic refrigeration system

tion technician with analytical troubleshooting skills, saving time and money for both technician and customer.

Compressor

One of the main functions of the compressor is to circulate refrigerant. Without the compres-

sor as a refrigerant pump, refrigerant could not reach other system components to perform its heat transfer functions. The compressor also separates the high pressure from the low pressure side of the refrigeration system. A difference in pressure is mandatory for fluid (gas or liquid) flow, and there could be no refrigerant flow without this pressure separation.

Another function of the compressor is to elevate or raise the temperature of the refrigerant vapor above the ambient (surrounding) temperature. This is accomplished by adding work, or heat of compression, to the refrigerant vapor during the compression cycle. The pressure of the refrigerant is raised, as well as its temperature. By elevating the refrigerant temperature above the ambient temperature, heat absorbed in the evaporator and suction line, and any heat of compression generated in the compression stroke can be rejected to this lower temperature ambient. Most of the heat is rejected in the discharge line and the condenser. Remember, heat flows from hot to cold, and there must be a temperature difference for any heat transfer to take place. The temperature rise of the refrigerant during the compression stroke is a measure of the increased internal kinetic energy added by the compressor.

The compressor also compresses the refrigerant vapors, which increases vapor density. This increase in density helps pack the refrigerant gas molecules together, which helps in the condensation or liquification of the refrigerant gas molecules in the condenser once the right amount of heat is rejected to the ambient. The compression of the vapors during the compression stroke is actually preparing the vapors for condensation or liquification.

Discharge Line
One function of the discharge line is to carry the high pressure superheated vapor from the compressor discharge valve to the entrance of the condenser. The discharge line also acts as a desuperheater, cooling the superheated vapors that the compressor has compressed and giving that heat up to the ambient (surroundings). These compressed vapors contain all of the heat that the evaporator and suction line have absorbed, along with the heat of compression of the compression stroke. Any generated motor winding heat may also be contained in the discharge line refrigerant, which is why the beginning of the discharge line is the hottest part of the refrigeration system. On hot days when the system is under a high load and may have a dirty condenser, the discharge line can

reach over 400°F. By desuperheating the refrigerant, the vapors will be cooled to the saturation temperature of the condenser. Once the vapors reach the condensing saturation temperature for that pressure, condensation of vapor to liquid will take place as more heat is lost.

Condenser
The first passes of the condenser desuperheat the discharge line gases. This prepares the high pressure superheated vapors coming from the discharge line for condensation, or the phase change from gas to liquid. Remember, these superheated gases must lose all of their superheat before reaching the condensing temperature for a certain condensing pressure. Once the initial passes of the condenser have rejected enough superheat and the condensing temperature or saturation temperature has been reached, these gases are referred to as 100% saturated vapor. The refrigerant is then said to have reached the 100% saturated vapor point, Figure 1-4.

One of the main functions of the condenser is to condense the refrigerant vapor to liquid. Condensing is system dependent and usually takes place in the lower two-thirds of the condenser. Once the saturation or condensing temperature is reached in the condenser and the refrigerant gas has reached 100% saturated vapor, condensation can take place if more heat is removed. As more heat is taken away from the 100% saturated vapor, it will force the vapor to become a liquid or to condense. When condensing, the vapor will gradually phase change to liquid until 100% liquid is all that remains. This phase change, or change of state, is an example of a latent heat rejection process, as the heat removed is latent heat not sensible heat. The phase change will happen at one temperature even though heat is being removed. *Note: An exception to this is a near-azeotropic blend of refrigerants where there is a temperature glide or range of temperatures when phase changing (see Chapter Eight on blend temperature glide).* This one temperature is the saturation temperature corresponding to the saturation pressure in the condenser. As mentioned before, this pressure can be measured anywhere

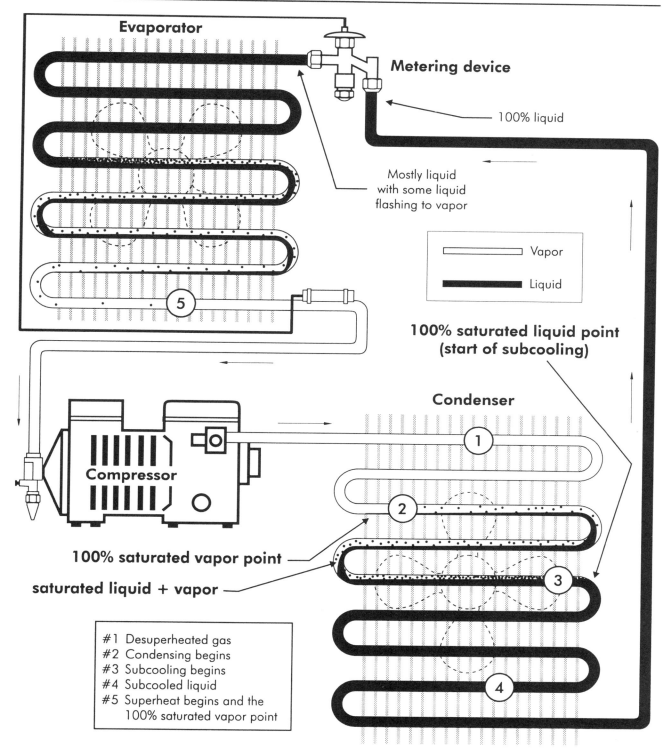

Evaporator

Metering device

100% liquid

Mostly liquid
with some liquid
flashing to vapor

☐ Vapor

■ Liquid

**100% saturated liquid point
(start of subcooling)**

Condenser

100% saturated vapor point

saturated liquid + vapor

Compressor

#1 Desuperheated gas
#2 Condensing begins
#3 Subcooling begins
#4 Subcooled liquid
#5 Superheat begins and the
    100% saturated vapor point

Figure 1-4. Basic refrigeration system showing 100% saturated vapor and liquid points

on the high side of the refrigeration system as long as line and valve pressure drops and losses are negligible.

The last function of the condenser is to subcool the liquid refrigerant. Subcooling is defined as any sensible heat taken away from 100% saturated liquid. Technically, subcooling is defined as the difference between the measured liquid temperature and the liquid saturation temperature at a given pressure. Once the saturated vapor in the condenser has phase changed to saturated liquid, the 100% saturated liquid point has been reached. If any more heat is removed, the liquid will go through a sensible heat rejection process and lose temperature as it loses heat. The liquid that is cooler than the saturated liquid in the condenser is subcooled liquid. Subcooling is an important process, because it starts to lower the liquid temperature to the evaporator temperature. This will reduce flash loss in the evaporator so more of the vaporization of the liquid in the evaporator can be used for useful cooling of the product load (see Chapter Two on the importance of liquid subcooling).

Receiver

The receiver acts as a surge tank. Once the subcooled liquid exits the condenser, the receiver receives and stores the liquid. The liquid level in the receiver varies depending on whether the metering device is throttling opened or closed. Receivers are usually used on systems in which a thermostatic expansion valve (TXV or TEV) is used as the metering device. The subcooled liquid in the receiver may lose or gain subcooling depending on the surrounding temperature of the receiver. If the subcooled liquid is warmer than receiver surroundings, the liquid will reject heat to the surroundings and subcool even more. If the subcooled liquid is cooler than receiver surroundings, heat will be gained by the liquid and subcooling will be lost.

A receiver bypass is often used to bypass liquid around the receiver and route it directly to the liquid line and filter drier. This bypass prevents subcooled liquid from sitting in the receiver and losing its subcooling. A thermostat with a sensing bulb on the condenser outlet controls the bypass solenoid valve by sensing liquid temperature coming to the receiver, Figure 1-5. If the liquid is subcooled to a predetermined temperature, it will bypass the receiver and go to the filter drier.

Liquid Line

The liquid line transports high pressure subcooled liquid to the metering device. In transport, the liquid may either lose or gain subcooling depending on the surrounding temperature. Liquid lines may be wrapped around suction lines to help them gain more subcooling, Figure 1-6. Liquid/suction line heat exchangers can be purchased and installed in existing systems to gain subcooling. The importance of liquid subcooling will be covered more extensively in Chapter Two.

Metering Device

The metering device meters liquid refrigerant from the liquid line to the evaporator. There are several different styles and kinds of metering devices on the market with different functions. Some metering devices control evaporator superheat and pressure, and some even have pressure limiting devices to protect compressors at heavy loads.

The metering device is a restriction that separates the high pressure side from the low pressure side in a refrigeration system. The compressor and the metering device are the two components that separate pressures in a refrigeration system. The restriction in the metering device causes liquid refrigerant to flash to a lower temperature in the evaporator because of its lower pressure and temperature.

Evaporator

The evaporator, like the condenser, acts as a heat exchanger. Heat gains from the product load and outside ambient travel through the sidewalls of the evaporator to vaporize any liquid refrigerant. The pressure drop through the metering device causes vaporization of some

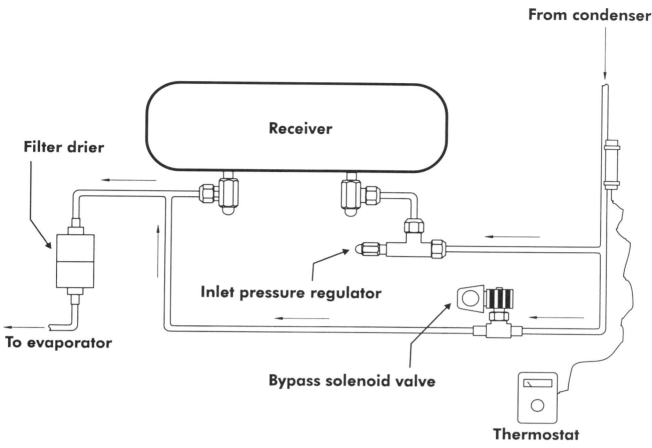

Figure 1-5. Receiver with thermostatically controlled liquid bypass

Figure 1-6. Liquid/suction line heat exchanger (Courtesy, Refrigeration Research, Inc.)

of the refrigerant and causes a lower saturation temperature in the evaporator. This temperature difference between the lower pressure refrigerant and the product load is the driving potential for heat transfer to take place.

The last pass of the evaporator coil acts as a superheater to ensure all liquid refrigerant has been vaporized. This protects the compressor from any liquid slopover, which may result in valve damage or diluted oil in the crankcase. The amount of superheat in the evaporator is usually controlled by a thermostatic expansion type of metering device.

Suction Line

The suction line transports low pressure superheated vapor from the evaporator to the compressor. There may be other components in the suction line such as suction accumulators, crankcase pressure regulators, p-traps, filters, and screens. Liquid/suction line heat exchangers are often mounted in the suction line to transfer heat away from the liquid line (subcool) and into the suction line, Figure 1-7.

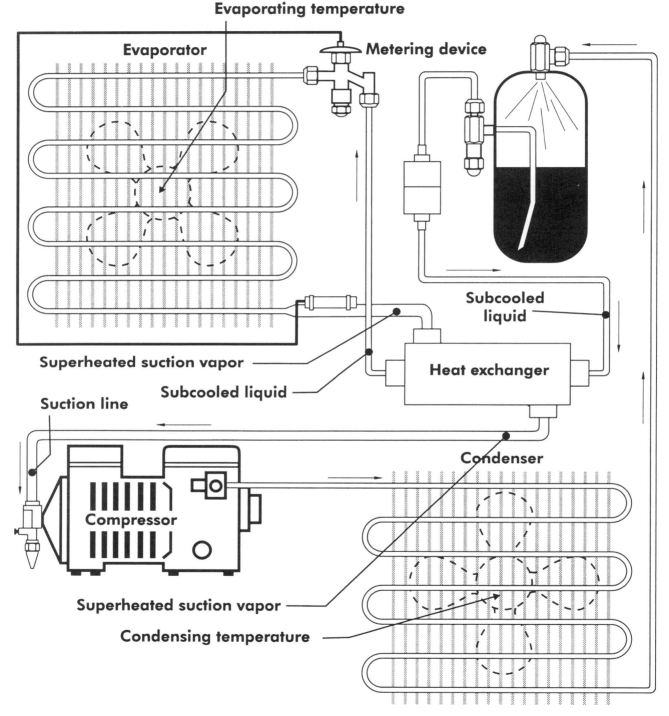

Figure 1-7. Refrigeration system showing liquid/suction line heat exchanger

Another function of the suction line is to superheat the vapor as it approaches the compressor. Even though suction lines are usually insulated, sensible heat stills penetrates the refrigerant vapor and adds more superheat. This additional superheat decreases the density of the refrigerant vapor to prevent compressor overload, resulting in lower amp draws. This additional superheat also helps ensure that the compressor will see vapor only under low loading conditions. Many metering devices have a tendency to lose control of evaporator superheat at low loads. It is recommended that systems should have at least 20°F of total superheat at the compressor to prevent liquid slugging and/or flooding of the compressor at low loadings. This topic will be covered extensively in Chapter Two.

APPLICATION OF PRESSURES, STATES, AND CONDITIONS

Systematic troubleshooting requires mastering the function of all refrigeration system components. It is also important to be able to recognize the pressure, state, and condition of the working fluid (refrigerant) in the refrigeration system components. Figure 1-8 illustrates the basic refrigeration system. The legend lists refrigerant pressures, conditions, and states for the points shown in Figure 1-8. An explanation of the pressure, condition, and state of each point should clarify any system weaknesses.

Assume the following conditions for Figure 1-8:

- Refrigerant = R-134a

- Discharge (condensing) pressure = 124 psig (100°F)

- Suction (evaporating) pressure = 6 psig (0°F)

- Discharge temperature = 180°F

- Condenser outlet temperature = 90°F

- TXV inlet temperature = 80°F

- Evaporator outlet temperature = 10°F

- Compressor inlet temperature = 40°F

See Table 1-1 for pressure/temperature relationships.

Compressor Discharge (Point #1)
The refrigeration compressor is a vapor pump, not a liquid pump. The vapor leaving the compressor will be high pressure superheated vapor. The compressor is one of the two components in the system that separates the high pressure side from the low pressure side. The compressor discharge is high pressure, and the compressor suction is low pressure. The compressor discharge vapor receives its superheat from sensible heat coming from the evaporator, suction line, motor windings, friction, and internal heat of compression from the compression stroke. Since the vapor is superheated, no pressure/temperature relationship exists. Its temperature is well above the saturation temperature of 100°F for the given saturation (condensing) pressure of 124 psig.

Condenser Inlet (Point #2)
As high pressure superheated refrigerant leaves the compressor, it instantly begins to lose superheat and cool in temperature. Its heat is usually given up to the surroundings. As mentioned before, this process is called desuperheating. Even though this refrigerant vapor is going through a desuperheating process, it is still superheated vapor. Pressure acts equally in all directions, so the vapor will be high pressure, or the same pressure as the compressor discharge, assuming that any line and valve pressure drops are ignored. Remember, the refrigerant is superheated and not saturated, so there is no pressure/temperature relationship. This high pressure superheated vapor is also above its saturation temperature of 100°F for the given discharge or condensing pressure of 124 psig. This point can be referred to as high pressure superheated vapor. This process of desuperheating will continue until the 100% saturated vapor point in the condenser is reached.

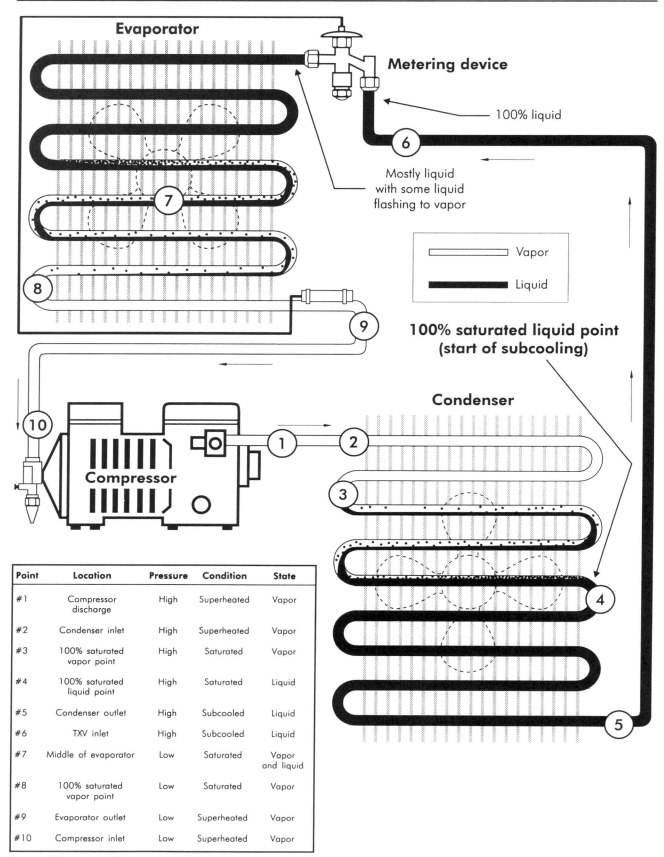

Point	Location	Pressure	Condition	State
#1	Compressor discharge	High	Superheated	Vapor
#2	Condenser inlet	High	Superheated	Vapor
#3	100% saturated vapor point	High	Saturated	Vapor
#4	100% saturated liquid point	High	Saturated	Liquid
#5	Condenser outlet	High	Subcooled	Liquid
#6	TXV inlet	High	Subcooled	Liquid
#7	Middle of evaporator	Low	Saturated	Vapor and liquid
#8	100% saturated vapor point	Low	Saturated	Vapor
#9	Evaporator outlet	Low	Superheated	Vapor
#10	Compressor inlet	Low	Superheated	Vapor

Figure 1-8. Basic refrigeration system showing refrigerant pressures, states, and condition locations

100% Saturated Vapor Point (Point #3)

Once all the superheat is rejected from the refrigerant gas, the saturation temperature of 100°F is finally reached for the condensing pressure of 124 psig (see Table 1-1 for pressure/temperature relationship). The vapor has now reached the 100% saturated vapor point and is at the lowest temperature it can be and still remain a vapor. This temperature is referred to as both its saturation and condensing temperature, and a pressure/temperature relationship exists. Any heat lost past the 100% saturated vapor point will gradually phase change the vapor to liquid (condensing). The heat removed from the vapor turning to liquid is referred to as latent heat and happens at a constant temperature of 100°F. As the vapor condenses to liquid, refrigerant molecules actually become more dense and get closer together. This molecular joining is what gives up most of the latent heat energy. This point is on the high side of the refrigeration system and is referred to as high pressure saturated vapor.

100% Saturated Liquid Point (Point #4)

Soon all of the vapor will give up its latent heat and turn to saturated liquid at a constant condensing temperature of 100°F. Any more heat given up by the refrigerant after this point will be sensible heat, because the phase change from vapor to liquid is complete. This point is still on the high side of the refrigeration system and can be referred to as high pressure saturated liquid. The entire condensing process takes place between the 100% saturated vapor point and the 100% saturated liquid point. Any sensible heat lost past the 100% saturated liquid point is referred to as subcooling.

Condenser Outlet (Point #5)

Once the 100% saturated liquid point is reached in the condenser, liquid subcooling occurs. Remember, any heat lost in the liquid past its 100% saturated liquid point is subcooling. Liquid subcooling can continue all the way to the entrance of the metering device if conditions are right. This point is on the high side of the system and is all subcooled liquid, so it is referred to as high pressure subcooled liquid. There is no pressure/temperature relationship at the subcooled condition, only at saturation. At a pressure of 124 psig, the corresponding temperature of the 100% saturated liquid point in the condenser is 100°F. The difference between 100°F and the condenser outlet temperature of 90°F is 10°F of condenser subcooling. Subcooling calculations will be covered much more extensively in Chapter Two.

TXV Inlet (Point #6)

The inlet to the TXV is on the high side of the system and consists of subcooled liquid. This subcooling should continue from the 100% saturated liquid point in the condenser. The tubing from the condenser outlet to the TXV inlet is often referred to as the liquid line. The liquid line may be exposed to very high or low roof temperatures depending on the time of year. This will seriously affect whether or not subcooling takes place and to what magnitude. If the liquid line is exposed to hot temperatures, liquid line flashing may occur (covered in Chapter Two). Since this point is on the high side of the system and is subcooled liquid, it is referred to as high pressure subcooled liquid.

Middle of Evaporator (Point #7)

When the subcooled liquid enters the TXV, flashing of the liquid occurs. Once in the evaporator, the liquid refrigerant experiences a severe drop in pressure to the new saturation (evaporator) pressure of 6 psig. This pressure decrease causes some of the liquid to flash to vapor in order to reach the new saturation temperature in the evaporator of 0°F. Once this new evaporator temperature is reached, the liquid/vapor mixture starts absorbing heat from the product load and continues to change from liquid to vapor. This process happens at the new saturation temperature of 0°F, corresponding to the saturation (evaporator) pressure of 6 psig. This is a classic example of heat absorbed by the refrigerant without increasing in temperature, which is called the latent heat of vaporization. The heat energy absorbed in the refrigerant breaks the liquid molecules into vapor molecules instead of increasing its temperature. Since the refrigerant is both saturated liquid and vapor and is on the low side of the refrigeration system, it is referred to as a low pressure saturated liquid and vapor.

100% Saturated Vapor Point (Point #8)

Once all of the liquid changes to vapor in the evaporator, the 100% saturated vapor point is reached. This point is still at the evaporator saturation temperature of 0°F. Any more heat absorbed by the refrigerant vapor will result in a temperature rise of the refrigerant. This heat energy goes into increasing the velocity and spacing of the vapor molecules, because there is no more liquid to be vaporized. This increase in molecular velocity can be measured in degrees. Any heat added past this 100% saturated vapor point is superheat. Since this 100% saturated vapor point is in the low side of the system and is saturated vapor, it is referred to as low pressure saturated vapor.

Evaporator Outlet (Point #9)

The evaporator outlet temperature is used for evaporator superheat calculations. This point is located at the evaporator outlet next to the TXV remote bulb. Because it is located downstream of the 100% saturated vapor point, it is superheated. This point is in the low side of the refrigeration system and is referred to as low pressure superheated vapor. The difference between the 100% saturated vapor temperature of 0°F and the evaporator outlet temperature of 10°F is called evaporator superheat. In this example, there are 10°F of evaporator superheat.

Compressor Inlet (Point #10)

The compressor inlet consists of low pressure superheated vapor. This vapor feeds the compressor. As the refrigerant travels from the evaporator outlet down the suction line to the compressor, more superheat is gained. Superheat ensures that no liquid refrigerant enters the compressor at low evaporator loadings when TXV valves are known to lose control of superheat settings. Because this point is superheated, no refrigerant pressure/temperature relationship exists.

CHAPTER TWO

Subcooling and Superheating

The concepts of subcooling and superheating are probably the two most important principles that the service technician must understand before attempting to systematically troubleshoot hvac/r systems. This chapter covers not only basic principles of subcooling and superheat, but also topics for the more advanced service technician. Examples dealing with superheat and subcooling amounts are also included for any clarifications that may be needed.

LIQUID SUBCOOLING

In today's competitive service market, every conscientious service technician should understand why a refrigeration or air conditioning system must have the proper amount of liquid subcooling. The amount of liquid subcooling affects system capacity, as well as the effectiveness and capacity of expansion type metering devices.

Subcooling is defined as the difference between the measured liquid temperature and the liquid saturation temperature at a given pressure. Any sensible heat taken away from the 100% saturated liquid point in the condenser can be defined as liquid subcooling. Liquid subcooling may occur from the start of the 100% saturated liquid point in the condenser to the metering device, Figure 2-1. The saturated liquid temperature can be obtained from a pressure/temperature relationship using the condensing pressure. This means that as soon as all of the saturated vapor in the condenser changes to saturated liquid, subcooling will start to occur if further heat is removed. This is a temperature change or sensible heat change, so any drop in temperature of the liquid below the saturation temperature for the pressure at that point is considered liquid subcooling.

SUBCOOLING CATEGORIES

There are two subcooling categories: *condenser subcooling* and *total subcooling*.

Condenser Subcooling

Condenser subcooling is the liquid subcooling present leaving the condenser. It can be measured by subtracting the actual condenser liquid out temperature from the saturation temperature measured at the condenser outlet:

$$\text{Condensing temperature (saturated)} - \text{Condenser liquid out temperature} = \text{Condenser subcooling}$$

[Formula 2-1]

When subcooled, the refrigerant is not generating or losing any vapor pressure, so there is no pressure/temperature relationship and a pressure/temperature chart cannot be used. The condenser outlet temperature must be measured

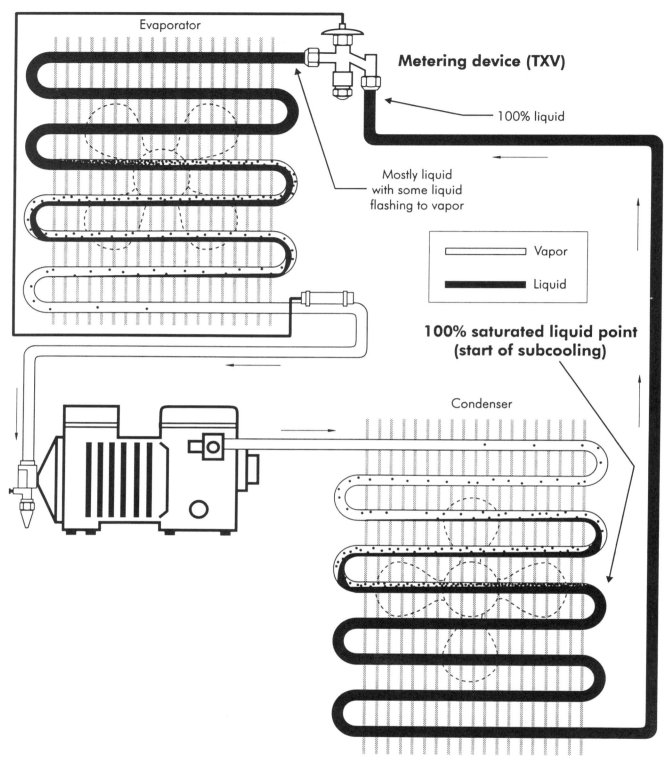

Figure 2-1. Start of liquid subcooling

with either a thermistor or a thermocouple fastened to the condenser outlet, Figure 2-2. The saturated temperature may be acquired from the condensing pressure read from a gauge on the condenser outlet. This is because a pressure/temperature relationship does exist in a saturated condition. If pressure drops exist in the system, the pressure must be measured where the temperature was taken to obtain an accurate liquid subcooling amount.

Example 2-1

What is the condenser subcooling if the head pressure on an R-22 system is 211 psig and a thermistor at the condenser outlet measures 95°F? (Assume no pressure drops through the condenser.) *Note: Thermistors or thermocouples must have good thermal contact onto the refrigerant line. The line must be straight and lightly sanded to ensure good thermal contact. Insulation must be placed around the line to insulate against outside ambient heat losses or gains.*

Solution 2-1

Convert 211 psig for R-22 to a 105°F condensing temperature using the pressure/temperature chart shown in Table 2-1:

| 105°F (condensing saturation temperature) | − | 95°F (condenser out temperature) | = | 10°F (condenser subcooling) |

This means that there are 10°F of liquid subcooling leaving the condenser to feed the liquid line or receiver, depending on the type of system employed. However, if the liquid goes to a flow-through receiver, subcooling could be lost.

Example 2-2

This example incorporates pressure drop through the condenser into the equation for condenser subcooling. In this case, the pressure must be measured where the condenser outlet temperature is taken.

What is the condenser subcooling if the head pressure on an R-22 system is 211 psig, the thermistor at the condenser outlet measures 95°F, and the pressure reads 205 psig at that

Temperature (°F)	Pressure (psig)
65	111.23
66	113.22
67	115.24
68	117.28
69	119.34
70	121.43
71	123.54
72	125.67
73	127.83
74	130.01
75	132.22
76	134.45
77	136.71
78	138.99
79	141.30
80	143.63
81	145.99
82	148.37
83	150.78
84	153.22
85	155.68
86	158.17
87	160.69
88	163.23
89	165.80
90	168.40
91	171.02
92	173.67
93	176.35
94	179.06
95	181.80
96	184.56
97	187.36
98	190.18
99	193.03
100	195.91
101	198.82
102	201.76
103	204.72
104	207.72
105	210.75
106	213.81
107	216.90
108	220.02
109	223.17

Table 2-1. Pressure/temperature relationship for R-22

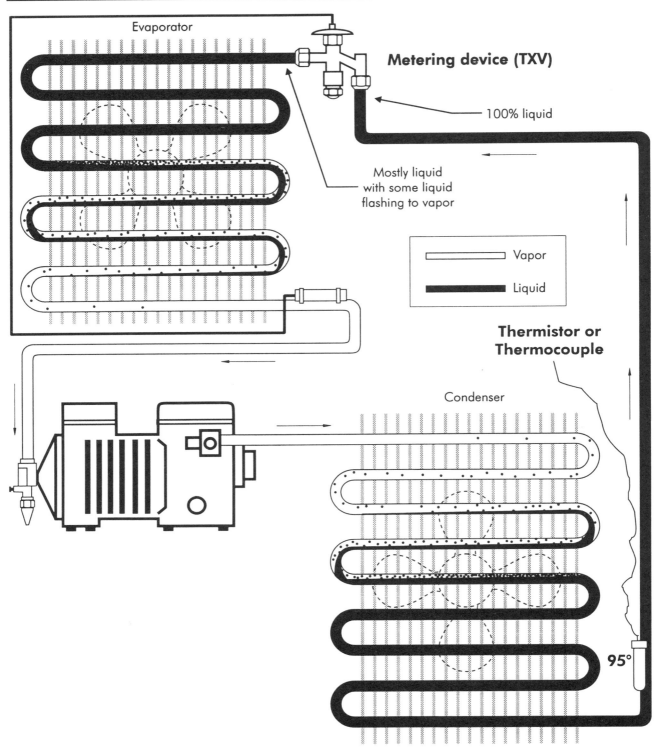

Figure 2-2. Thermistor or thermocouple at the condenser outlet

point? There is a 6 psi pressure drop through the condenser from line losses. The condenser outlet pressure can be read at the king valve, receiver charging valve, or simply by installing a line tap valve.

Solution 2-2
Convert 205 psig for R-22 to 103°F using Table 2-1:

| 103°F (saturation temperature at condenser outlet) | − | 95°F (condenser out temperature) | = | 8°F (condenser subcooling) |

Total Subcooling

Total subcooling encompasses liquid subcooling from the 100% saturated liquid point to the expansion valve or metering device, Figure 2-3. This means that total subcooling includes condenser subcooling and any subcooling that takes place after the condenser. Liquid can lose heat and subcool in the receiver, filter drier, and liquid line before it gets to the metering device. However, if there is any refrigerant vapor in the receiver, the subcooled liquid will not subcool further. The vapor will re-condense to a liquid in the receiver, and both the liquid and vapor will reach a new saturation temperature. As the vapor re-condenses it rejects heat, which will be absorbed by the subcooled liquid. Subcooling can be lost in this situation if the liquid stays in the receiver too long.

Subcooling of liquid rarely occurs when a capillary tube metering device is used because of the severe pressure drop of the liquid as it travels the length of the capillary tube. Due to this reduction in pressure through the capillary tube, the liquid may actually be saturated with flash gas present. Because of the difficulty measuring the temperature at the capillary tube outlet in a service environment, assume that total subcooling ends at the capillary tube entrance.

Total subcooling can be calculated by subtracting the liquid line temperature at the entrance

of the metering device from the saturation temperature at that point:

| Saturation temperature from pressure at TXV inlet | − | TXV inlet temperature | = | Total subcooling |

[Formula 2-2]

The concept of total subcooling is meaningless without incorporating pressure drops into the equation. Example 2-3 is a total subcooling calculation that does not take pressure drop into consideration.

Example 2-3
What is the total subcooling if the head pressure of an R-22 system is 223 psig, and a thermistor on the liquid line at the entrance of the metering device reads 90°F (assume no pressure drop)?

Solution 2-3
Convert 223 psig for R-22 to a 109°F condensing temperature using Table 2-1:

| 109°F (condensing saturation temperature) | − | 90°F (liquid at metering device temperature) | = | 19°F (total subcooling) |

There are 19°F of total liquid subcooling before the liquid enters the metering device. *Note: When measuring the liquid line temperature at the entrance of a TXV or AXV metering device, place the thermistor or thermocouple on the liquid line about 3 to 4 inches in front of the expansion valve and insulate it from ambient heat gains. Measurement instruments placed too close to the metering device will measure colder temperatures due to heat conduction to the colder expansion device.*

Example 2-4
In this example, pressure drop is incorporated into the system as the liquid travels through the condenser to the metering device. This is typical of most medium and large systems.

What is the total subcooling if the head pressure of an R-22 system is 223 psig, a thermistor on the liquid line at the entrance of the metering device reads 90°F, and the pressure is 216 psig at that point? There is a 7 psi pressure

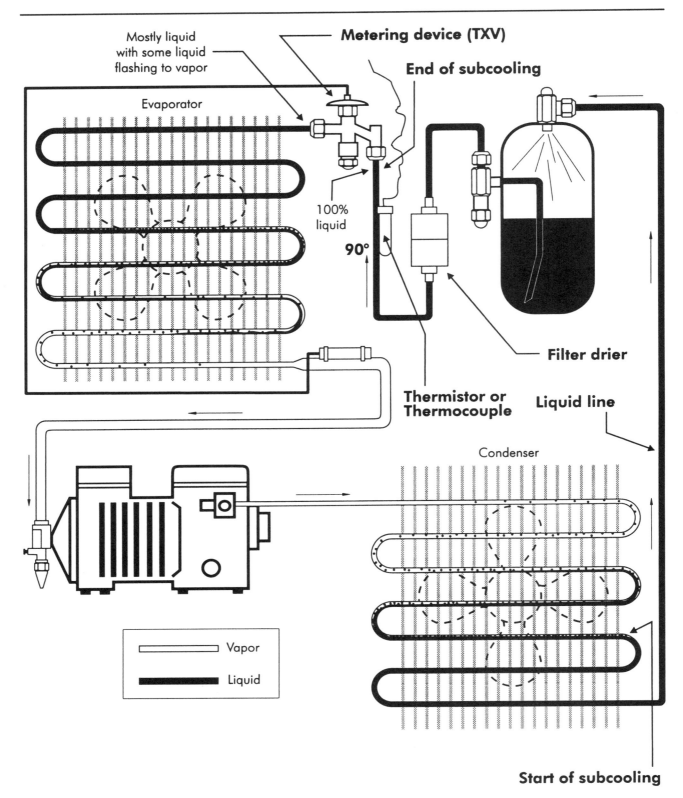

Mostly liquid
with some liquid
flashing to vapor

Metering device (TXV)

End of subcooling

Evaporator

100%
liquid

90°

Filter drier

**Thermistor or
Thermocouple**

Liquid line

Condenser

Vapor

Liquid

Start of subcooling

Figure 2-3. Thermistor or thermocouple at the TXV inlet

drop before the liquid reaches the metering device. This pressure drop could be static, friction, or both. The pressure of the liquid is measured just before the metering device where the 90°F temperature is taken, Figure 2-4. Line tap valves are often needed for this procedure.

Solution 2-4
Convert 216 psig for R-22 to a 107°F saturated temperature using Table 2-1:

107°F (saturated temperature for 216 psig at TXV inlet)		90°F (liquid at metering device temperature)		17°F (total subcooling)
	−		=	

SUBCOOLING NEEDED TO PREVENT LIQUID LINE FLASH

Condenser subcooling ensures there is a liquid seal at the condenser bottom so the liquid line or receiver is not fed with vapor. This condition prevents any refrigerant vapor or noncondensibles (e.g., air) from leaving the condenser. On receiverless systems, condenser subcooling can be an indicator of how much refrigerant charge is in a system. See manufacturer specifications for condenser subcooling amounts when charging a system using this criteria.

Another important reason why condenser subcooling must be present is to prevent liquid line flash gas (vapor bubbles) from forming in the liquid line. Liquid line flash gas is primarily caused from a pressure drop in the liquid line. Pressure drop in the liquid line can be caused by any restrictive device, including filter driers, sightglasses, solenoid valves, shutoff valves, or kinked or undersized liquid lines. There is also pressure drop associated with the length of run and the vertical lift of liquid in refrigerant lines. The longer the run and the more turns and bends associated with the liquid line, the larger the associated pressure drop. There is also more pressure drop with more vertical lift of liquid.

If there were no subcooling, saturated liquid from the condenser (while experiencing these

pressure drops) would try to cool itself down to a lower saturation temperature to match the progressively lower pressure in the liquid line. The only way the saturated liquid could cool itself down is for some of its own liquid to flash into vapor. Whenever a liquid flashes into vapor, heat energy is needed for the vaporization process. Almost all of the heat supplied to the flashing process is from the heat in the saturated liquid itself. This cools the remaining liquid down to the new saturation temperature associated with the new lower pressure caused from the pressure drop. This type of cooling is referred to as *adiabatic cooling*, because there is no net loss or gain of energy in the liquid. Whenever saturated liquid experiences liquid line pressure drop, liquid line flash gas results and is detrimental to system performance. Metering devices will then experience a mixture of liquid/vapor instead of a solid column of liquid at their entrances. The liquid/vapor mixture decreases the capacity of the valves, because TXV capacity tables are based on vapor-free liquid at their inlets.

The balanced port TXV is designed to handle a certain amount of flash gas due to its oversized port. It is designed to maintain better control under such conditions and at light loads. The balanced port TXV can operate with its valve pin very close to the seat. This allows a large port to handle small loads. When liquid flashes in the liquid line, it also leaves less liquid to flash in the evaporator. This takes away from the refrigeration effect and adversely affects system capacity. Excessive liquid line flash gas may even erode the needle and seat areas of some TXVs. The erosion of needle valves usually occurs only with ammonia valves.

PRESSURE DROP
Total pressure losses that take place in the liquid line are the result of either friction pressure losses or static pressure losses.

Friction Pressure Losses
Friction pressure losses were discussed earlier. They are caused from refrigerant flowing

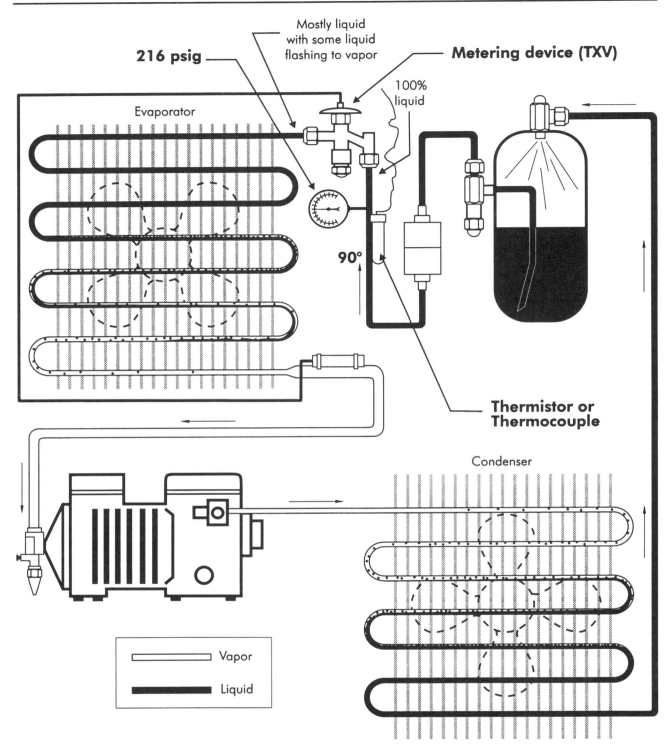

Figure 2-4. Thermistor or thermocouple and gauge at TXV inlet

either through tubing, driers, elbows, or even a sightglass on its way to the metering device. Some of the fluid energy is used to overcome friction from the side walls of the pipe or from an obstruction such as the filter drier. A velocity profile is set up in the pipe with the slowest moving fluid at the sidewalls and the fastest moving fluid in the center of the pipe, Figure 2-5. This lost energy in the fluid causes a loss in pressure, which is why liquids must be subcooled before experiencing these pressure drops to prevent liquid flashing. Proper line sizing, short runs, and unobstructed accessories like clean filter driers can prevent excessive friction pressure drops.

Static Pressure Losses

Static pressure losses occur in the liquid line whenever liquid has to flow uphill through vertical lifts. A vertical column of liquid naturally has more pressure at its bottom than its top due to the weight of the entire column of liquid at the bottom, Figure 2-6. Whenever liquid from a condenser or receiver must be piped uphill, there will be static pressure drops as the liquid climbs higher in the piping. This is in addition to the friction pressure losses from the pipe sidewalls as the liquid climbs. See Table 2-2 for static pressure losses corresponding to vertical lift for five different refrigerants. Refrigerant line sizing does not reduce or increase static pressure losses from vertical lift.

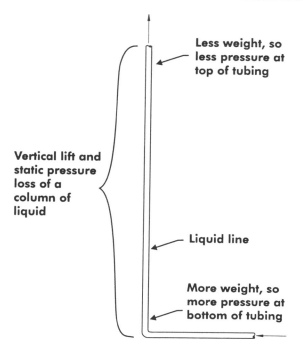

Figure 2-6. Vertical column of liquid showing static pressure losses

REFRIGERANT	VERTICAL LIFT — FEET				
	20	40	60	80	100
	STATIC PRESSURE LOSS — psi				
12	11	22	33	44	55
22	10	20	30	39	49
134a	10	20	30	40	50
502	10	20	30	40	50
717 (Ammonia)	5	10	15	20	25

Table 2-2. Static pressure losses as a function of vertical lift for five refrigerants (Courtesy, Sporlan Valve Company)

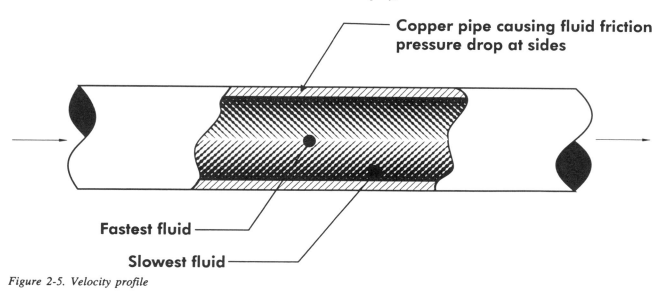

Copper pipe causing fluid friction pressure drop at sides

Fastest fluid

Slowest fluid

Figure 2-5. Velocity profile

How Much Subcooling Is Needed?

Subcooling can be accomplished by numerous methods. Common methods include liquid/suction line heat exchangers, ambient air, and direct expansion mechanical subcooler heat exchangers. On receiverless systems, simply adding a little more refrigerant to the system can increase subcooling, but this is usually not recommended because of added inefficiencies that may occur. Increasing the pressure of the liquid line is another popular method to subcool liquid and prevent flash gas. This method will be covered later in this chapter.

The amount of total subcooling needed is system dependent. The more pressure drops (friction and static) associated with the lines and accessories that carry the liquid in the system, the more need there is for liquid subcooling to prevent liquid line flash gas. For example, assume an R-12 refrigeration system is operating at a condensing pressure of 127 psig (105°F), Table 2-3. Also assume that all of the vapor in the condenser has turned to liquid and that it has reached the 100% saturated liquid point at 127 psig (105°F). For this example, assume no liquid subcooling. Now assume that this same saturated liquid experiences total pressure drops of 10 psig before it reaches the TXV. Because the liquid is saturated, it must eventually reach a new saturation pressure of 117 psig (127 psig - 10 psig). There is a saturated temperature associated with all saturated pressures, and for the saturated pressure of 117 psig, the new saturated temperature is 100°F (Table 2-3). However, for liquid to drop in temperature from 105° to 100°F, some heat must be removed from the liquid, which is where liquid line flash gas appears. As the liquid gradually travels through the 10 psig pressure drop, a portion of the liquid will continually flash and cool the column of liquid immediately behind it to the new saturation temperature of 100°F. This flashing cools the saturated liquid to 100°F, which corresponds to 117 psig at the entrance of the metering device. As mentioned earlier, premature liquid flashing that does not take place in the evaporator is lost refrigeration effect and can also derate the tonnage of expansion valves. The only sure way to safeguard against liquid

Temperature (°F)	Pressure (psig)
70	70.192
71	71.520
72	72.863
73	74.222
74	75.596
75	76.986
76	78.391
77	79.813
78	81.250
79	82.704
80	84.174
81	85.66
82	87.16
83	88.68
84	90.22
85	91.77
86	93.34
87	94.93
88	96.53
89	98.15
90	99.79
91	101.45
92	103.12
93	104.81
94	106.52
95	108.25
96	110.00
97	111.76
98	113.54
99	115.34
100	117.16
101	119.00
102	120.86
103	122.74
104	124.63
105	126.55
106	128.48
107	130.43
108	132.41
109	134.40
110	136.41
111	138.44
112	140.49
113	142.57
114	144.66
115	146.77
116	148.91
117	151.06
118	153.24
119	155.43
120	157.65
121	159.89
122	162.15
123	164.43
124	166.73
125	169.06

Table 2-3. Pressure/temperature relationship for R-12

line flash gas is to subcool the liquid before it experiences any pressure drop.

Example 2-5
Based on the R-12 example just discussed, figure out how much liquid subcooling needs to be present to prevent liquid line flash gas.

Solution 2-5
Refer to Table 2-3 to find the temperature span of 5°F that covers the pressure drop of 10 psig (127 psig - 117 psig). If there is at least 5°F of subcooling in the system, it acts as a sensible heat cushion to prevent liquid line flash gas. If the liquid is subcooled 5°F and experiences the gradual 10 psig pressure drop, flashing does not occur, because the liquid is already subcooled to accommodate the new lower pressures. When subcooled, the liquid is already at or below the saturation temperature for the new pressures. The *minimal amount* of subcooling required to accommodate the 10 psig pressure drop without liquid flashing is 5°F.

You can always use a pressure/temperature chart to figure the minimal amount of subcooling for any known pressure drop and refrigerant. However, different refrigerants require different amounts of liquid subcooling for the same pressure drops, Table 2-4. If charts such as Table 2-4 do not exist, calculations must be performed to find the amount of subcooling needed to handle whatever pressure drops exist. If the pressure drop is not known, make sure there is enough subcooling to handle an extreme pressure drop. Do not overcharge the system by adding too much refrigerant on receiverless systems. Overcharging increases condenser subcooling, but the head pressure will elevate and give unwanted inefficiencies from higher compression ratios. Always consult the system manufacturer for the proper amount of condenser subcooling if charging to this specification.

Example 2-6
Apply the principles just learned to an R-22 system piping layout, Figure 2-7. Assume the following conditions:

- Design pressure drop for friction losses in liquid line is 2 psig for copper pipes only

- Filter drier pressure drop is 1 psig (friction pressure drop)

- Vertical lift of liquid is 40 feet (static pressure drop)

- Static pressure drop is 20 psig (Table 2-2)

- System has a condensing pressure of 211 psig (105°F)

What is the *minimal* amount of liquid subcooling required to prevent liquid line flash gas in this R-22 refrigeration system?

Solution 2-6
The total pressure drop for this system is 23 psig (2 psig + 1 psig + 20 psig). The 23 psig pressure drop puts the new pressure at 188 psig (211 psig - 23 psig). According to Table 2-1, the saturation temperature that corresponds to 188 psig is about 97°F. Therefore, the liquid must be subcooled to at least 97°F to prevent liquid line flash gas. This means the system must have at least 8°F (105° - 97°F) of subcooling to prevent flash gas.

REFRIG-ERANT	100° F. Condensing					
	PRESSURE LOSS — psi					
	5	10	20	30	40	50
	REQUIRED SUBCOOLING — °F.					
12	3	6	12	18	25	33
22	2	4	8	11	15	19
500	3	5	10	15	21	27
502	2	3	7	10	14	18
717 (Ammonia)	2	4	7	10	14	17
REFRIG-ERANT	130° F. Condensing					
	PRESSURE LOSS — psi					
	5	10	20	30	40	50
	REQUIRED SUBCOOLING — °F.					
12	3	5	9	14	18	23
22	2	4	6	9	12	14
500	2	4	8	11	15	19
502	1	3	5	8	11	13
717 (Ammonia)	2	3	5	7	10	12

Table 2-4. Required subcooled liquid amounts to compensate for pressure drops (Courtesy, Sporlan Valve Company)

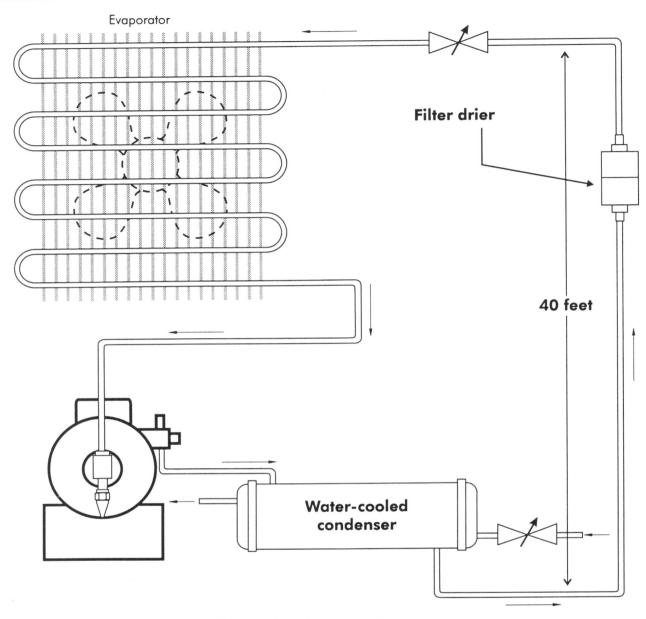

Figure 2-7. Refrigeration system with a 40-ft vertical rise in the liquid line

LIQUID PRESSURE AMPLIFICATION™ (LPA) AND SUPERHEAT SUPPRESSION

Liquid pressure amplification technology[1] has given refrigeration systems capacity boosts while saving electrical energy. Using this technology, liquid refrigerant entering the liquid line is pressurized by a small centrifugal pump, Figure 2-8. The pressurized amount is equivalent to the pressure loss between the condenser outlet and the TXV inlet on receiverless systems, or the receiver and the TXV inlet on TXV/receiver systems. By increasing the pressure of the liquid refrigerant, the associated saturation temperature is raised, while the actual liquid temperature remains the same. The liquid becomes subcooled and will not flash if exposed to pressure drops in the liquid line. Because the liquid centrifugal pump motor is external to the refrigeration system and the impeller is driven by a revolving magnetic field, negligible energy and heat are added to the system. The

Figure 2-8. Centrifugal pump for liquid pressure amplification™ (Courtesy, Hy-Save, Inc.)

liquid is pressurized with a negligible addition of temperature or heat and allows for a completely sealed system, because there is no drive shaft protruding from the motor to the pump. The LPA centrifugal pump can increase the pressure of the liquid by approximately 8 to 20 psi. The liquid pressure amplification system is shown in Figure 2-9 and can be compared to a more common system shown in Figure 2-10.

Since subcooling exists in any liquid below its saturation temperature for a given pressure, there are really several ways to subcool liquid. One way is to sensible cool the liquid in the bottom of the condenser, giving it a sensible heat reduction to prevent flashing from liquid line pressure drops. However, subcooling this way will take up valuable condenser volume with subcooled liquid at its bottom, since condensing cannot occur in this area. This will cause higher head pressures and compression

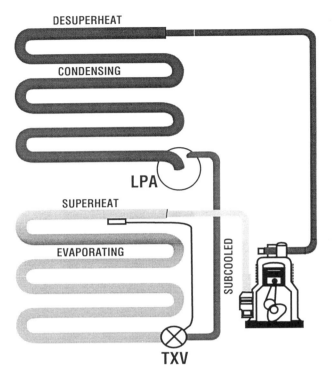

Figure 2-9. Normal refrigeration system with centrifugal pump for liquid pressure amplification (Courtesy, Hy-Save, Inc.)

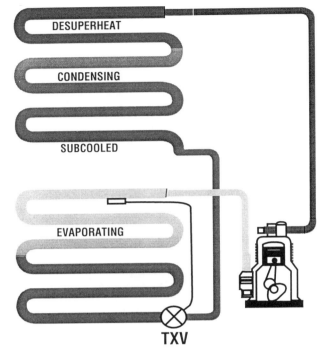

Figure 2-10. Typical refrigeration system (Courtesy, Hy-Save, Inc.)

ratios, thus lower efficiencies. To only consider the subcooling of the liquid without regard to decreasing condenser surface area, there will be a gain of 1/2% of capacity for every degree of liquid subcooling. Also, 7°F of ambient subcooling costs .167 brakehorsepower per ton because of elevated head pressures. However, if we consider the reduction of effective condensing surface area due to the condenser flooding necessary to achieve liquid subcooling, there is a net loss in capacity due to increased condensing pressures and temperatures.

This type of subcooling, often called ambient subcooling, has been practiced for years and was thought to be a "free" method of subcooling. This simply is not the case with receiverless systems having no separate subcooling coil. Ambient subcooling is usually accomplished at the cost of increased head pressures. It was used in refrigeration systems simply as a liquid seal in the condenser bottom and to prevent liquid line flash gas. This kept

a solid column of liquid supplied to the metering device.

Ambient subcooling cannot be maintained at a given level with only air side controls. Condensing pressures are directly related to the temperature of the condenser cooling medium and the useful condensing area in the condenser. Useful or effective condensing area is the total condensing area minus the area used for desuperheating and the area used for subcooling:

Total condenser area	–	Condenser area used for desuperheating and subcooling	=	Useful or effective condensing area

[Formula 2-3]

As one can see, the more desuperheating and liquid subcooling that is done by the condenser, the less useful condenser area there is. This will raise condensing pressures and compression ratios and cause inefficiencies with higher power draws.

A more efficient way to subcool liquid usually is to increase the pressure of the liquid without raising the temperature. This puts the liquid at a higher pressure, thus it has a higher associated saturation temperature but does not change its actual temperature. As the subcooling definition states, this liquid is subcooled in an amount equal to the difference between the saturation temperature and the actual temperature, and the liquid is below its saturation temperature for that new pressure. By increasing the pressure of the subcooled liquid to overcome any pressure losses that occur in the liquid line, condensing pressures can be allowed to fall to their lowest pressures attainable without flash gas developing in the liquid line.

Another term for attaining the lowest possible head pressure is *floating the head pressure.* Condensing temperatures of 20°F are not uncommon in low temperature systems incorporating LPA. However, if one tries to float the head pressure with the ambient, these lower head pressures will require more subcooling for the same pressure drops in the liquid line in order to prevent flashing. This phenomenon happens because the refrigerant pressure/temperature graph is non-linear. The pressure/temperature graph is much flatter at the lower pressures, meaning that the same amount of liquid subcooling is needed to overcome less of a pressure drop at these lower pressures and temperatures, Figure 2-11. This is one of the reasons why LPA is incorporated in the system when the head pressure is floated with the ambient. It subcools the liquid by increasing the pressure of the liquid and forces the liquid to have a new higher saturation temperature. Thus, flash gas is prevented when head pressures are allowed to float, because the LPA ensures that the liquid line pressure (and saturation temperature) are always higher than the actual liquid temperature.

If head pressure is reduced without the LPA and the liquid experiences the same pressure drop through the liquid line as it did at the higher condensing pressures, the flash gas will

occupy more volume in the liquid line because of the higher specific volume of the flashed vapors. The TXV will begin to hunt, alternately overfeeding and underfeeding the evaporator. The system capacity will be reduced, and damage to the compressor may result.

For example, consider the curve for a CFC refrigerant, Figure 2-12. As the pressure in the liquid line drops, more liquid progressively flashes into vapor to cool the remaining liquid to the saturation temperature corresponding to the progressively lower pressure. With an 8 psig pressure drop, the flash gas by weight is 2% with a 100°F condensing temperature (214 psig). The vapor bubbles in the liquid line are very compressed and occupy only 20% of the volume in the liquid line. However, reduce the pressure to 97 psig (50°F), and the flashing vapor will occupy 38% of the liquid line volume. This vapor reduces the flow through the expansion valve, has little refrigeration effect, and must be recompressed after doing work. Again, system capacity will suffer, the evaporator will starve, and the TXV will begin to hunt. This is the primary value of the LPA system - to ensure a solid column of liquid to the TXV, so the TXV can supply adequate liquid to the evaporator.

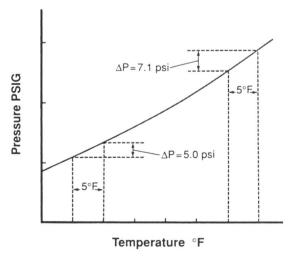

Figure 2-11. R-22 pressure/temperature curve (Courtesy, Sporlan Valve Company)

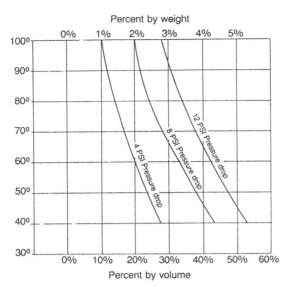

Figure 2-12. Effects of condensing pressure on flash gas (Courtesy, Hy-Save, Inc.)

Floating Head Pressures and LPA Advantages

In the past, designers of air conditioning and refrigeration systems picked an outdoor design condition for the system. This outdoor design condition typically was a temperature that would not be reached any more than 2% of the time in the life of the system. This design condition would also occur only a couple of hours at a time when reached. The selection of the condenser was then made based on this seldom-reached condition.

This was acceptable several years ago when energy was much cheaper, and designers could select condensing temperatures at 20° to 30°F above the ambient. This was done because it was thought the higher condensing temperatures and pressures would enhance the flow through the metering device to outweigh any inefficiencies from the high compression ratios and lower efficiencies. With today's escalating energy costs, designers are specifying larger condensers with condensing temperatures 10° to 15°F above the ambient. The significant energy savings from lower compres-

sion ratios and possibly increased subcooling of liquid negate the higher costs of the larger condenser.

After much research with metering valve suppliers, it was discovered that TXVs would work with much less pressure drop across them than expected in the past, as long as pure liquid was supplied to them. The balanced port TXV design today is noted for its low pressure drop performance. With this new knowledge, condensing pressures and temperatures were allowed to "float" downward with the ambient temperature. In fact, a majority of the outdoor temperatures in the USA are below 70°F more than they are above. The compressor capacity increases about 6% for every 10°F drop in condensing temperature.

Pressure drops of less than 30 psig across the expansion valve should be avoided for proper evaporator feeding of liquid. For an evaporator to operate at peak efficiency, it must have a high percent of liquid to vapor ratio entering the evaporator. To accomplish this, the expansion valve must allow liquid refrigerant to enter the evaporator at the same rate that it evaporates. With an LPA pump, subcooling and pressure can be maintained at a constant at the metering device. Overfeeding and underfeeding by the expansion valve, which dramatically affect the efficiency of the evaporator, will then be minimized.

Historically, high head pressures and temperatures were artificially maintained in a refrigeration system so it would function well at low outdoor temperatures. These higher pressures were considered mandatory in order for the TXV to feed the evaporator adequately. Power costs 50 years ago were not the primary consideration, so the added cost of lower efficiency did not matter. At today's power costs, inefficiency is an unacceptable part of a company's overhead. LPA allows lowering of the head pressure and reduced power consumption along with higher efficiencies. Here are four advantages to having an LPA system:

- Eliminates liquid line flashing by overcoming line pressure losses

- Reduces energy costs because pumping liquid refrigerant is up to 40 times more efficient than using head pressure from the compressor to do the same work

- Increases evaporator capacity along with the net refrigeration effect

- Lowers compression ratios and reduces stress on compressors, resulting in longer compressor life

The days of the fixed elevated head pressures are fading. No longer are consumers willing to pay for inefficiencies. Customers are looking at life cycle costs, which are the equipment costs plus the operational costs for the life of the equipment.

Superheat Suppression

Superheat suppression can be used in conjunction with LPA. The superheat suppression process injects liquid refrigerant into the compressor discharge line or inlet of the condenser, Figure 2-13. This liquid usually comes from the same centrifugal pump used in the LPA

process (see Figure 2-9). This liquid flashes to vapor while cooling the superheated discharge gas closer to its condensing temperature. As a result, less surface area is required for desuperheating. This leaves a more efficient condenser because of the increase in useful condensing surface area, which increases the overall performance of the system. Savings from 6% to 12% can be realized with superheat suppression.

Superheat suppression processes have been used in large ammonia plants for years. However, the process was not usually feasible on smaller systems. This spurred the use of LPA and superheat suppression on the same system. A small portion of the pressurized liquid refrigerant provided by LPA (centrifugal pump) is diverted to the compressor discharge line to cool the superheated vapors coming from the compressor. Reduced superheat of the gas entering the condenser results in higher condenser efficiency, a lower condensing temperature, and greater compressor efficiency. Although some efficiency gains are seen at low ambients, the greatest gain with superheat suppression is realized at the higher ambient temperatures.

Suppression or cooling of superheated vapors in a refrigeration system has four main advantages:

- Reduced superheated vapor heat intensity (temperature), so the pressure and volume of these superheated vapors will decrease

- Faster saturation of superheated vapors, which condense more quickly

- Condensing of the vapors occurs closer to the inlet of the condenser, resulting in lower condensing temperatures and possibly more ambient subcooling

- Higher overall condenser heat transfer because of the increased liquid/vapor mixture heat transfer and increased effective condensing area

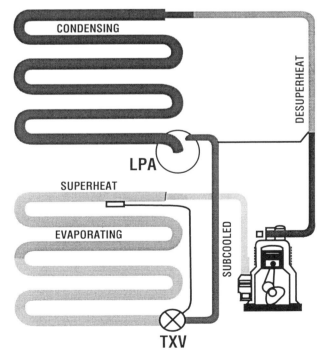

Figure 2-13. System with LPA and superheat suppression (Courtesy, Hy-Save, Inc.)

LPA/Superheat Suppression with Subcooling/Reheat Coil

In an air conditioning system, the evaporator removes moisture from the air. Therefore, it would seem reasonable that if the efficiency of the evaporator were increased, more moisture could be removed. Part of dehumidification is reheating the air that comes out of the evaporator to a lower relative humidity and raising the temperature to an acceptable delivery level. Accomplishing both of these functions greatly enhances the desired operation. Figure 2-14 shows a combination LPA/superheat suppression system incorporating a subcooling/reheat coil at the evaporator air outlet.[2] The coil allows subcooled liquid to be further subcooled, while at the same time providing reheat to the air leaving the evaporator. The relative humidity of the air is thus lowered. Depending on the size of the coil, the liquid should be able to be subcooled to within 8°F of the air temperature leaving the evaporator. With 60°F leaving air, the liquid should be subcooled to approximately 68°F. Remember, subcooling reduces liquid flash loss in the evaporator creating a greater net refrigeration effect.

With the rapid transition to alternative refrigerants and escalating energy costs facing the refrigeration industry today, more attention is being paid to operating costs and system efficiencies. Inattention to system pressures and liquid subcooling can make systems very inefficient. Because of this, it is very important that every technician understands the principles behind subcooling and system pressures.

ENVIROGUARD™ REFRIGERANT CONTROL SYSTEM

Enviroguard is a refrigerant control system that allows head pressures to float with the ambient temperature, while reducing system refrigerant charge up to one-third when compared to conventional systems.[3] Key features and benefits of this system include the following:

- Reduces system refrigerant charge up to one-third

- Reduces refrigerant leak liabilities compared to conventional systems

- Provides early warning of dirty condenser or failed condenser fans

- Provides early warning of a system refrigerant leak

- Increases compressor life because of lowered compression ratio and discharge temperatures due to lower condensing pressures when operated in cooler ambients

- Increases volumetric efficiencies due to lowered compression ratios

Principles of Operation

The amount of liquid refrigerant used in the Enviroguard system is controlled by a system pressure regulator (SPR), Figure 2-15. The system refrigerant charge is reduced by taking the receiver out of the active refrigerant circuit and allowing liquid refrigerant to return directly to the liquid manifold, which feeds the branch refrigeration circuits. With the receiver out of the active refrigeration circuit, a minimum receiver charge is no longer required as with conventional system design. The receiver is now only used as a storage vessel to store the condenser charge variations between summer and winter operations.

The SPR is controlled by a pilot pressure from a remote mounted, refrigerant filled, ambient air sensor connected to the SPR by a pilot line. The SPR's remote sensing bulb is charged with the same refrigerant as the system. The remote mounted ambient air sensor is located under the air-cooled condenser to sense the ambient air temperature entering the condenser, Figure 2-16. Figure 2-17 illustrates the location of the remote sensor in the evaporative condenser sump if an evaporative condenser is used.

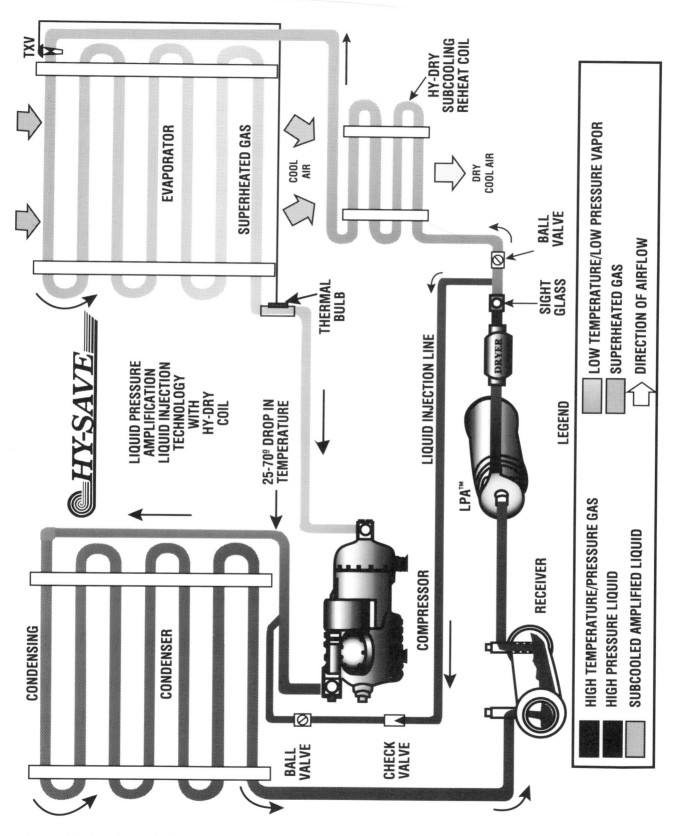

Figure 2-14. Combination LPA/superheat suppression system with a subcooling/reheat coil (Courtesy, Hy-Save, Inc.)

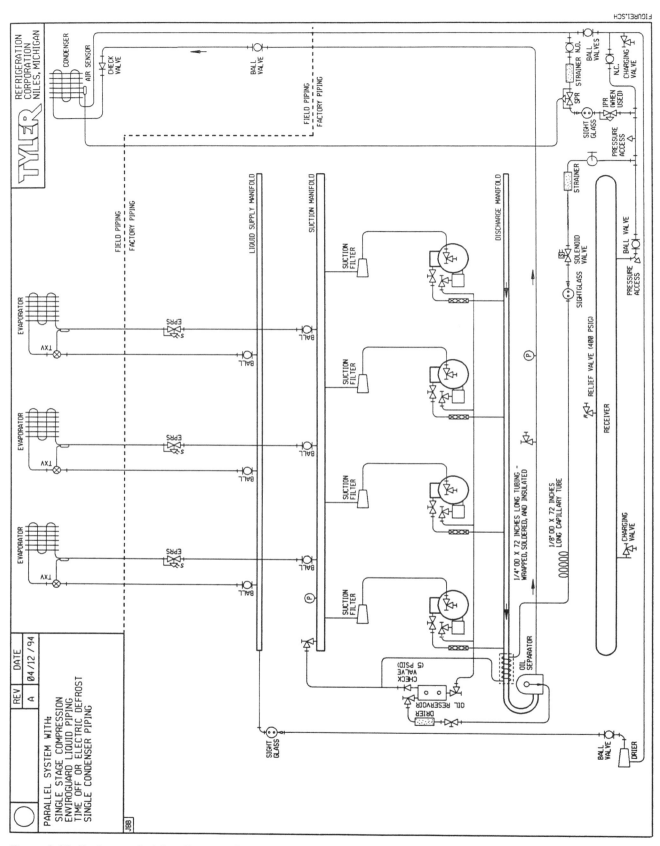

Figure 2-15. Enviroguard piping diagram (Courtesy, Tyler Refrigeration Corporation)

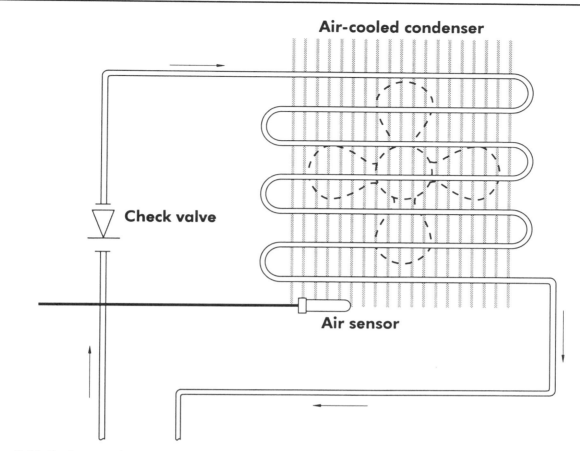

Air-cooled condenser

Check valve

Air sensor

Figure 2-16. Condenser with air sensor for SPR valve (Courtesy, Tyler Refrigeration Corporation)

When the ambient air temperature rises, the pressure inside the sensor and pilot line rises, exerting a higher pressure in the pilot line and diaphragm of the SPR valve. A spring with an adjustable resistance (differential) counteracts the SPR's diaphragm pressure. Diaphragm force against spring force determines the position of the SPR valve, and therefore, how much refrigerant will be bypassed to the receiver. When the ambient air temperature falls, the pressure inside the sensor and pilot line also falls, exerting a lower pressure on the pilot and diaphragm of the SPR. The SPR setting is adjusted to achieve a differential of approximately 45 psi for R-22 low temperature air-cooled applications and approximately 61 psi for R-22 medium temperature air-cooled applications. Since the SPR pilot pressure is equivalent to the saturated refrigerant pressure at the ambient temperature, the pressure at which the SPR begins to bypass refrigerant into the receiver on an R-22 low temperature system is the sum of the saturated refrigerant pressure corresponding to the actual ambient air temperature plus the appropriate differential pressure setting of either 45 or 61 psi.

Whenever the ambient air temperature drops, the pressure setting at which the SPR bypasses liquid refrigerant to the receiver also drops relative to the ambient air temperature. Anytime the condensing pressure rises 45 or 61 psi above the corresponding ambient air sensor pressure, the SPR begins to bypass refrigerant into the receiver. Condensing pressure changes occur at the same rate and time relative to changes in the ambient air temperature so liquid feed is always constant.

If the condenser should become damaged, fouled internally or externally, or should it lose a fan motor, an elevated condensing pressure

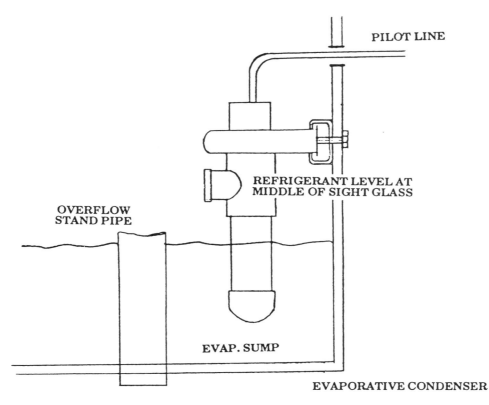

PILOT LINE

REFRIGERANT LEVEL AT
MIDDLE OF SIGHT GLASS

OVERFLOW
STAND PIPE

EVAP. SUMP

EVAPORATIVE CONDENSER

Figure 2-17. Remote sensor in the evaporative condenser sump (Courtesy, Tyler Refrigeration Corporation)

will occur, resulting in refrigerant being by-passed into the receiver by the SPR valve. This will prevent high compression ratios and high discharge temperatures with lower volumetric efficiencies. These occurrences often go unnoticed for long periods of time and stress compressor systems. Eventually, branch circuit evaporator temperatures will rise, because refrigerant is being bypassed out of the working part of the system into the receiver, simulating a refrigerant starved system. Rising evaporator and product temperatures will be noticed sooner, and the problem can be remedied. If a refrigerant leak should occur in the system, it will also be noticed earlier because of higher evaporator temperatures. Overall, less refrigerant is lost to the atmosphere than with conventional system design, because system problems are noticed much faster.

Whenever any or all of the compressors are running, there is a bleed circuit that is opened to the suction manifold, which bleeds from the receiver back to the system for use. This allows the refrigerant working charge in the system to seek its own level of equilibrium relative to the ambient temperature. Because of this bleed circuit, certain conditions may allow the receiver to be void of refrigerant and at suction pressure.

As refrigerant bleeds from the receiver to the suction manifold, a capillary restrictor tube and discharge line heat exchanger vaporize any liquid (see Figure 2-15). This prevents liquid from entering the suction manifold. This same refrigerant is also strained and filtered before entering the suction manifold. For example, on most systems, when the ambient air temperature is below 70°F, the receiver is empty because refrigerant is flooding the condenser. This is because less condenser surface is required during winter operation due to lower ambient air temperatures. During typical system operation when the ambient air temperature is above 70°F, part of the refrigerant charge normally flooding the condenser is stored in the receiver since more condenser surface is required to reject the total heat of rejection at the higher ambient temperatures.

Determining the SPR Valve Setting

For a forced air condenser, the proper steps to be taken in determining the proper SPR valve setting for a low temperature R-22 system application are as follows:

1. Choose a design ambient air temperature for the geographical region .... 95°F

2. Choose the condenser temperature difference (TD) .... 10°F

3. Design in a TD safety factor .... 5°F

4. Add the first three together to obtain the adjusted condensing temperature .... **110°F**

5. Corresponding saturated refrigerant pressure equal to the adjusted condensing temperature (110°F) .... 227 psig

6. Corresponding saturated refrigerant pressure equal to design ambient air temperature of 95°F (SPR bulb and pilot line pressure at design ambient) .... 182 psig

7. Subtract No. 6 from No. 5 for target SPR differential pressure .... **45 psig**

8. Actual ambient air temperature at time of system start-up .... 60°F

9. Corresponding saturated refrigerant pressure equal to actual ambient air temperature of 60°F .... 102 psig

10. Target SPR differential pressure .... 45 psig

11. Add No. 9 and No. 10 for the target SPR bypass pressure .... **147 psig**

The actual pressure at which the SPR valve will be set to bypass refrigerant to the receiver for this low temperature R-22 example will be 147 psig with the ambient air temperature measured at 60°F.

For an evaporative condenser, the proper steps to be taken in determining the proper SPR valve setting for a low temperature R-22 system application are as follows:[4]

1. Determine the area design ambient wet bulb temperature.

2. Determine the design wet bulb temperature to refrigerant condensing temperature differences. Typical condensing temperatures for low temperature systems are 90° to 95°F with a 20°F TD. The design condensing temperature for medium temperature applications is

from 95° to 100°F with a 25°F TD. Always refer to manufacturer guidelines.

a)	Design wet bulb temperature for the geographical area	75°F
b)	TD wet bulb to refrigerant	20°F
c)	Add a) and b) for saturated condensing temperature	——— **95°F**
d)	Corresponding saturation pressure at condensing temperature (95°F)	182 psig
e)	Corresponding saturation pressure at design wet bulb (75°F)	132 psig
f)	Subtract e) from d) to obtain SPR setting required	——— **50 psig**

SUPERHEAT

Superheat is any heat added to completely saturated vapor that results in a rise in temperature (sensible heat change) of the gas. Superheat is measured as the difference between the actual temperature of refrigerant vapor and the saturation temperature of the refrigerant at that same point. Superheat on the system's low side can be divided into two types: evaporator superheat and total (compressor) superheat.

Evaporator Superheat

Evaporator superheat starts at the 100% saturated vapor point in the evaporator and ends at the outlet of the evaporator, Figure 2-18. Technicians usually put a thermistor or thermocouple at the evaporator outlet to obtain the evaporator outlet temperature. A pressure gauge at the same point as the temperature reading will give the technician the saturated vapor temperature.

A Schrader tee can be used to obtain evaporator outlet pressure. Simply install the Schrader tee into the external equalizer access fitting on the side of the TXV once the system is pumped down, Figure 2-19. Once the pressure gauge and external equalizer line are reconnected to the TXV, the technician should wait from 10 to 15 minutes to let the system reach a running equilibrium again. The evaporator outlet pressure can then be read.

To determine evaporator superheat, consider an R-134a refrigeration system. In this system, the low side gauge reading at the evaporator outlet is 20 psig, or 23°F (see Table 1-1), and the evaporator outlet temperature (thermistor reading) is 30°F. The evaporator superheat calculation is as follows:

$$30°F \text{ (evaporator outlet temperature)} - 23°F \text{ (saturation temperature at evaporator outlet)} = 7°F \text{ (evaporator superheat)}$$

If a technician were to measure the pressure at the compressor instead of the evaporator outlet, a higher, fictitious superheat value would be read. As the refrigerant travels the length of the suction line, there is associated pressure drop from friction and/or restrictions. This causes the pressure at the compressor to be lower than the pressure at the evaporator outlet. This higher, fictitious superheat reading may lead the technician to open the TXV more to compensate for the fictitious reading. This could cause compressor damage from liquid flooding or slugging from a superheat setting that is too low.

To illustrate this point, consider the same R-134a refrigeration system just discussed with a 5 psi pressure drop from the evaporator outlet to the compressor. In this system, the low side gauge reading at the compressor inlet is 15 psig or 15°F, and the evaporator outlet temperature (thermistor reading) is 30°F. The evaporator superheat calculation is as follows:

$$30°F \text{ (evaporator outlet temperature)} - 15°F \text{ (saturation temperature at compressor inlet)} = 15°F \text{ (superheat)}$$

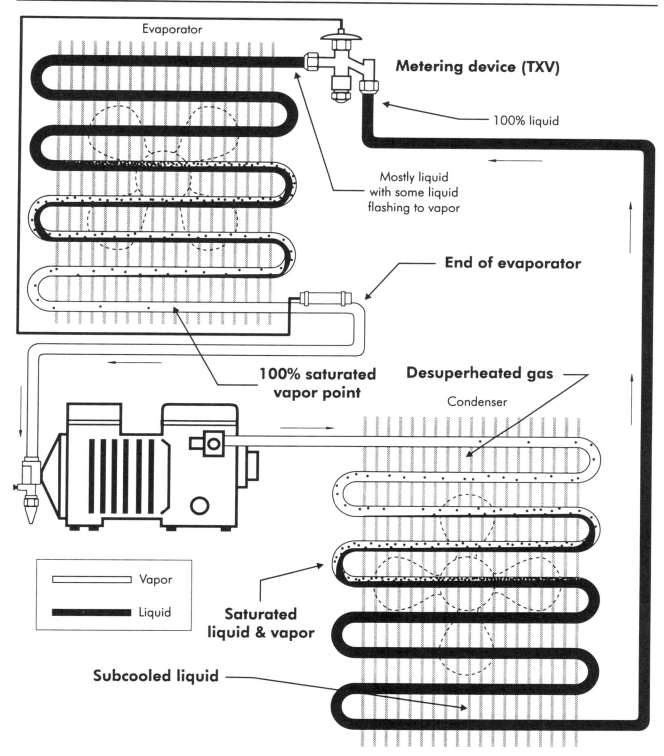

Figure 2-18. 100% saturated vapor point in evaporator and start of superheat

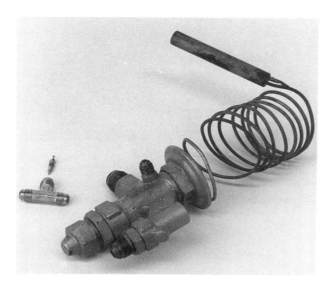

Figure 2-19. Externally equalized TXV with Schrader tee (Photo by Bill Bitzinger, Office of University Communication Services, Ferris State University)

The superheat changed from 7° to 15°F simply by reading the pressure at the compressor inlet instead of the evaporator outlet. The correct evaporator superheat is 7°F. It is best to measure the pressure at the same location as the temperature to exclude any system pressure drops. As an alternate, read the pressure at the compressor inlet and add estimated losses in the suction line to arrive at the approximate pressure at the evaporator outlet. (This alternate method is to be used on air conditioning systems only.)

The amount of evaporator superheat that is required for certain applications varies. Commercial ice makers call for 5° to 6°F of evaporator superheat to fill out their ice sheets; however, suction line accumulators are often em-

ployed on these systems for added protection. This helps ensure that all of the refrigerant entering the compressor is liquid free. This also helps keep a fully active evaporator. Lower temperature applications generally utilize lower evaporator superheat. Consult the case manufacturer if in doubt. In the absence of manufacturer data, Table 2-5 illustrates guidelines for superheat settings.

Total Superheat

There will always be times when the evaporator has a light load and the TXV may lose control of its evaporator superheat due to limitations of the valve and system instability. This is when total superheat comes into play. Total superheat, also called compressor superheat, is all the superheat in the low side of the refrigeration system. It starts at the 100% saturated vapor point in the evaporator and ends at the compressor inlet, Figure 2-20. Total superheat consists of evaporator superheat and suction line superheat. Total superheat can be measured by a technician by placing a thermistor or thermocouple at the compressor inlet and taking the temperature. A pressure reading will also be needed at this same location.

For example, consider an R-134a system with a low side pressure at the compressor of 20 psig or 23°F and a compressor inlet temperature of 50°F. The total superheat calculation is as follows:

50°F (compressor inlet temperature)	−	23°F (saturation temperature)	=	27°F (total superheat)

Applications	Air Conditioning and Heat Pump	Commercial Refrigeration	Low Temperature Refrigeration
Evaporator temperature	40° to 50°F	0° to 40°F	-40° to 0°F
Suggested superheat setting	8° to 12°F	6° to 8°F	4° to 6°F

Table 2-5. Evaporator superheat guidelines

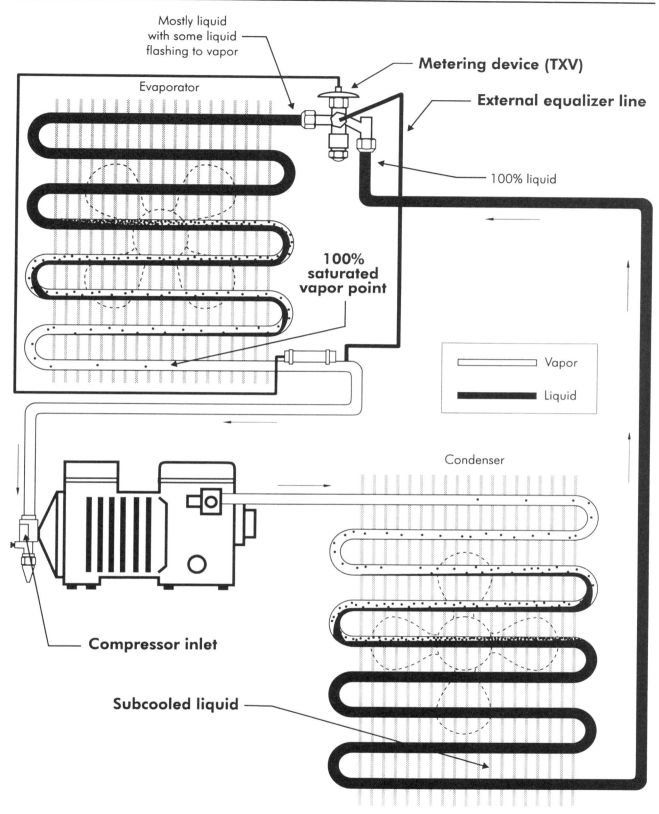

Mostly liquid with some liquid flashing to vapor

Metering device (TXV)

Evaporator

External equalizer line

100% liquid

100% saturated vapor point

Vapor

Liquid

Condenser

Compressor inlet

Subcooled liquid

Figure 2-20. Refrigeration system showing total superheat

In this example, the total superheat is 27°F. It is possible to have a TXV that is adjusted to control superheat at the coil (evaporator superheat) and still return liquid refrigerant to the compressor at certain low load conditions. If so, the conditions causing the floodback should be found and corrected. It is recommended that all TXV controlled refrigeration systems have at least 20°F of compressor superheat to ensure that the compressor does not see liquid refrigerant (flood or slug) at low evaporator loads. Total superheats from 20° to 30°F are recommended to ensure adequate compressor cooling and preventive liquid control to the compressor. The TXV, however, should be set to maintain proper superheat for the evaporator.

Air-cooled compressors are more vulnerable to slugging and valve damage, because the suction gases are not heated by the motor windings. The gases enter the sidewall of the compressor and go directly to the valves. The 20°F of compressor superheat is a buffer in case the TXV loses control of superheat at these low loads. However, the evaporator superheat must still be maintained by the guidelines in Table 2-5. A buffer of 20° to 30°F of compressor superheat makes sure that the refrigerant vapor entering the compressor is not too dense. Density of vapors that are too high entering the compressor will cause the compressor to have a higher than normal amp draw. This will overload the compressor motor in many instances and open thermal overloads. On the other hand, excess compressor or total superheat and/or long periods of low mass flow rate (e.g., an unloaded compressor) can result in insufficient cooling of the stator and opening of the internal protectors.

Remember, the TXV controls evaporator superheat. To obtain more total superheat, add a liquid/suction heat exchanger or even run a bit longer suction line to allow heat gains from the surrounding temperature to heat the suction line. Taking the insulation off the suction line to increase total superheat is not recommended, because this will cause the suction line to sweat from water vapor in the air reaching its dew point on the suction line. Freezing of this condensation may also occur if suction line temperatures are below 32°F. If at all possible, do not sacrifice (raise) evaporator superheat to obtain the amount of total superheat needed. This will not maintain an active evaporator and system capacity will suffer.

Low evaporator loads can be caused by many different situations, including the following:

- Evaporator fan motor not operating

- Iced up or dirty coil

- Defrost circuit malfunction causing coil icing

- End of the refrigeration cycle

- Low air flow across evaporator coil

- Low refrigerant charge

Any time the evaporator coil sees a reduced heat load, a TXV can lose control and hunt. Hunting is when the valve overfeeds then underfeeds while trying to find a controlling point. Hunting occurs during periods of system unbalance (e.g., low loads) when temperatures and pressures become unstable. The TXV tends to overfeed and underfeed in response to these rapidly changing values until the system conditions settle out and the TXV can stabilize. It is this overfeeding condition that may damage compressors. Evaporator superheat settings that are too low may also cause the TXV to hunt. Remember, a total superheat setting of at least 20°F can prevent the compressor from seeing any liquid refrigerant at most conditions.

NOTES

[1] Liquid Pressure Amplification™ is a trademark of Hy-Save, Inc. of Portland, Oregon.

[2] This type of subcooling/reheating has been patented by Hy-Save, Inc. of Portland, Oregon.

[3] Enviroguard™ is a trademark of Tyler Refrigeration Corporation, which holds the patent for this technology.

[4] This system was originally designed by Ron Wells and Dave Goodson, who were employees of Refrigeration Engineering Inc. (REI) of Grand Rapids, Michigan, and developed with the approval of the owner Bob Howieson. A patent was obtained and the rights sold to Tyler Refrigeration of Niles, Michigan.

CHAPTER THREE

Compression System

COMPRESSION RATIO

Compression ratio is the term used with compressors to describe the difference between the low and high sides of the compression cycle. It is calculated as follows:

$$\text{Compression ratio} = \frac{\text{Absolute discharge pressure}}{\text{Absolute suction pressure}}$$

Compression ratio is as important to a refrigeration and air conditioning technician as a thermometer and barometer are to a meteorologist. By knowing the compression ratio, a technician can determine the volumetric efficiency of the cooling system and thus determine how efficiently it is operating.

Most service technicians realize that their service gauges read zero when not connected to a system, even though there is a pressure of approximately 15 psi on the gauges exerted from atmospheric pressure. These gauges are calibrated to read zero at atmospheric pressure. In order to use the true or absolute discharge and suction pressure at zero gauge pressure or above, a technician must add 14.696 psi (approximately 15 psi) to the gauge reading. Absolute pressure is expressed in pounds per square inch absolute (psia), and gauge pressure is expressed in pounds per square inch gauge (psig).

Example 3-1
Determine the compression ratio for pressures above or equal to zero gauge pressure. The discharge pressure is 145 psig, and the suction pressure is 5 psig. Determine the compression ratio.

Solution 3-1
The compression ratio is as follows:

$$\text{Absolute discharge pressure} = \text{gauge reading} + 15\ \text{psi}$$

$$\text{Absolute suction pressure} = \text{gauge reading} + 15\ \text{psi}$$

$$\text{Compression ratio} = \frac{145\ \text{psig} + 15\ \text{psi}}{5\ \text{psig} + 15\ \text{psi}} = \frac{160\ \text{psig}}{20\ \text{psig}} = 8$$

The actual compression ratio for this problem is 8 to 1, which is expressed as 8:1. This simply means that the discharge pressure is 8 times the magnitude of the suction pressure.

Below Zero
There may be some confusion when computing the absolute suction pressure when the system is operating below zero gauge pressure (vacuum). This can be accomplished by subtracting the vacuum reading in inches of mercury from 30 inches of mercury (in. Hg) and dividing the results by two:

$$\text{Absolute suction pressure} = \frac{30\ \text{in. Hg} - \text{gauge reading in inches Hg}}{2}$$

This equation should only be used when the low side gauge is reading in the vacuum range.

The reason that absolute pressures rather than gauge pressures are used in the compression ratio equation is to keep the calculated compression ratio from becoming a negative number. One inch of vacuum can be represented as -1/2 psi, since 15 psi atmospheric pressure is equal to approximately 30 in. Hg on the barometer. If a negative number is used, a negative compression ratio would be calculated and have no real meaning. Absolute pressure keeps compression ratios positive and meaningful.

Example 3-2
Determine the compression ratio if the discharge pressure is 145 psig, and the suction pressure is 4 in. Hg (vacuum).

Solution 3-2
The compression ratio is as follows:

$$\text{Absolute discharge pressure} = \text{gauge reading} + 15 \text{ psi}$$

$$\text{Absolute suction pressure} = \frac{30 \text{ in. Hg} - \text{gauge reading in inches Hg}}{2}$$

$$\text{Compression ratio} = \frac{145 \text{ psig} + 15 \text{ psi}}{(30 \text{ in. Hg} - 4 \text{ in. Hg}) \div 2}$$

$$= \frac{160 \text{ psia}}{13 \text{ psia}} = 12.3$$

Again, the actual compression ratio for this problem is 12.3:1. This indicates to the technician that the absolute or true discharge pressure is 12.3 times as great as the absolute suction pressure.

Notice that even though the suction pressure gauge in this example read vacuum, the compression ratio came out positive because absolute pressures were used when calculating the compression ratio. This would not have been the case if gauge pressures were used in place of absolute pressures.

DISCHARGE AND SUCTION STROKE
So what does the compression ratio physically mean in a cooling system? In reciprocating compressors, there must be some clearance space between the piston at top dead center and the valve plate, otherwise there would be a collision of the two, Figure 3-1a. This intentionally designed clearance volume or clearance pocket traps a certain amount of refrigerant vapor after the discharge valve closes. Even though compressor manufacturers are reducing the amount of clearance volume found between the valve plate and piston head, some clearance will always remain. The clearance volume gas pressure is assumed to be at the discharge pressure if valve weight and valve spring forces are ignored. The vapor left in the clearance volume becomes compressed to the discharge pressure.

Once the downstroke of the piston starts, this same clearance volume vapor must be re-expanded to just below the suction pressure before the suction valve can open and let new vapors into the cylinder. The piston, however, will have already completed part of its suction stroke, and the cylinder will already have been filled with re-expanded clearance vapors from the clearance volume before new vapors enter Figure 3-1b. These re-expanded clearance volume vapors take up valuable space that new suction vapors coming from the suction line cannot occupy. Hence, suction vapors from the suction line will fill only part of the cylinder volume that is not already filled with re-expanded discharge gases. Therefore, the total volume of the piston's cylinder is not completely utilized in taking in new refrigerant gases, and the system is said to have a volumetric efficiency.

Volumetric Efficiency
Volumetric efficiency is defined as the ratio of the actual volume of the refrigerant gas pumped by the compressor to the volume displaced by the compressor pistons. The volumetric efficiency is expressed as a percentage from 0% to 100%, depending on the system. A high volumetric efficiency means that more of the piston's cylinder volume is being filled with

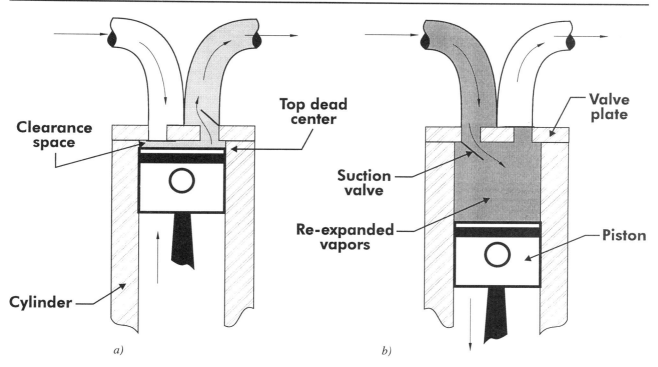

Figure 3-1. Piston and cylinder arrangements showing clearance space and re-expanded vapors

new refrigerant from the suction line and not re-expanded clearance volume gases. The higher the volumetric efficiency, the greater the amount of new refrigerant that will be introduced into the cylinder with each downstroke of the piston, and thus more refrigerant will be circulated with each revolution of the crankshaft. The system will have better capacity and a higher efficiency. The lower the discharge pressure, the less re-expansion of discharge gases to the suction pressure. The higher the suction pressure, the less re-expansion of discharge gases, because of the discharge gases experiencing less re-expansion to the higher suction pressure and the suction valve opening sooner.

Volumetric efficiency depends mainly on system pressures. In fact, the farther apart the discharge pressure magnitude is from the suction pressure magnitude, the lower the volumetric efficiency, because of more re-expansion of discharge gases to the suction pressure before the suction valve opens. Compression ratio measures how many times greater the discharge pressure is than the suction pressure; in other words, their relative magnitudes.

Remember, a compression ratio of 10:1 indicates that the discharge pressure is 10 times greater than the suction pressure, and a certain amount of re-expansion of vapors occurs in the cylinder before new suction gases enter. A relationship between compression ratio and volumetric efficiency can be seen in Figure 3-2. This graph is a typical multipurpose compressor curve. Deviations of this graph apply to multistage and air conditioning compressors. Figure 3-2 can only be used with halocarbon refrigerants. Halocarbons, or halogenated hydrocarbons, have had some or all of their hydrogen atoms replaced with fluorine, chlorine, bromine, or iodine atoms, Figure 3-3. R-11, R-12, R-13, R-14, R-22, R-32, R-123, R-502, R-134a, R-113, R-114, R-124, R-125, R-152a, and many other refrigerants, including most refrigerant blends, are halocarbons and can accurately be used with Figure 3-2.

To use Figure 3-2, consider that the compression ratio of a system is found to be 10:1. On the horizontal axis of Figure 3-2, simply draw a vertical line from the 10:1 compression ratio until it intersects the volumetric efficiency curve. At that point, draw a straight horizontal line to the left until it crosses the vertical axis,

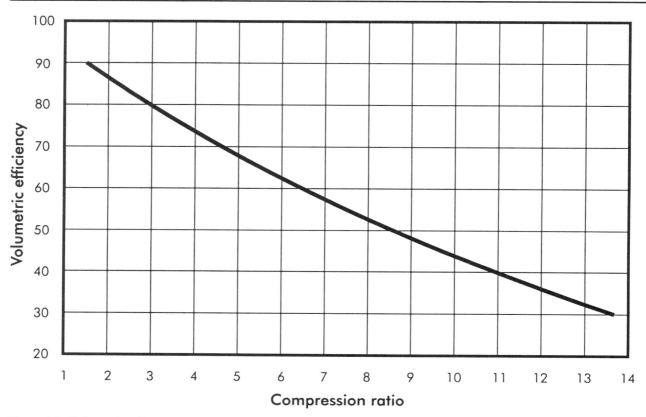

Figure 3-2. Volumetric efficiency versus compression ratio for halocarbon refrigerants

which is at 45%, Figure 3-4. This is the volumetric efficiency of the system when running at a 10:1 compression ratio. A 45% volumetric efficiency means that only 45% of the total cylinder volume is letting in new suction gases each downstroke. Compression ratios should be kept as low as possible in order to keep volumetric efficiencies as high as possible. If the compression ratios exceed the 14:1 curve limit in Figure 3-2, some way of lowering the discharge pressure or raising the suction pressure should be investigated. Otherwise, compound compression designs, where appropriate, should be incorporated to lower compression ratios.

With energy efficiencies becoming a national priority, it is going to be the service technician's responsibility to keep energy consumption to a minimum once the hvac/r equipment is up and running. Most hvac/r equipment runs on electricity, and most electricity is produced by the combustion of fossil fuels. Inefficient systems with low volumetric efficiencies cause more

electricity to be consumed and more fossil fuels to be burned. As more fossil fuels are burned, more carbon dioxide is emitted into the atmosphere. This contributes to the greenhouse effect (global warming) and environmental pollution. Try to keep those compression ratios low and volumetric efficiencies high for the best energy efficiency possible.

Compressor Flooding, Migration, and Slugging

Hvac/r field service terminology is often confusing and misused by even the most seasoned hvac/r service veterans. Clarification of such terminology is of utmost importance in order to determine the real problem and efficiently find the correct remedy. Clear, concise, and accurate communication between service technicians, part suppliers, customers, and the home shop is rapidly gaining importance as the hvac/r field becomes more technically oriented.

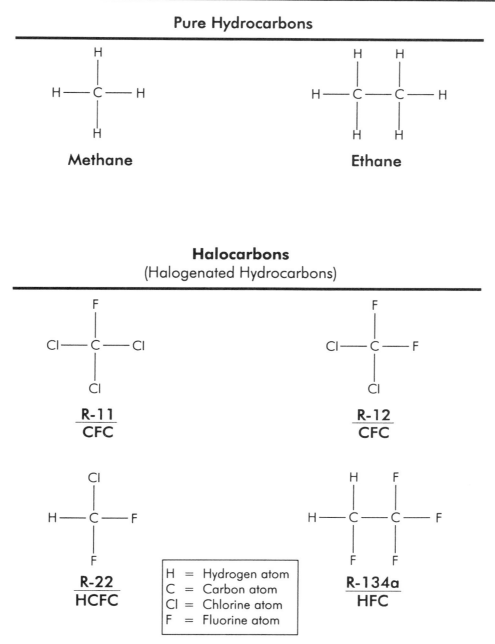

Pure Hydrocarbons

Methane

Ethane

Halocarbons
(Halogenated Hydrocarbons)

R-11
CFC

R-12
CFC

R-22
HCFC

H = Hydrogen atom
C = Carbon atom
Cl = Chlorine atom
F = Fluorine atom

R-134a
HFC

Figure 3-3. Halocarbon refrigerants

Flooding, *migration*, and *slugging* are three important service terms that are often misunderstood and misused by service technicians. These terms apply to refrigeration and air conditioning compressors and should be thoroughly understood by service technicians.

Flooding

Flooding is when liquid refrigerant enters the compressor crankcase while the compressor is running. Flooding of a compressor occurs only during the on cycle. The reasons why flooding might occur include the following:

- Wrong TXV setting (no compressor superheat)

- Overcharge

- Evaporator fan out

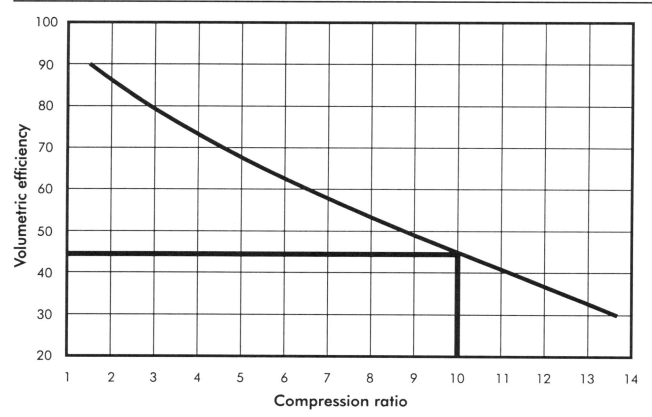

Figure 3-4. Volumetric efficiency of 45% with compression ratio of 10:1

- Low load on evaporator

- End of cycle (lowest load)

- Defrost clock or heater out (iced coil)

- Dirty or blocked evaporator coil

- Capillary tube overfeeding

- Capillary tube system overcharged

- Expansion bulb loose on evaporator outlet

- Oversized expansion valve

- Flooding after hot gas termination

- Heat pump changeover

- Defrost termination

Since liquid refrigerants are heavier than refrigeration oils, liquid refrigerant returning to the compressor will settle under the oil in the bottom of the compressor crankcase. This liquid refrigerant will gradually boil off due to low pressures in the crankcase. However, since the liquid refrigerant being boiled off is under the oil in the crankcase, very small oil particles will be entrained in this vaporization process. The oil level in the crankcase will then drop and rob mechanical parts of vital lubrication. Refrigerant-cooled semi-hermetic compressors often have check valves located on a partition between the crankcase and motor barrel to prevent oil and liquid refrigerant from mixing. Air-cooled semi-hermetic compressors and hermetic compressors are often more prone to flooding. Suction accumulators can help a flooding condition, but if the situation is severe, accumulators can also flood, especially if sized improperly.

Crankcase pressures can become excessively high from liquid refrigerant boiling in the crankcase. These high crankcase pressures can cause refrigerant and entrained oil particles to escape around the rings of the pistons during the downstroke. Once in the compressor cylinders, these particles are pumped by the compressor into the discharge line. The compressor then pumps oil and refrigerant and robs the crankcase of lubrication.

Oil in the system and not in the crankcase will coat the inner walls of the tubing and valves and cause unwanted inefficiencies. These inefficiencies include higher than normal crankcase pressures caused from the higher density refrigerant and oil mixture being pumped through the compressor cylinders, which will cause high compressor current draws. This may overheat and even trip the compressor. Broken valves may also occur.

A telltale sign of flooding is a cold, frosted, or sweaty crankcase. A foaming oil sightglass with a low oil level and/or higher than normal current draws are also signs of flooding.

Migration

Migration occurs when refrigerant liquid or vapor returns to the compressor crankcase or suction line during the off cycle. Migration only occurs during the off cycle of the compressor. Migration can occur when it is not an automatic pumpdown system or when there is a leaky liquid line (pumpdown) solenoid.

During the compressor off cycle, both liquid and vapor refrigerant have a tendency to migrate to the compressor crankcase and settle under the oil. Refrigeration oil has a much lower vapor pressure than the liquid or vapor refrigerant; therefore, refrigerant migrates or flows to the compressor crankcase where the lowest pressure exists. If the compressor is located in a cold ambient, migration takes place much faster, because cold causes an even lower vapor pressure in the compressor crankcase. Migration can even take place from a suction line accumulator to a compressor because of the difference in vapor pressure.

Migration can take place with liquid or vapor refrigerant, and the refrigerant vapor can flow uphill or downhill. Once the refrigerant vapor

reaches the crankcase, it will condense and settle to the bottom of the crankcase under the oil if the compressor remains off long enough. Oil and refrigerant mix readily. On short off cycles, the refrigerant does not have a chance to settle under the oil but will mix with the oil. When the compressor turns on, the sudden crankcase pressure drop causes the oil refrigerant mixture in the crankcase to flash. The oil level in the crankcase then drops and mechanical parts can become scored. Oil foaming appears and a combination of oil and refrigerant can be forced around piston rings to be pumped by the compressor. High current draws, motor overheating, and broken valves can occur.

The only sure remedy for compressor migration is an automatic pumpdown system, which clears all the refrigerant (liquid and vapor) from the evaporator and suction line before every off cycle. Automatic pumpdown is accomplished with a thermostat-controlled liquid line solenoid in combination with a low pressure controller terminating the on cycle once the evaporator and suction lines are void of any refrigerant. This ensures that there is no refrigerant in the evaporator or suction line to migrate towards the compressor. Automatic pumpdown systems are covered later in this chapter.

It is often thought that a crankcase heater prevents migration. Crankcase heaters keep the compressor crankcase warm and prevent migration to the compressor oil. However, condensed migrated refrigerant will be driven from the compressor and collect in the suction line near the compressor waiting for the next on cycle. If excessive liquid refrigerant is driven to the suction line, severe liquid slugging may occur during start-ups. Frequently, compressor damage such as broken valves and damaged pistons will occur. Crankcase heaters can be effective in combating migration, but they will not remedy slugging at start-ups from liquid floodback unless used in conjunction with a properly sized suction line accumulator.

Slugging

Slugging is when liquid refrigerant or liquid refrigerant and oil enters the compressor cylinder during an on cycle. The reasons why slugging might occur include the following:

- No compressor superheat
- Migration (off cycle)
- Bad TXV
- TXV hunting
- Low load
- End of cycle (lowest load)
- Evaporator fan out
- Iced evaporator coil
- Defrost timer or heater out
- Dirty evaporator
- Capillary tube overfeeding
- Overcharge

Air-cooled semi-hermetic compressors are more prone to slugging than refrigerant-cooled semi-hermetic compressors. This is because refrigerant is often drawn directly into an air-cooled semi-hermetic compressor cylinder without passing through the motor barrel. Slugging can result in broken valves, broken head gaskets, broken connecting rods, and other major compressor damage.

Refrigerant-cooled semi-hermetic compressors often draw liquid from the suction line through hot motor windings in the motor barrel, which assists in vaporizing any liquid. Even if liquid refrigerant gets past the motor windings, the check valve in the partition between the crankcase and the motor barrel will prevent any liquid refrigerant from entering the crankcase. High current draws will be noticed from dense refrigerant vapors entering the compressor cylinder.

Most hermetic compressor suction lines end at the shell of the compressor. If liquid refrigerant enters the compressor, it will fall directly into the crankcase oil and eventually be flashed. This is referred to as flooding, which causes oil foaming and excessively high crankcase pressures. Refrigerant and oil droplets will then reach the compressor cylinder and slugging will occur.

Slugging in hermetic compressors can also be caused by a migration problem. Foaming oil and refrigerant in the crankcase due to migration generate excessive crankcase pressures when the on cycle occurs. These oil and refrigerant droplets can then get past piston rings and other small openings and enter the compressor cylinder. The end result is slugging of refrigerant and oil. Slugging can damage reed valves, piston rods, bearings, and many more mechanical parts.

AUTOMATIC PUMPDOWN SYSTEMS

As mentioned earlier, the automatic pumpdown cycle is the only sure remedy for refrigerant migration to a compressor during the off cycle. These systems also prevent slugging of liquid refrigerant or oil caused by a severe migration problem and at cycle start-ups.

The automatic pumpdown system consists of a normally closed (NC) liquid line solenoid valve installed in the liquid line of a refrigeration system. The solenoid is a normally closed electric shut-off valve that is controlled by a thermostat, Figure 3-5. Solenoids come in different types and series for different applications, Figures 3-6 and 3-7. The thermostat is located somewhere in the refrigerated space. When the desired temperature is reached in the refrigerated space, the thermostat opens and de-energizes the liquid line solenoid valve, Figure 3-8, which closes the solenoid valve. The compressor continues to run and evacuate any refrigerant from the solenoid valve outlet to and including the compressor, as well as part of the liquid line, evaporator, suction line, and crankcase. This refrigerant is stored in the condenser and receiver on the high side of the refrigeration system. The compressor is then shut off by the low pressure switch, which is set to open at about 0 psig, Figure 3-9. This ensures that no refrigerant will migrate during the off cycle to the compressor.

Figure 3-5. Typical solenoid valve (Courtesy, Sporlan Valve Company)

On a call for cooling, the thermostat closes and energizes the liquid line solenoid valve. This action sends liquid refrigerant to the expansion valve and into the empty evaporator, thus increasing evaporator pressure. Once the cut-in pressure of the low pressure control is reached, the compressor starts easily and resumes a normal refrigeration cycle. Automatic pumpdown systems do not need a larger receiver, because the receiver is designed to hold all of the refrigerant in the entire system and still have a 20% vapor head for safety.

The cut-in pressure of the low pressure control should be set at a high enough pressure to ensure that the system will not short cycle if residual pressure remains in the low side of the system once pumped down. Compressor short cycling results in overheating, which can be devastating to motor windings and starting controls. However, the cut-in pressure has to be low enough to ensure the system will cut in once the liquid line solenoid is energized by the thermostat to start the next on cycle. These pressures are dependent upon refrigerant type and box temperature applications desired. Consult the case manufacturer for specific cut-in pressure settings.

It is good practice to install automatic pumpdown systems on all refrigeration systems employing large amounts of refrigerant and oil. Automatic pumpdown systems are not seen on systems with small amounts of oil or refrigerant, because there is not enough refrigerant to cause any damage if migration does occur. A good example of this is a domestic refrigerator or freezer. These refrigerators or freezers often have a refrigerant charge of only 8 to 16 ounces of refrigerant, while their oil charge is from 12

for Refrigerants 12-22-134a-502 ■ Types B19 and E19 Series ■

Figure 3-6. Typical solenoid valves (Courtesy, Sporlan Valve Company)

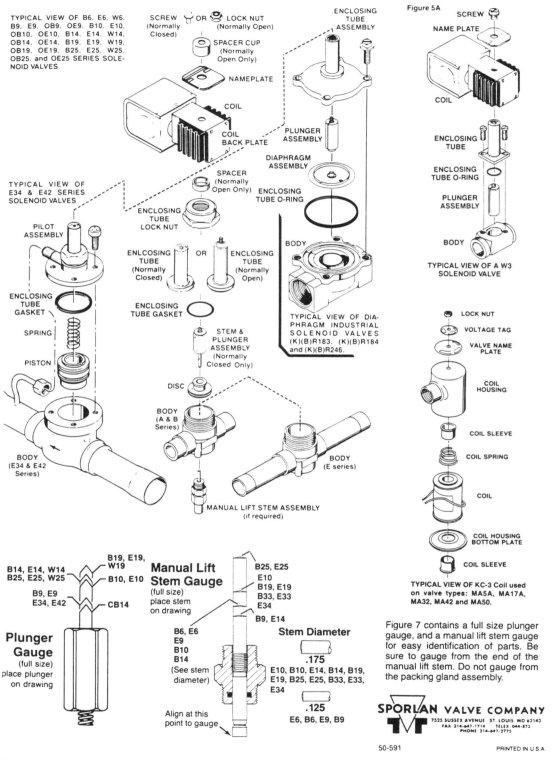

Figure 3-7. Exploded views of solenoid valves (Courtesy, Sporlan Valve Company)

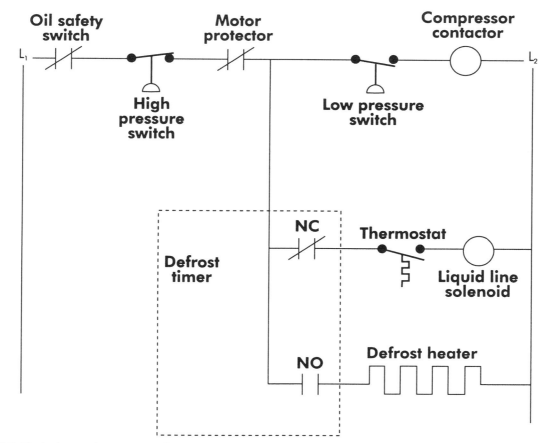

Figure 3-8. Typical pumpdown system showing thermostat in series with liquid line solenoid valve

to 20 ounces, resulting in more oil than refrigerant in these systems. Even if all of the refrigerant migrates to the crankcase during an off cycle, no real damage will be done, because the refrigerant flashes off during the next on cycle. Automatic pumpdown systems should be employed if the refrigerant to oil ratio is large. Automatic pumpdown systems are also used because they:

- rid the evaporator, suction line, and crankcase of refrigerant before the off cycle so migration of refrigerant to the crankcase cannot occur.

- prevent surges of liquid refrigerant from entering the suction port of the compressor (slugging) during start-ups.

- rid the crankcase of refrigerant to prevent foaming of oil on start-ups and robbing compressor mechanical parts of lubrication.

If an automatic pumpdown system is not employed, certain evaporator piping practices can be employed to help keep the residual refrigerant in the evaporator from migrating to the compressor during the off cycle, Figure 3-10. In severe situations, even piping arrangements cannot prevent migration to a compressor.

Electric Defrost Cycle

An automatic pumpdown can be used on electrical defrost systems in order to prevent the refrigerant from migrating to the suction line, compressor, and oil while in defrost. During defrost, the evaporator is full of liquid and vapor refrigerant. Once the defrost heaters energize, this refrigerant looks for a lower pressure and heads for the compressor. When the defrost terminates and the compressor starts back up, there is enough refrigerant in the suction line and compressor crankcase to damage the compressor. By pumping all of the refrigerant out

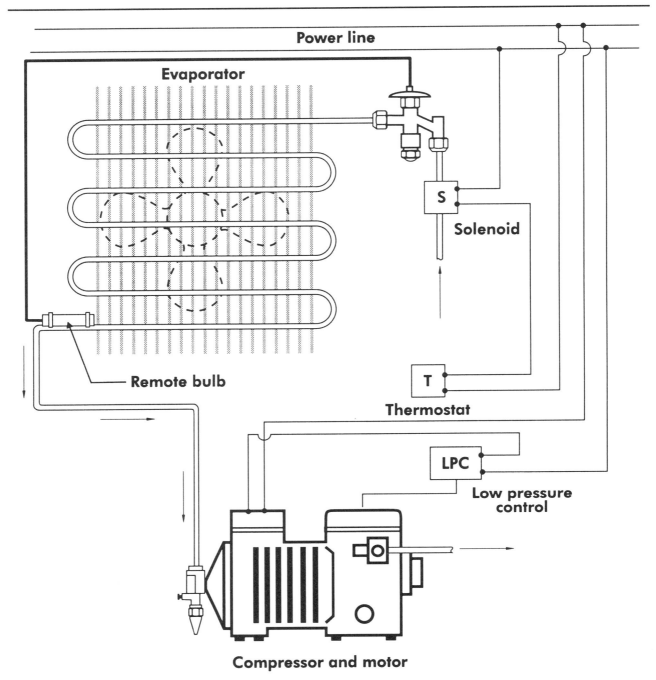

Figure 3-9. Electrical hook-up of a pumpdown system

of the evaporator before or during defrost, the compressor is protected from slugging refrigerant and oil. This process prevents oil foaming due to refrigerant migration to the crankcase and slugging of refrigerant from the suction line at start-up after the defrost cycle. Pumpdown during defrost also shortens the defrost time, because the heat for defrost melts ice and does not vaporize refrigerant left in the evaporator. This system also performs an automatic pumpdown before every off cycle.

Figure 3-11 is an electrical diagram of an automatic pumpdown system that initiates during defrost or when the thermostat opens. Note the normally closed contacts of the defrost timer in series with the control thermostat and the liquid line solenoid. When defrost calls, the

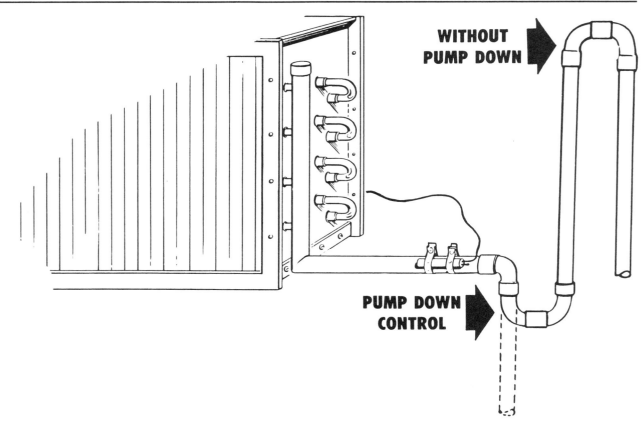

Figure 3-10. Good piping practice for compressor below evaporator configurations (Courtesy, Sporlan Valve Company)

normally closed contacts open. This action de-energizes the solenoid valve and starts a pumpdown. At the same time, the normally open (NO) contacts of the defrost timer close and energize the defrost heaters. Once pumpdown during defrost is complete, a low pressure switch opens the compressor circuit.

There is a short time during defrost when the compressor and defrost heaters are on at the same time. It is of utmost importance that the fuses are sized to carry the current of both the compressor and defrost heaters. They will both operate until the low pressure switch opens and shuts off the compressor.

Another option for pumpdown during defrost is to use a double acting low pressure switch. This type of pumpdown operates the same as the aforementioned system except that the double acting low pressure switch shuts off the compressor and turns on the defrost heaters at the same time. This is accomplished by a set of normally closed and normally open contacts

within the low pressure switch, Figure 3-12. This type of set-up eliminates the high amp draw when both the compressor and defrost heaters are on at the same time. This system also performs an automatic pumpdown before every off cycle.

Non-Recycling Automatic Pumpdown System

Once a system is pumped down, it should stay pumped down until the thermostat calls for cooling and energizes the liquid line solenoid. This action causes the evaporator to receive refrigerant from the liquid line and increase in pressure. The low pressure control then closes and starts the compressor.

In time, all mechanical parts fail. Liquid line solenoid valves can leak, causing compressors to short cycle on their low pressure control while in pumpdown. Discharge valves can also leak and cause refrigerant pressure to build in the crankcase, which can short cycle compres-

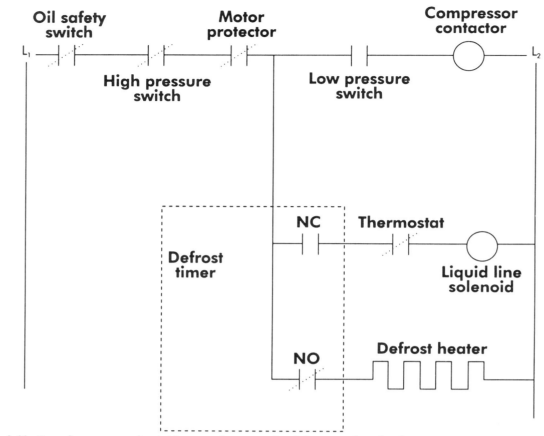

Figure 3-11. Pumpdown system that initiates at the onset of a defrost or when the thermostat opens

sors on their low pressure controls while in pumpdown. Cylinder unloading devices are also known to leak and short cycle compressors while in pumpdown. Serious compressor motor damage and contactor damage can result if the short cycling continues unchecked for any period of time.

One compressor manufacturer has solved the problem of short cycling during pumpdown caused by leaking components by designing a non-short cycling pumpdown circuit using a latching relay. Figure 3-13 shows the system in a pumpdown with the compressor off.

On a call for cooling, the thermostat closes and energizes the liquid line solenoid. This allows the low side of the system to experience refrigerant pressure, which allows the low pressure switch to close. The latching relay coil (LRC) is energized through the normally closed (NC) contacts A and B. Once the LRC is energized,

it closes contacts B and C and contacts C and D, while opening contacts A and B. With the LRC energized, the compressor contactor (CC) is energized through contacts C and D and the low pressure switch. This starts the compressor, and the system is in a normal cooling cycle.

When the thermostat opens, the liquid line solenoid is de-energized. This physically closes off the liquid line and initiates an automatic pumpdown. The low pressure switch opens, de-energizing the compressor contactor (CC), which shuts off the compressor. The system is pumped down, and the compressor remains off until there is another call for cooling from the thermostat. However, if the low pressure control tries to close from a leaking component while the thermostat is still open and not calling for cooling, the compressor contactor (CC) will not be energized because of the normally open (NO) contacts C and D of the latching relay. This prevents short cycling. The LRC

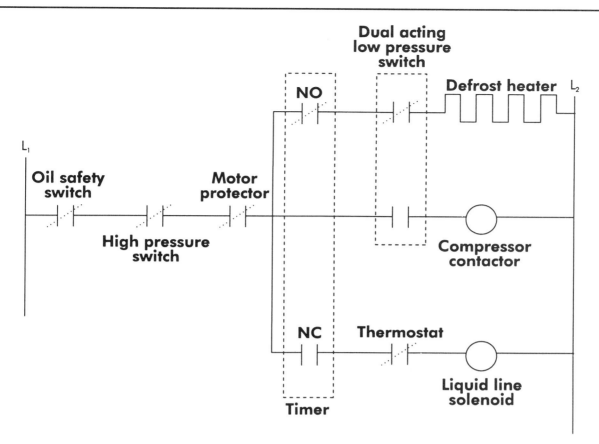

Figure 3-12. Pumpdown system using a double acting low pressure switch

will also not be energized because of the NO contacts B and C of the latching relay. These latching relay contacts prevent compressor short cycling during pumpdown when components leak, which causes low side pressures to increase and low pressure switches to close prematurely.

CRANKCASE HEATERS

As mentioned earlier, refrigerant migration to the compressor crankcase during an off cycle can cause serious compressor damage if the problem is not remedied. The automatic pumpdown system is the most effective remedy for migration, but a crankcase heater may help the problem.

The crankcase heater is a resistance heater that is wrapped around, strapped on, inserted, or

connected in some way to the compressor crankcase, Figure 3-14. It combats refrigerant migration by holding the oil in the compressor at a temperature higher than the coldest part of the system. Refrigerant entering the crankcase is vaporized and driven back into the suction line; however, during the next on cycle, refrigerant from the suction line may be drawn causing slugging of the compressor. Again, crankcase heaters do help in combating migration but do not prevent slugging at start-ups or liquid floodback to compressors.

Crankcase heaters may be energized continuously, Figure 3-15, or during the on cycle, Figure 3-16. However, to avoid carbonizing of the oil from excessive heat, the wattage input of the crankcase heater must be limited. In ambients approaching 0°F, or when exposed suction lines and cold winds impose an added cold load, the crankcase heater may be overpowered and migration may still occur.

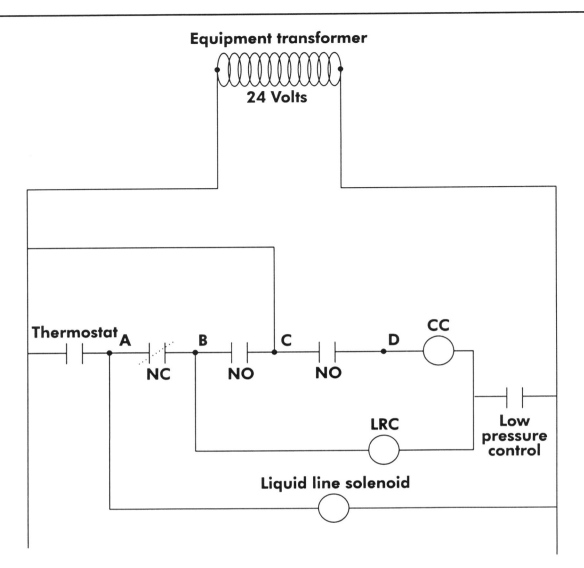

Figure 3-13. Non-short cycling automatic pumpdown system using a latching relay

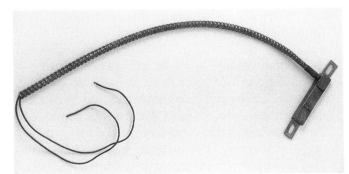

Figure 3-14. Crankcase heater for large compressor (Photo by Bill Bitzinger, Office of University Communication Services, Ferris State University)

One major compressor manufacturer pumps down the evaporator and suction line before each off cycle and at the same time energizes a crankcase heater during the off cycle. The crankcase heater is a safety precaution in case the liquid line solenoid leaks refrigerant. The heater prevents refrigerant from getting to the crankcase and causing oil flash at start-up; however, it will not prevent slugging or flooding of liquid refrigerant from the suction line or evaporator at start-ups. Other manufacturers employ both a crankcase heater and a properly sized suction line accumulator to protect the compressor from liquid returning to the compressor.

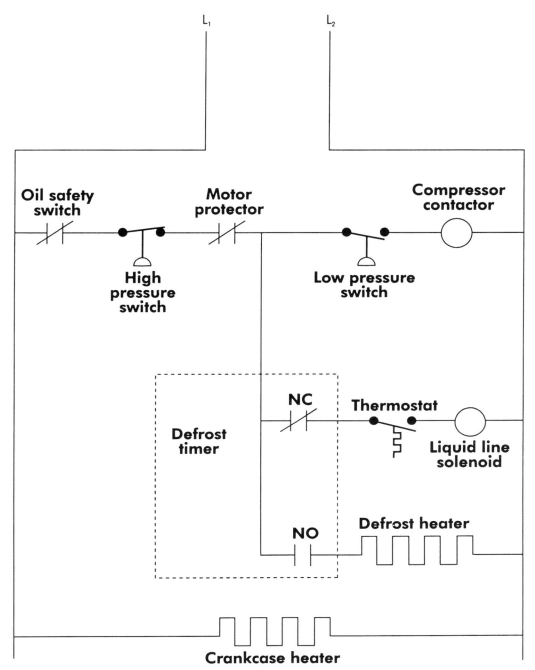

Figure 3-15. Continuously energized crankcase heater

Off Cycle Heat

The use of off cycle motor heat is often considered instead of a crankcase heater to help prevent migration, because it is more economical than a crankcase heater.

Off cycle heat employs a run capacitor connected directly to the power line beyond the unit disconnect instead of the run terminal of the compressor, Figure 3-17. When the compressor is cycled off, the start winding is still connected across the 20 mfd (microfarad) capacitor. The frequency of alternating current in the line charges and discharges the 20 mfd capacitor. This causes a heating effect and keeps the crankcase warm. Both the 15 mfd and 20 mfd capacitors are really run capacitors, but they have been connected in parallel for the off cycle

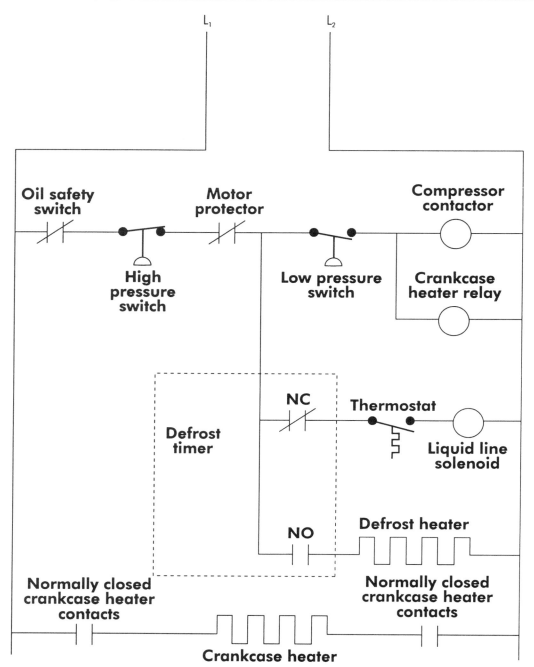

Figure 3-16. Crankcase heater energized only during off cycle

heat application. When capacitors are connected in parallel, their capacitance is added together. When the thermostat is closed and the compressor is running, the motor sees a 35 mfd capacitor wired across run and start. This makes it a permanent split capacitance motor when running.

SUCTION LINE ACCUMULATORS

As discussed previously, liquid refrigerant floodback is a compressor's worst nightmare. Floodback dilutes the compressor oil with liquid refrigerant and causes foaming in the compressor crankcase, resulting in bearing wash. Even if the compressor superheat is kept over 20°F, there are times when evaporator heat loads are low, defrost periods are ending, mechanical

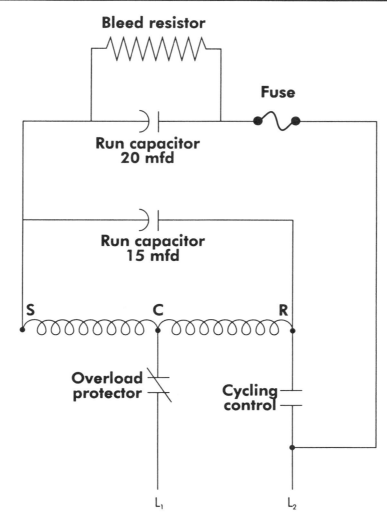

Bleed resistor

Fuse

**Run capacitor
20 mfd**

**Run capacitor
15 mfd**

S   C   R

**Overload
protector**

**Cycling
control**

L₁   L₂

Figure 3-17. Off cycle motor heat utilizing an existing run capacitor

valves fail or are misadjusted, or lack of system maintenance can cause compressor floodback. If there is no way to prevent these short periods of liquid floodback to the compressor, a suction line accumulator is needed on the system.

Suction line accumulators are designed as compressor protection devices when flooding and migration occur. They are usually installed in the suction line between the evaporator and compressor, as close to the compressor as possible. In reverse cycle systems such as heat pumps, they should be installed in the suction line between the compressor and the reversing valve. Accumulators can also act as suction line mufflers to quiet compressor pulsation noises.

Accumulators act as a temporary reservoir for liquid refrigerant and/or oil. They collect liquids from the suction line and hold them until they evaporate and return to the compressor naturally. Their outlet tubes to the compressor are located at the highest point in the accumulator so that only refrigerant vapor enters the compressor, Figure 3-18. The refrigerant liquid level hopefully never reaches this highest point.

Most accumulators are designed to meter both the liquid refrigerant and oil back to the compressor at an acceptable rate that will not damage compressor parts or cause oil foaming in the crankcase. This is done with a small metering orifice at the bottom of the outlet tube while the compressor is running, Figure 3-19. This small orifice is mainly designed for oil return;

Figure 3-18. Suction line accumulator (Courtesy, AC&R Components, Inc.)

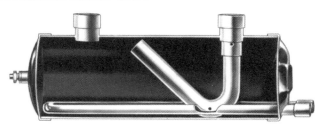

Figure 3-19. Horizontal suction line accumulator showing metering orifice (Courtesy, AC&R Components, Inc.)

however, liquid refrigerant can be slowly metered and vaporized through it also.

Some accumulators use electric heaters or liquid line heat sources to evaporate liquid refrigerant in the accumulator, Figure 3-20. However, any heat source that boils off liquid refrigerant causes the refrigerant gas to be saturated. This means the gas contains no superheat as it leaves the accumulator. Compressors often experience high ampere draws from near saturated vapors being compressed. The den-

sity of saturated or near-saturated vapors are much greater than superheated vapors and will often cause a higher mass flow rate than the compressor can handle. The result is high amp draw and often overheating of the compressor. This happens because the vapor must be removed from contact with the vaporizing liquid before it can be superheated. If there is no liquid being vaporized in the accumulator, however, the gases can become superheated. If the gases are superheated too much, inefficient compressor cooling can result. In refrigerant-cooled compressors, it is the returning refrigerant gas that cools the compressor.

Many indoor compressor installations have problems with suction accumulators that sweat and drip. The only way around this problem is to insulate the accumulator. The accumulator must be insulated completely and vapor sealed to prevent condensate from forming under the insulation. Rusting problems can result if the accumulator is exposed to moisture for any long period of time. Even though manufacturers do supply accumulators with rust preventative paints, these paints can be burned off during the welding process, leaving exposed metals.

Suction accumulators can assist in flooding and migration conditions; however, if flooding or migration problems are severe, suction line accumulators may flood and cause compressor damage. This is why the only safe accumulator is one that can hold 100% of the entire system refrigerant charge.

Accumulators should be selected with three basic considerations in mind (consult the manufac-

Figure 3-20. Suction line accumulators with liquid line heat exchangers (Courtesy, Refrigeration Research)

turer for specific catalog numbers and tonnage ratings):

- It should have adequate liquid holding capacity (not to be less than 50% of the entire system charge).

- It should not add excessive pressure drop to the system.

- Never base the accumulator on the line size of the suction line. The accumulator may not have the same line size as the suction line.

OIL SEPARATORS

Because refrigerants and refrigeration oils can mix together (are miscible), there will always be some oil that leaves the compressor with the refrigerant being circulated. Any time flooding or migration occurs, the crankcase oil is also sure to be diluted with refrigerant. This will cause oil foaming at start-ups and crankcase pressures to build, often forcing oil and refrigerant around the rings of the compressor's cylinders to be pumped into the discharge line.

Oil separators remove oil from the compressor discharge gas, temporarily store the oil, and then return it to the compressor crankcase. Oil separators are located close to the compressor in the discharge line, Figure 3-21. Even though most oil separators are designed to be mounted vertically, there are some horizontal models available on the market. Oil separators are essential on low and ultra-low temperature refrigeration systems and on large air conditioning systems up to 150 tons. Most compressor manufacturers require oil separators on all two-stage compressors. Oil separators can also act as discharge mufflers to quiet compressor pulsation and vibration noises.

Oil from the compressor crankcase may be rapidly removed due to some unusual condition that is beyond the control of both the designer and installer. The velocity of the refrigerant flowing through the system should return oil to the compressor crankcase. Even though proper refrigerant system piping designs maintain enough refrigerant velocity to ensure good oil return, sometimes this added pressure drop, which assists in getting the right refrigerant velocity for oil return, hampers system efficiency. Many times a higher than normal pressure drop is intentionally designed into a system

for better oil return. This causes higher compression ratios and lower volumetric efficiencies, which lead to lower capacities.

Oil that gets past the compressor and into the system not only robs the compressor crankcase of vital lubrication, but it also coats the walls of the condenser and evaporator. Oil films on the walls of these important heat exchangers will reduce heat transfer. An oil-coated condenser will not be able to reject heat efficiently. Even though this oil film will be hotter and thinner than in the evaporator, system efficiencies will suffer. Head pressures will rise, causing higher compression ratios and lower volumetric efficiencies with lower than normal system capacities.

Oil that coats the walls of the evaporator will decrease heat transfer to the refrigerant in the evaporator. A film of oil bubbles, which acts as a very good insulator, will form on the inside of the evaporator. The evaporator will see a reduced heat load, which will cause the suction pressure to be lower. Lower suction pressures cause higher compression ratios and lower

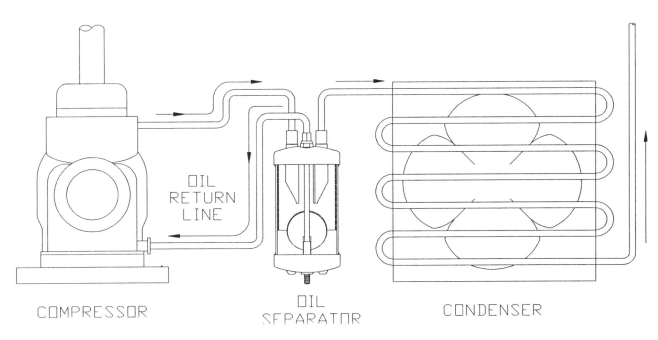

Figure 3-21. Oil separator and location (Courtesy, AC&R Components, Inc.)

volumetric efficiencies. The result is lower system capacity with much longer running times.

Most control valving, including TXV and capillary tubes, will also experience inefficient performance due to the presence of oil filming. Capillary tubes may experience a wide variation in flow rates. Reduced refrigerant flow rate with higher head pressures and lower suction pressures will usually occur. TXV remote bulbs may not sense the correct refrigerant temperature at the evaporator outlet causing improper superheat control, and TXV hunting may occur.

If an oil separator is not employed, the compressor may see slugs of oil that are returning from the evaporator. Compressor pistons can momentarily pump slugs of liquid oil, which can build tremendous hydraulic forces because of the incompressibility of most liquids. Serious compressor valve and drive gear damage can result.

How Oil Separators Work

As oil laden discharge gas enters the oil separator, its velocity is immediately slowed. This low velocity is the key to good oil separation. The oil mixes with the discharge gas and forms a fog. This refrigerant/oil fog runs into internal baffling, which forces the fog mixture to change direction; at the same time, the baffles cause the fog mixture to slow down rapidly. Very fine oil particles collide with one another and form heavier particles. Fine mesh screens separate the oil and refrigerant even more, causing larger oil droplets to form and drop to the bottom of the separator. The oil collects in a sump at the bottom of the separator. A magnet is often connected to the bottom of the oil sump to collect any metallic particles.

When the level of oil becomes high enough to raise a float, an oil return needle is opened, and the oil is returned to the compressor crankcase through a small return line. The pressure difference between the high and low sides of the refrigeration or air conditioning system is the driving force for the oil to travel from the oil separator to the crankcase. The oil separator is

in the high side of the system and the compressor crankcase in the low side. The float-operated oil return needle valve is located high enough in the oil sump to allow clean oil to automatically return to the compressor crankcase. Only a small amount of oil is needed to actuate the float mechanism, which ensures that only a small amount of oil is ever absent from the compressor crankcase at any given time. When the oil level in the sump of the oil separator drops to a certain level, the float forces the needle valve closed.

Helical Oil Separators

Helical oil separators, Figure 3-22, offer 99% to 100% efficiency in oil separation with low pressure drop. Upon entering the oil separator, the refrigerant gas and oil fog mixture encounters the leading edge of a helical flighting, Figure 3-23. The gas/oil mixture is centrifugally forced along the spiral path of the helix, causing heavier oil particles to spin to the perimeter where impingement with a screen layer occurs.

Figure 3-22. Helical oil separator (Courtesy, AC&R Components, Inc.)

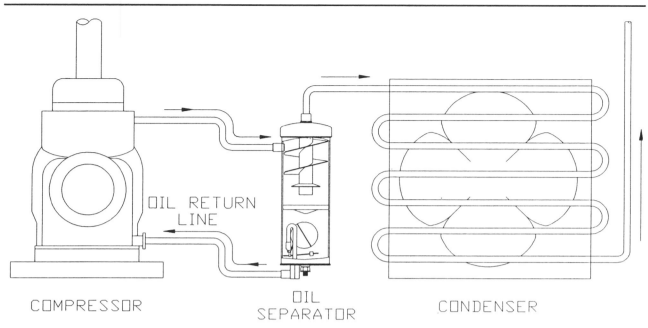

Figure 3-23. Helical oil separator showing helical flightings (Courtesy, AC&R Components, Inc.)

This screen layer serves as an oil stripping and draining medium. The separated oil now flows downward along the boundary of the shell through a baffle and into an oil collection area at the bottom of the separator. The specially designed baffle isolates the oil collection and eliminates oil re-entrainment by preventing turbulence. Virtually oil-free refrigerant gas exits the separator through a fitting just below the lower edge of the helical flighting. A float activated oil return valve allows the captured oil to return to the compressor crankcase or oil reservoir.

Selection

Although oil separator catalogs show capacity in tons or horsepower, the actual tonnage or Btu capacity may vary widely from the horsepower size of the compressor. Actual capacity of compressors is dependent on suction pressures, discharge pressures, liquid temperatures, rpm, and density of the suction gases.

The larger the capacity of the compressor, the larger the separator volume must be, regardless of piping size connections of the separator. The separator must be large enough to match the compressor, and the connection sizes must be the same, or larger, than the discharge line size of the system. This allows the discharge gases in the separator to be near the same pressure as the discharge line because of minimal pressure drop within the oil separator. Do not ever undersize an oil separator; it will lose its ability to return oil and cause high pressure drop resulting in system inefficiencies.

When adding an oil separator to an existing system, care must be taken to check the system frequently after start-up. Gallons of oil may be trapped in a system and gradually return to the compressor crankcase and overfill it.

Installation

The oil separator should be rigidly supported by a solid surface and never supported by the discharge line alone. Its own weight plus the weight of the oil in the separator will stress the discharge line and may cause failure. A vibration eliminator should always be installed in the discharge line before the oil separator. The separator should then be supported by a solid surface such as a concrete wall or floor, steel post, or to the same base as the compressor. Simple lead anchors and bolts are not good for vibrating equipment, because they usually work loose. Studs should be anchored in the floor, and special anchors guaranteed and designed to handle vibration should be used on an existing floor.

On new installations, the compressor and separator should be mounted in their final position, then the interconnecting piping installed while making sure alignment is perfect. There should not be any long, vertical discharge line. If there is a vertical discharge line of more than 10 to 12 feet, it is a good practice to install a drop leg trap to the floor just before the riser. This catches any oil left in the vertical discharge line and prevents it from draining back to the compressor discharge.

The oil return line, which runs from the oil separator to the compressor crankcase, is usually 3/8 SAE flare. There is a chance that some

liquid refrigerant may condense in the separator during the off cycle. Do not return this mixture directly back to the compressor crankcase. If the float and needle assembly in the separator sticks open, hot gas will blow back directly into the crankcase and cause severe overheating of the oil. In semi-hermetic compressors, oil normally returns to the compressor with the suction gas through the end bell of the compressor, Figure 3-24. This gas must flow through the motor compartment and an oil return check valve in the crankcase. This helps vaporize any condensed refrigerant in the oil before it reaches the crankcase. The tap on the compressor suction housing makes an excellent return point for oil coming from an oil separator. Any liquid refrigerant mixed with the oil will be vaporized by the motor heat, and any hot gas blown back will mix with cool suction gas. This reduces the

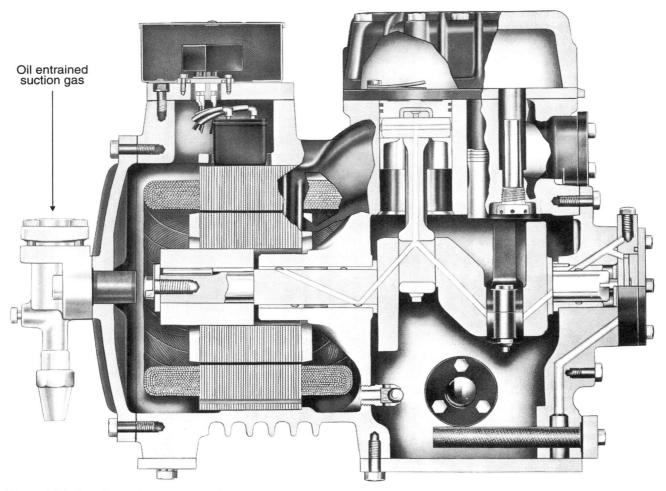

Oil entrained suction gas

Figure 3-24. Semi-hermetic compressor showing suction gas entering end bell of compressor (Courtesy, Copeland Corporation)

chance of harming the compressor more so than if an oil fill hole is used directly to the crankcase.

Some oil separator manufacturers can provide a wrap-around heater of about 50 watts to help prevent refrigerant from condensing in the oil separator during the off cycle. Most oil separators must also be insulated to keep them hot during the on and off cycles. This prevents refrigerant from condensing and mixing with the oil in their sumps.

Parallel Compressor Systems

Parallel compressor systems apply two or more compressors to a common suction header and a common discharge header. Parallel compressor systems provide diversification, flexibility, and higher efficiencies, which lead to lower operating costs. However, parallel compressors have always been a problem when it comes to maintaining correct oil levels in each of the compressors.

As the load varies, the compressors are turned on and off in response to the ever changing load. Because of this, oil return is not equal to each compressor; some compressors are overfilled with oil and some are underfilled. Even a slight difference in pumping rate between likesized compressors or significant differences in pumping rates caused by uneven parallel systems (compressors of different sizes) will cause uneven oil return rates to the compressors involved.

An answer to the problem is the use of a large oil separator in the common discharge line to trap the pumped oil, Figure 3-25. The oil is not returned directly from the separator to the compressors, but pumped by differential pressure to a reservoir. From the reservoir, the oil passes to a float valve, which is mounted on each compressor at its sightglass location, Figure 3-26. The float valve opens and closes as necessary to ensure an adequate oil level in the compressor crankcase.

The oil system consists of three major components: the *oil separator*, *oil reservoir*, and *oil level regulator*. Figure 3-27 shows a reservoir pressure valve located on top of the oil reservoir. This valve maintains a set pressure difference between reservoir pressure and suction pressure. Depending on the system, the differential pressure is usually about 5 psi. This allows the float valve orifices to be calibrated to the 5 psi pressure differential over suction pressure and not have excessive inlet pressures to overfill the compressors before the float valve can react. This also prevents spraying oil, oil foam, and agitation in the crankcase.

One manufacturer has an electronic oil level controller that functions with transducers on each compressor sightglass and offers low level control and low level alarm, as well as a panel mounting as an option, Figure 3-28.

Oil Traps

Systems without good oil separators or with no separator at all will often have oil build-up points throughout the piping. The fine oil mist that travels with the refrigerant usually builds up and collects at low points in the system piping, Figure 3-29. Most of the time, the refrigerant vapor pushes this slug of oil to the next low spot until the oil finally returns to the compressor. Large oil slugs returning to the compressor can damage valve plates and internal running gear if hydraulic pressure is too great in the cylinders. Oil traps prevent large oil slugs from forming by trapping a small amount of oil in the trap and returning it to the compressor gradually.

Refrigerant velocity carries oil throughout a refrigeration and air conditioning system. While oil clings to the sidewalls of the piping, refrigerant gas velocity sweeps small oil particles away in suspension. As oil collects in the trap, Figure 3-30, there is a restriction of the internal pipe area. This causes the refrigerant gas to travel faster as it passes by the oil in the trap. It is this higher velocity refrigerant that picks up minute surface droplets of oil out of the

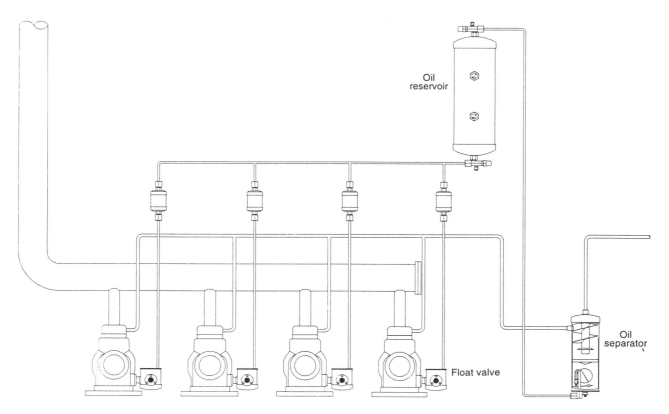

Figure 3-25. Typical parallel compressor system (Courtesy, AC&R Components, Inc.)

Figure 3-26. Oil float valve (Courtesy, AC&R Components, Inc.)

trap. The small oil droplets are then carried either to the compressor or to the next trap. Oil traps are constantly being fed and depleted of oil at a gradual rate. It is this gradual rate that protects compressors from large oil slugs. It is recommended that oil traps be installed in all vapor-carrying lines in the system to ensure proper oil return.

Risers

Risers are upward runs of piping, which cause the refrigerant to flow upward. If there is oil mixed in with the refrigerant, it will have a tendency to fall back to the bottom of the riser because of the long climb to the top of the riser. This is why there should be a trap at the bottom of the riser. There should also be an oil trap every 15 to 20 feet of riser to help temporarily store the oil as it makes its way to the top of the riser, Figure 3-31. This is because the refrigerant/oil mixture gradually loses velocity as it climbs the riser.

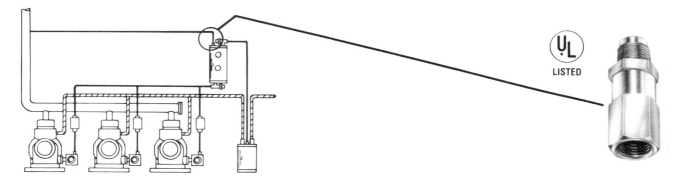

Figure 3-27. Oil reservoir pressure valve (Courtesy, AC&R Components, Inc.)

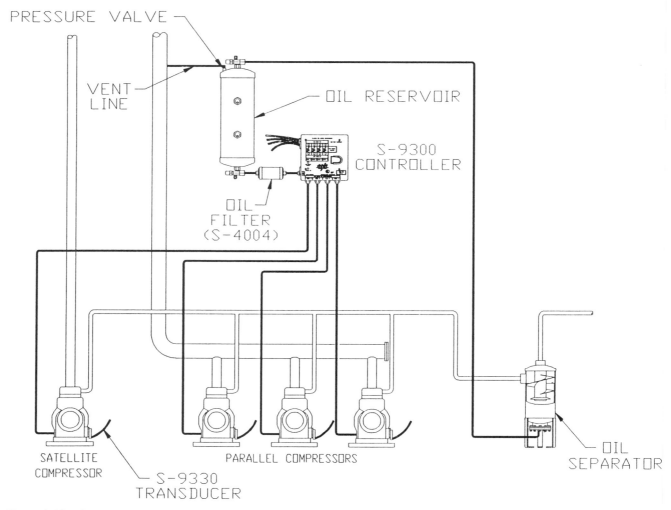

Figure 3-28. Electronic oil level controller (Courtesy, AC&R Components, Inc.)

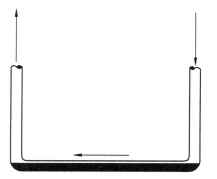

Figure 3-29. Oil collecting at system low points

Figure 3-30. Typical oil trap

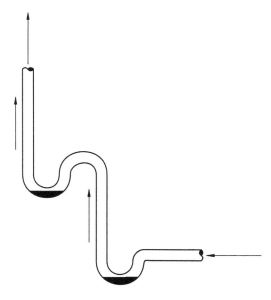

Figure 3-31. Riser with oil traps

COMPRESSOR PERFORMANCE CURVES

A compressor performance curve provides the service technician with the Btu pumping rate (capacity), operating amperage, and operating wattage, when the suction pressure, condensing pressures, total subcooling, and total superheat are known for the system. Compressor curves are available from the compressor manufacturer for each of their compressors. Compressor curves may be either in graph type or table type, Figures 3-32 and 3-33.

Graph Compressor Curve

Figure 3-32 is a graph-type compressor curve. Its left side axis is the capacity values in Btu per hour (Btuh) multiplied by 1000. Any horizontal lines drawn from the left side capacity axis are lines of constant capacity. On the right side vertical axis are condensing pressures in psig and the equivalent saturated condensing temperatures in °F. The sloping lines going down from the right side axis are lines of constant condensing pressure.

Example 3-3

A technician measures the head pressure to be 283 psig and the suction pressure to be 15.3 psig. Using Figure 3-32, what capacity should the R-502 Copeland compressor (9RS3-0765-TFC) be putting out at these two pressures?

Solution 3-3

Using Figure 3-32, draw a line straight up from the bottom suction pressure axis at 15.3 psig (-20°F) until it intersects the 283 psig (120°F) condensing pressure curve. From this intersection, draw a horizontal line to the left until it intersects the vertical (capacity) axis. The capacity of the compressor is 33,600 Btuh. The compressor is circulating enough R-502 through the system for about 33,600 Btuh of cooling.

Curve Ratings

The answer obtained in Solution 3-3 is only a ballpark figure. Looking at the curve closer, notice the set of conditions or ratings that accompany the capacity curves. The ratings are 65°F return gas, 0°F total liquid subcooling, 95°F ambient, and 60 Hz electrical power. The

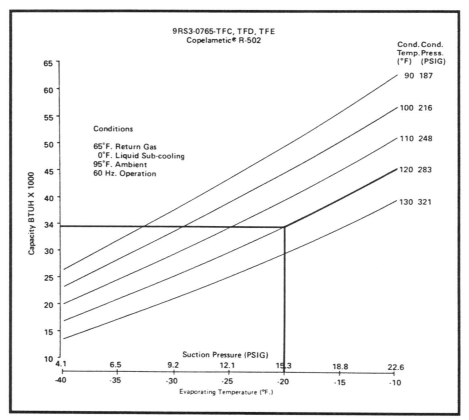

Figure 3-32. *Graph-type compressor performance curve (Courtesy, Copeland Corporation)*

exact capacities on the curves can only be used if the system has these ratings. These ratings are called the *nominal* or *rated* capacities. The Air Conditioning and Refrigeration Institute (ARI) makes these ratings so that compressors can be compared to one another easily and fairly.

In the field, it would be harmful to system capacity to have 0°F of subcooling because of the flash gas that would form in the liquid line as soon as any pressure drop occurred. A 65°F return gas temperature with a -20°F evaporating temperature would give 85°F of compressor (total) superheat, which would overheat compressors and decrease capacity. More realistic values of compressor superheat are 20° to 30°F to let the compressor enjoy a long healthy life. It is necessary to know the total liquid subcooling and total superheat to obtain the capacity, as well as two "rules of thumb:"

1. There is an approximate 1/2% change in capacity for every 1°F of liquid subcooling change. As subcooling increases, capacity increases.

2. There is an approximate 1% change in capacity for every 10°F of total superheat change. As total superheat increases, capacity decreases.

Assume 20°F of total subcooling and a return gas temperature of 15°F. The system would then have the following measurements:

*   Condensing pressure = 283 psig (120°F)

*   Suction pressure = 15.3 psig (-20°F)

*   Total subcooling = 20°F

*   Return gas temperature = 15°F

*   Total superheat = (15° - -20°F) = 35°F

- Nominal or rated capacity from curves = 33,600 Btuh

The actual capacity can now be calculated using the rules of thumb for subcooling and superheat. The curves are based on 0°F of subcooling and there are 20°F of subcooling, so there will be a 10% (1/2% x 20°F) increase in capacity with the added subcooling. Using Rule #1, this increases the capacity to 36,960 Btuh (33,600 increased by 10%).

The curves are also based on a 65°F return gas temperature, which would give 85°F of total superheat at a -20°F evaporator temperature. The 15°F return gas temperature will give 35°F of total superheat at a -20°F evaporating temperature. This causes a 50°F (85° - 35°F) decrease in total superheat. Using Rule #2, this increases the capacity 5% (50°F ÷ 10°F). The new capacity is now 38,808 Btuh (36,960 increased by 5%). The more correct answer for Example 3-3 is an actual compressor capacity of 38,808 Btuh. Note that this is much higher than the nominal or rated capacity of 33,600 Btuh because of an increase in liquid subcooling and a decrease in total superheat. In addition, any increase in head pressure will decrease capacity and any lowering of suction pressure will decrease capacity. (For a quick review of how head pressure and suction pressure affect compressor capacity, review the compression ratio section.)

These "rules of thumb" are accurate enough for the service technician to use as a tool when servicing compressor systems. However, if more accurate data is to be used in an engineering atmosphere, pressure/enthalpy diagrams must be plotted to figure the difference in capacity when subcooling and superheat values do not match nominal capacities.

Table Compressor Curve

Compressor capacity data can also be expressed in table form. The compressor shown in Figure 3-33 is a 10 hp, semi-hermetic, low temperature, Discus model using R-22. The upper left corner of the table gives the rated conditions. The table gives capacity data, power (watts),

current (amps), and evaporator mass flow rate. To use Figure 3-33, determine the nominal capacity for this compressor, which is running at a 90°F condensing temperature and a 0°F evaporating temperature. Simply draw a horizontal line from the vertical condensing temperature column at 90°F until it intersects a line drawn down from the horizontal evaporating temperature column at 0°F. The intersection of these lines in Figure 3-33 shows a nominal capacity of 93,400 Btuh. This figure can now be corrected for subcooling and superheat values if they are known.

The same method can be used for watts and amps. For example, using Figure 3-33, determine the amp draw for the same compressor with a 0°F evaporating temperature and a 90°F condensing temperature. The intersection of 90°F condensing temperature with a 0°F evaporating temperature results in an amp draw of 31.6 amps. Using this same method for watts, the watt draw is 10,000 watts at these same temperatures.

Compressor Specifications

Included with compressor curves are compressor specifications and system application data, Figure 3-34. This specification sheet includes: evaporator temperature range applications, refrigerant, compressor cooling, capacity, watts, amps, rated conditions, bore and stroke, displacement, rpm, suction and discharge connections, voltage ranges, nameplate ratings, oil charges, viscosity and grade of oil, and compressor weight.

Scaled drawings are also included with a set of compressor specifications and performance curves, as shown in Figure 3-34. Figure 3-35 is an example of compressor capacity, amps, and watts in curve form instead of table form for an R-502 compressor model.

COMPRESSOR EFFICIENCY

Many service technicians believe that front seating the suction service valve on a compres-

CONDITIONS
65°F Return Gas
0°F Liquid Sub Cooling
95°F Ambient Air Over
60 Hz Operation
Use only with Demand Cooling ™ Installed

LOW TEMPERATURE

TENTATIVE DATA
2380 CFH   39.2 CIR
Outline Drawing 90-58
Wiring Diagram 90-60

4DA3-1000-TSK/TSE
COPELAMETIC® R-22
DISCUS-COMPRESSOR

208/230/460-3-60
575-3-60

Cond. Temp.	EVAPORATOR CAPACITY (BTU/HR)					Evaporating Temperature			
Degrees F	-40	-35	-30	-25	-20	-15	-10	-5	0
(Degrees C)	(-40)	(-37.2)	(-34.4)	(-31.7)	(-28.9)	(-26.1)	(-23.3)	(-20.6)	(-17.8)
70 (21.1)	32800	39200	46200	53800	62100	71100	80800	91400	103000
80 (26.7)	29100	35400	42300	49800	58000	66800	76500	87000	98300
90 (32.2)	25400	31600	38400	45700	53700	62500	71900	82200	93400
100 (37.8)	21500	27600	34200	41400	49200	57700	67000	77100	88100
110 (43.3)	17400	23200	29600	36600	44200	52500	61500	71400	82100
120 (48.9)	12800	18400	24500	31200	38600	46600	55400	64900	75400
130 (54.4)	7490	12800	18700	25100	32100	39900	48300	57600	67700

Cond. Temp.	POWER (WATTS)					Evaporating Temperature			
Degrees F	-40	-35	-30	-25	-20	-15	-10	-5	0
(Degrees C)	(-40)	(-37.2)	(-34.4)	(-31.7)	(-28.9)	(-26.1)	(-23.3)	(-20.6)	(-17.8)
70 (21.1)	5520	6000	6420	6800	7140	7480	7800	8140	8510
80 (26.7)	5630	6200	6710	7170	7590	7990	8390	8780	9200
90 (32.2)	5720	6400	7010	7570	8090	8570	9050	9520	10000
100 (37.8)	5730	6530	7260	7930	8550	9140	9720	10300	10900
110 (43.3)	5570	6510	7380	8170	8920	9630	10300	11000	11700
120 (48.9)	5160	6260	7280	8220	9100	9940	10800	11500	12300
130 (54.4)	4440	5710	6880	7990	9020	10000	11000	11900	12800

Cond. Temp.	CURRENT (AMPS)					@ 230 Volts (Multiply Amps by 0.5 for 460 Volts)			
Degrees F	-40	-35	-30	-25	-20	-15	-10	-5	0
(Degrees C)	(-40)	(-37.2)	(-34.4)	(-31.7)	(-28.9)	(-26.1)	(-23.3)	(-20.6)	(-17.8)
70 (21.1)	22.0	22.8	23.8	24.4	25.2	25.8	26.6	27.4	28.0
80 (26.7)	22.2	23.2	24.2	25.2	26.0	27.0	27.8	28.8	29.6
90 (32.2)	22.4	23.6	24.8	26.0	27.2	28.2	29.2	30.4	31.6
100 (37.8)	22.4	24.0	25.4	26.8	28.2	29.4	30.8	32.2	33.4
110 (43.3)	22.2	24.0	25.6	27.4	29.0	30.6	32.2	33.8	35.4
120 (48.9)	21.4	23.4	25.6	27.4	29.4	31.2	33.2	35.0	37.0
130 (54.4)	19.9	22.4	24.8	27.0	29.2	31.6	33.8	36.0	38.4

Cond. Temp.	EVAPORATOR MASS FLOW (LBS/HR)					Evaporating Temperature			
Degrees F	-40	-35	-30	-25	-20	-15	-10	-5	0
(Degrees C)	(-40)	(-37.2)	(-34.4)	(-31.7)	(-28.9)	(-26.1)	(-23.3)	(-20.6)	(-17.8)
70 (21.1)	382	458	540	629	727	834	950	1080	1210
80 (26.7)	351	428	512	604	704	813	932	1060	1200
90 (32.2)	318	397	482	576	677	789	910	1040	1190
100 (37.8)	281	361	447	542	646	759	883	1020	1170
110 (43.3)	237	317	404	500	605	720	846	984	1130
120 (48.9)	182	262	350	447	553	669	797	937	1090
130 (54.4)	112	192	280	377	484	601	731	873	1030

Nominal performance values (±5%) based on 1750 RPM on 60 Hz models. Subject to change without notice. Refer to AE-4-1287 for additional information.

Figure 3-33. Table-type compressor performance curve (Courtesy, Copeland Corporation)

sor to see how far into a vacuum the compressor will pull is a valid check for compressor efficiency. This is simply not true! When the compressor suction valve is closed while running, the technician is actually unloading the compressor, not loading it. While in a vacuum, the compressor sees very little refrigerant, and its amp draw is very low. There is no load on the compressor. Compressors are not designed to run in a vacuum. With the exception of a few low temperature R-12 systems, no single stage compressor used in refrigeration applications operates in a vacuum. This includes applications operating as low as -40°F evaporating temperature. This is why many compressor manufacturers will not guarantee their compressors will pull down to a deep vacuum. Pulling the compressor into a deep vacuum may cause damage from insufficient cooling if they are refrigerant cooled. These deep vacuums will put the compressors right off of the suction pressure operating range of their curves.

COPELAMETIC® R-12 MOTOR-COMPRESSOR

9RA1-0500-TFC
9RA1-0500-TFD
9RA1-0500-TFE

208/230-3-60	200/220-3-50
460-3-60	380/400-3-50
575-3-60	500-3-50

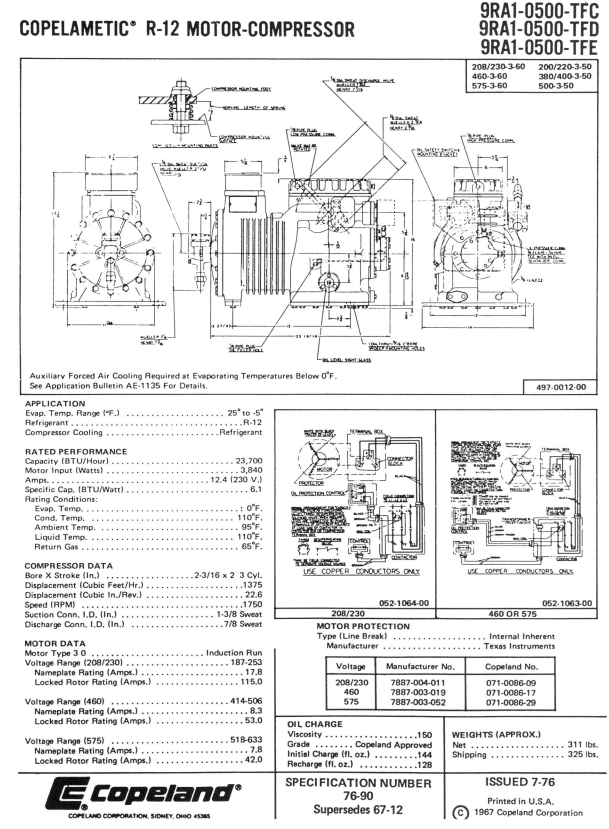

Auxiliary Forced Air Cooling Required at Evaporating Temperatures Below 0°F.
See Application Bulletin AE-1135 For Details.

497-0012-00

APPLICATION
Evap. Temp. Range (°F.) . . . . 25° to -5°
Refrigerant R-12
Compressor Cooling . . . . Refrigerant

RATED PERFORMANCE
Capacity (BTU/Hour) 23,700
Motor Input (Watts) 3,840
Amps. . . . . 12.4 (230 V.)
Specific Cap. (BTU/Watt) 6.1
Rating Conditions:
  Evap. Temp. . . . . : 0°F.
  Cond. Temp. . . . . 110°F.
  Ambient Temp. . . . . 95°F.
  Liquid Temp. . . . . 110°F.
  Return Gas . . . . 65°F.

COMPRESSOR DATA
Bore X Stroke (In.) . . . . 2-3/16 x 2 3 Cyl.
Displacement (Cubic Feet/Hr.) 1375
Displacement (Cubic In./Rev.) 22.6
Speed (RPM) 1750
Suction Conn. I.D. (In.) . . . . 1-3/8 Sweat
Discharge Conn. I.D. (In.) . . . . 7/8 Sweat

MOTOR DATA
Motor Type 30 . . . . Induction Run
Voltage Range (208/230) 187-253
  Nameplate Rating (Amps.) 17.8
  Locked Rotor Rating (Amps.) 115.0

Voltage Range (460) 414-506
  Nameplate Rating (Amps.) 8.3
  Locked Rotor Rating (Amps.) 53.0

Voltage Range (575) 518-633
  Nameplate Rating (Amps.) 7.8
  Locked Rotor Rating (Amps.) 42.0

052-1064-00	052-1063-00
208/230	460 OR 575

USE COPPER CONDUCTORS ONLY

MOTOR PROTECTION
Type (Line Break) . . . . Internal Inherent
Manufacturer . . . . Texas Instruments

Voltage	Manufacturer No.	Copeland No.
208/230	7887-004-011	071-0086-09
460	7887-003-019	071-0086-17
575	7887-003-052	071-0086-29

OIL CHARGE
Viscosity 150
Grade . . . . Copeland Approved
Initial Charge (fl. oz.) 144
Recharge (fl. oz.) 128

WEIGHTS (APPROX.)
Net . . . . 311 lbs.
Shipping . . . . 325 lbs.

COPELAND®
COPELAND CORPORATION, SIDNEY, OHIO 45365

SPECIFICATION NUMBER
76-90
Supersedes 67-12

ISSUED 7-76
Printed in U.S.A.
© 1967 Copeland Corporation

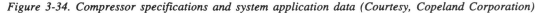

Figure 3-34. Compressor specifications and system application data (Courtesy, Copeland Corporation)

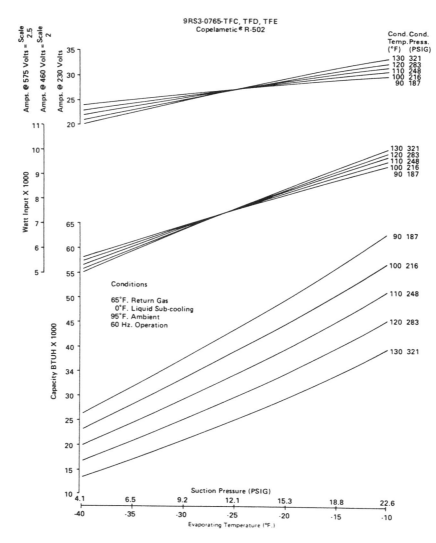

9RS3-0765-TFC, TFD, TFE
Copelametic® R-502

Cond. Cond.
Temp. Press.
(°F) (PSIG)

Conditions

65°F. Return Gas
0°F. Liquid Sub-cooling
95°F. Ambient
60 Hz. Operation

Figure 3-35. Compressor capacity, amp, and wattage curve (Courtesy, Copeland Corporation)

Front seating the suction service valve can be used to see if the discharge valves are seating properly; however, pulling the compressor to 4 or 5 psig will suffice for these tests. If at least one piston is operating and the valves are in place, the compressor will pull down. No person and no compressor curve can tell the technician how long it will take. If the discharge valves are in place and properly seated, the pressures will hold. Pulling the compressor suction pressure down to some level does have value for a basic valve test, but it is not a test for compressor efficiency.

The only sure way to check the efficiency of an operating compressor is to check its operating amp draw against the compressor amp draw curve, Figure 3-36. The horizontal or bottom axis of this curve is the suction pressure in psig or evaporating temperature in °F. The top vertical axis is condensing pressure (discharge pressure) in psig or condensing temperature in °F. The lines of constant discharge pressure slope down and to the left. The left vertical axis is the amp scale. These amps are rated at a specific voltage. The curves' amp values must be corrected for the actual voltage measured at the compressor terminals.

Assume again that a compressor has a suction pressure of 15.3 psig (-20°F) and a condensing pressure of 283 psig (120°F). In Figure 3-36,

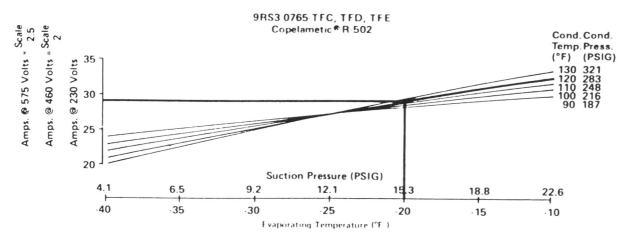

Figure 3-36. Compressor amp curve (Courtesy, Copeland Corporation)

enter the curve at 15.3 psig and draw a straight vertical line to where it intersects the condensing pressure of 283 psig. At this intersection, move horizontally to the left vertical amp axis. This amp reading will read 29 amps at 230 volts. If the voltage measured at the compressor terminals is 230 volts, then an ammeter should measure 29 amps (±5%). If the measured value is between 27.5 and 30.5 amps, the compressor is running efficiently. Using compressor amp curves or tables is the only method to check if a compressor is running efficiently.

If valves or motor parts are malfunctioning or the system is overcharged or undercharged, the amp reading will be higher or lower depending on the situation. If, for example, the discharge and/or suction valves of this compressor were not seated properly, there would be a very low head pressure and a high suction pressure. The only way a compressor could have low head pressure and high suction pressure is if either the suction and/or discharge valves were bad or not seating or if the piston rings were badly worn. This situation would cause a short cycling of refrigerant within the compressor cylinders, contributing to a low amp draw. The flow of refrigerant through the compressor and the system would be reduced, reducing both compressor work and amp draw. If suction pressure rose to 22.6 psig and the condensing pressure

dropped to 187 psig, the amp draw should be about 30 amps, according to Figure 3-36. However, if the amp draw of the actual compressor is only 20 amps, the amp draw of the compressor is less than the compressor amp curve, indicating an inefficient compressor.

Motor problems may also cause amp draws that do not match curve amp ratings. Motor related problems will often cause the amp draw of the compressor to be higher than the compressor curve amp draw. In either case, there is a problem.

Overcharging a system will also cause the actual measured compressor amps to be higher than that of the compressor curve amps. Undercharging will cause the actual measured compressor amps to be lower than that of the compressor curve amps.

Voltage Variations

In the examples given earlier, if the voltages measured at the compressor terminals were not exactly 230 volts, the following formula would correct for voltage differences:

$$(Vc)\ (Ac)\ =\ (Vm)\ (Am)$$

Where: V_c = voltage from the curve (230 volts)
        A_c = amperage from the curve
        V_m = volts measured
        A_m = amps measured

If the voltage measured at the compressor is 245 volts and the ammeter reads 26 amps, is the compressor running efficiently? Solve for amps measured (A_m):

$$A_m = \frac{(V_c)(A_c)}{V_m}$$

$$A_m = \frac{(230)(29)}{245} = 27.22$$

The ammeter measures 26 amps, which is well within the (±5%) of 27.22 amps. Therefore, the compressor is running within specification.

CHAPTER FOUR

Metering Devices

Metering devices separate the high and low side pressures of a refrigeration and/or air conditioning system by adding restriction to the system. Metering devices meter liquid refrigerant from the liquid line to the evaporator, and some metering devices control evaporator superheat, and/or pressures, and even prevent evaporator pressures from exceding maximum set pressures. The most common types of metering devices are capillary tubes, thermostatic expansion valves, and automatic expansion valves.

CAPILLARY TUBES

A capillary tube is a long fixed-length tube with a very small diameter, which varies from .025" to .060". The capillary tube is installed between the condenser and the evaporator, Figure 4-1, and meters the refrigerant from the condenser to the evaporator. Because of its long length and small diameter, there is an associated large friction loss as refrigerant flows through the tube. As subcooled liquid travels through the capillary tube, it may flash because pressure drops may bring it lower than the saturation pressure for its temperature. This flashing is caused by the expansion of the liquid as it experiences pressure drop. Liquid flashing amounts, if any, depend on the amounts of liquid subcooling coming from the condenser and in the capillary tube itself. If liquid flashing does occur, it is desirable to keep the flashing as close to the evaporator as possible to ensure better system performance. To prevent liquid flashing, capillary tubes are usually twisted around, placed inside, or soldered to suction lines, Figure 4-2. Because the capillary tube restricts and meters the flow of liquid to the evaporator, it helps maintain the needed pressure difference for proper system operation.

Manufacturers use capillary tubes as metering devices because of their simplicity and low cost. Systems employing capillary tubes as metering devices also do not require receivers, which is another cost savings. Capillary tube systems are found mainly in domestic and small commercial applications in which hermetic compressors are employed, and somewhat constant loads and small refrigerant flow rates are experienced.

The capillary tube has no moving parts and does not control evaporator superheat at all loading conditions, Figure 4-3. Even with no moving parts, the capillary tube varies the flow rate as system pressures change in the evaporator and condenser, or both. In fact, it can only reach its best efficiency at one set of high and low side pressures. This is because the capillary tube works off of the pressure difference between the high and low sides of the refrigeration system. As the pressure difference between the high and low side of the system becomes greater, the flow rate of refrigerant increases. The capillary tube operates satisfactorily over quite a large range of pressure differences but not very efficiently.

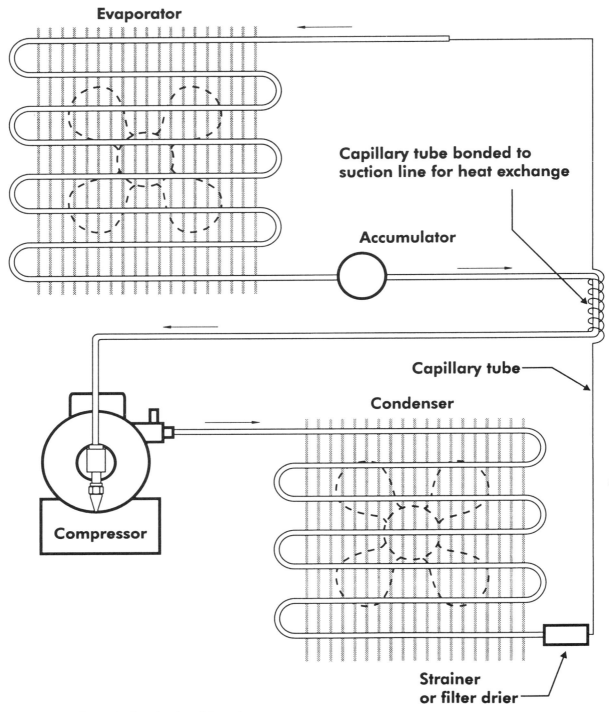

Figure 4-1. Capillary tube location in refrigeration system

Sizing

Since the capillary tube, evaporator, compressor, and condenser are in series, the flow rate of the capillary tube must be equal to the compressor pumping rate. This is why the designed length and diameter of the capillary tube at designed vaporizing and condensing pressures is critical and must be equal to the compressor pumping capacity at these same design conditions. Too many turns in the capillary tube will affect resistance of flow and system balance.

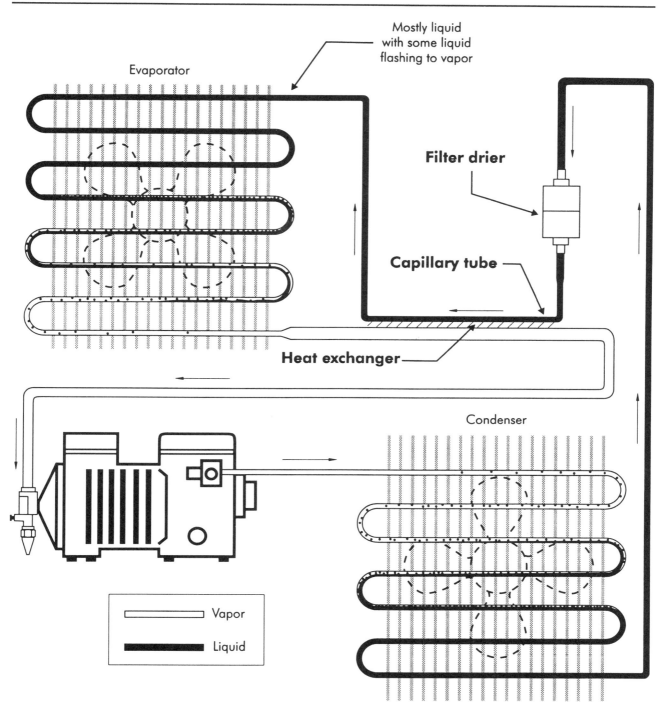

Figure 4-2. Capillary tube/suction line heat exchanger

If the resistance of the capillary tube is too great because it is too long, a partial restriction exists, the diameter is too small, or there are too many turns as it is coiled, the capacity of the tube will be less than that of the compressor. The evaporator will then be starved, causing low suction pressure and high superheats.

At the same time, subcooled liquid will back up in the condenser, causing higher head pressure, because there is no receiver in the system to hold the refrigerant. With the higher head pressure and lower evaporator pressure, the refrigerant flow rate will be increased because of a greater pressure difference across the capil-

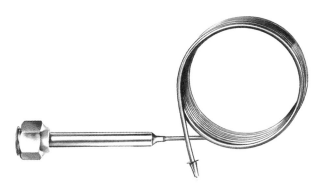

Figure 4-3. Capillary tube with built-in strainer (Courtesy, Refrigeration Research)

lary tube. At the same time, the compressor capacity will decrease because of higher compression ratios and lower volumetric efficiencies. This will cause the system to establish a balance but at higher head pressures and lower evaporating pressures, causing unwanted inefficiencies.

If the resistance of the capillary tube is less than called for because it is too short or its diameter is too large, refrigerant flow will be greater than the pumping capacity of the compressors. This will cause high evaporator pressures, low superheats, and possible flooding of the compressor due to overfeeding of the evaporator. In addition, subcooling will drop in the condenser, causing low head pressures and even a loss of liquid seal at the condenser bottom. This low head pressure and higher than normal evaporator pressures will decrease the compression ratio of the compressor, causing high volumetric efficiencies. This will increase the pumping rate of the compressor, and a balance may be reached if the compressor can keep up with the high refrigerant flow in the evaporator. Usually flooding occurs and the compressor fails to keep up. It is for these reasons that the capillary tube should be sized properly by calling the manufacturer. It is also necessary that capillary tube systems have an exact (critical) charge of refrigerant in their systems. Too much or too little refrigerant will cause a severe imbalance to occur and seriously damage compressors from slugging or flooding.

If a capillary tube system is overcharged, it will back up the excess liquid in the condenser,

causing high head pressures. This is because there is no receiver in the system. The pressure difference between the low and high side of the system will also increase, causing an increased flow rate to the evaporator and overfeeding of the evaporator, which may result in low superheats, flooding, or slugging of the compressor. This is another reason why a capillary tube system must be critically charged with a specified amount of refrigerant.

Capillary tube systems do not stop the flow of refrigerant after the compressor shuts off. Refrigerant will continue to flow from the condenser to the evaporator until high and low pressures equalize. In fact, most of the subcooled liquid in the condenser will flow through the capillary tube to the evaporator during the off cycle, which is why the capillary tube system must be critically charged with refrigerant. Too much liquid in the evaporator during the off cycle can cause compressor damage at start-ups. The refrigerant charge should satisfy the requirements of the evaporator and maintain a subcooled liquid seal at the condenser bottom and capillary tube entrance. If any excess refrigerant is added past the critical charge, subcooled liquid will back up in the condenser, causing high head pressures and unwanted inefficiencies. Because of the pressure equalization during off cycles, capillary tube systems can employ low starting torque motors. These motors usually use current starting relays sometimes without starting capacitors. The critical charge is usually stamped on the manufacturer nameplate, Figure 4-4.

Model #	AC-16345-227
Serial #	L-2091C
Volts 115	HZ = 60
Volts 100	HZ = 50
PH1	LRA 22.0
	Made in USA
Refrigerant	12
Charge	8.6 oz

Figure 4-4. System nameplate showing critical charge amounts

It is important that larger commercial capillary tube systems have accumulators at their evaporator outlets to catch any liquid that may escape vaporization in the evaporator during start-ups. Liquid coming from the evaporator enters the bottom of the accumulator, and the vapors are drawn from the top into the compressor.

High Heat Loads

Under high loads, capillary tube systems will often cause high superheat in the evaporator of 40° or 50°F. This is because the refrigerant in the evaporator will vaporize rapidly and drive the 100% saturated vapor point back up in the evaporator to give the system a high superheat reading, Figure 4-5. The capillary tube does not have a feedback mechanism to tell it there is high superheat, which makes the system run inefficiently at higher loads. This is one of the main disadvantages of a capillary tube system. Many technicians have the urge to add more refrigerant to the system at these high loads because of the high superheat readings. Adding refrigerant at this point in time will only over-charge the system. Before adding refrigerant, check the superheat reading at the low evaporator load and see if it is normal. If unsure, recover the refrigerant, evacuate the system, and add the specified nameplate critical charge of refrigerant.

Low Heat Loads

Once the high heat load on the evaporator drops off and the system is under a low load, the 100% saturated vapor point in the evaporator will climb down the last passes of the evaporator. This is caused by the decreased rate of vaporization of refrigerant in the evaporator from the low heat load. The system will then have a normal evaporator superheat of about 10° to 12°F, Figure 4-6. These normal evaporator superheat readings only occur with low heat loads on the evaporator.

Service

One of the major problems with capillary tube systems is that they can become restricted or completely blocked due to foreign materials, freeze-up if water exists in the system, or kink-

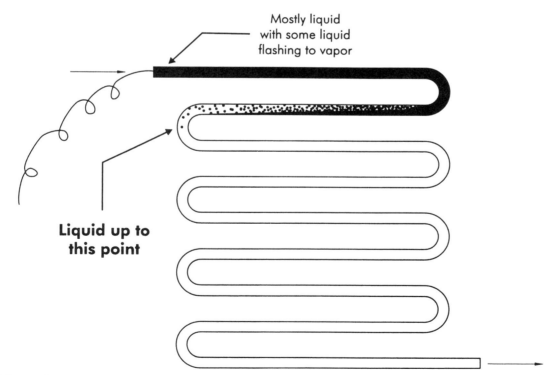

Mostly liquid with some liquid flashing to vapor

Liquid up to this point

Figure 4-5. Capillary tube feeding evaporator under a high heat load

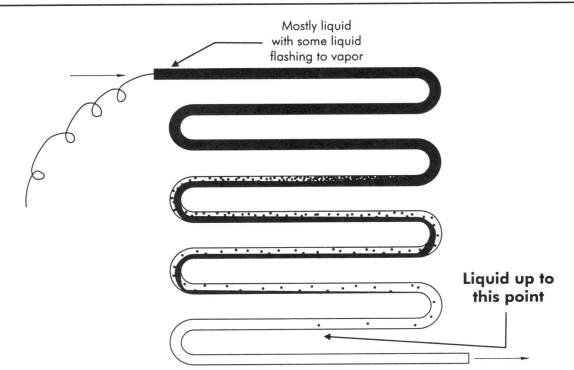

Mostly liquid
with some liquid
flashing to vapor

**Liquid up to
this point**

Figure 4-6. Capillary tube feeding evaporator with proper amount of superheat for low heat load

ing. Dirt, sandpaper grit, steel wool, and sand can all restrict or completely block capillary tubes.

It is extremely important to install a filter drier before the capillary tube to catch any foreign objects. Filter driers also help catch moisture so freeze-up will not occur at the evaporator entrance. High head pressure from backed-up subcooled liquid in the condenser, which will starve the evaporator and cause low pressures in the low side of the system, is one symptom of a partially blocked capillary tube. The symptoms will vary depending on how severe the capillary tube is restricted. A completely plugged capillary tube system will have dangerously high head pressures with evaporator pressures in the deep vacuum range.

Moisture in the system will pass through the condenser and the capillary tube. Once the moisture experiences the sudden pressure drop of the evaporator, it will gradually freeze. This gradual freezing around the inside diameter of the capillary tube will cause a slight restriction and hinder system performance. Head pressures will elevate and evaporator pressures will drop

from the partial restriction. If the moisture problem is severe, this frozen layer will soon cause an entire restriction and shut down the compressor by thermal overload. Once the system is off, the frozen layers will melt and system pressures will equalize and wait for the thermal overload to reset. The system will have to be dehydrated and recharged.

There are hydraulic capillary tube cleaners on the market that can be very successful in removing foreign objects from capillary tubes if used properly, Figure 4-7a. Hydraulic pressure is applied to one end of the capillary tube while the other end of the tube is removed from the system. Some capillary tubes may require a small piece of wire to be hydraulically blown through them in order to clean foreign objects and sludge from their internal diameter. There is a kit available that comes with coils of small wire pistons to clear capillary tubes. This kit also includes a hydraulic cleaner, a file, and a capillary tube diameter gauge, Figure 4-7b. The wire is just undersized enough to act as a piston to clear the capillary tube when used with the hydraulic cleaner.

Some technicians may replace a plugged or restricted capillary tube with a new one. The replacement capillary tube must have the same diameter, length, and number of coils or turns as the old one. There are capillary tube gauges on the market for measuring the right diameter of capillary tube, Figure 4-8. When cutting a capillary tube, filing a small V-notch in the side of the capillary tube with an appropriate file and bending the tube back and forth where the V-notch is cut will make for a nice clean break. Burrs at the ends of capillary tubes will hinder refrigerant flow and cause inefficient system performance.

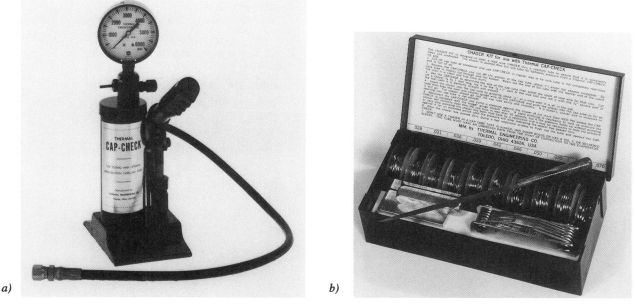

a)   b)

Figure 4-7. a) Hydraulic capillary tube cleaner; b) accessories included in capillary tube cleaner kit (Courtesy, Thermal Engineering Company)

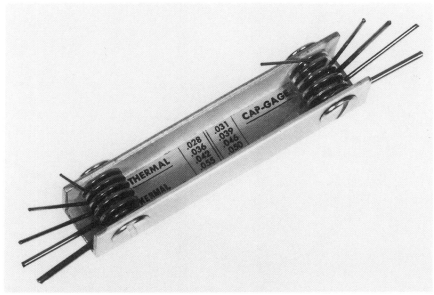

Figure 4-8. Capillary tube feeler gauges (Courtesy, Thermal Engineering Company)

Soldering the capillary tube in place requires certain skills. The capillary tube must be inserted about one or two inches past where the solder joint is to be made, Figure 4-9. This ensures solder will not creep up or down by capillary action and block the inlet or outlet of the capillary tube. A little bit of refrigeration oil smeared on the end of the capillary tube may help stop solder from bonding to the capillary tube end. When crimping the line where the capillary tube is inserted, care must be taken not to crimp or harm the capillary tube itself.

THERMOSTATIC EXPANSION VALVE

The thermostatic expansion valve (TXV or TEV), is probably the most popular and widely used valve in the hvac/r industry today, Figure 4-10. Chapters of material could be written about TXV applications, types, and functions, but this section will cover TXV basics. Figure 4-11 shows an exploded view of a TXV.

The TXV is a metering device located between the liquid line and the evaporator, Figure 4-12. The TXV is a dynamic valve that constantly modulates and meters just enough refrigerant into the evaporator to satisfy all heat loads on the evaporator. The basic TXV controls a set amount of evaporator superheat under varying heat loads. The three main functions of a TXV include:

- providing a constant amount of evaporator superheat under varying load conditions, provided the range and capacity of the valve are not exceeded.

- keeping the entire evaporator full of refrigerant under all heat load conditions.

- preventing floodback of refrigerant to the compressor under varying evaporator heat loads.

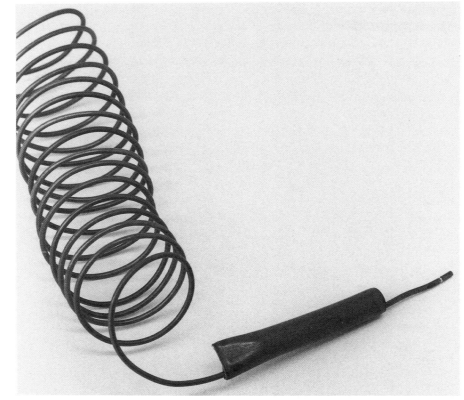

Figure 4-9. Capillary tube inserted two to three inches past solder joint to prevent solder from plugging capillary tube (Photo by Bill Bitzinger, Office of University Communication Services, Ferris State University)

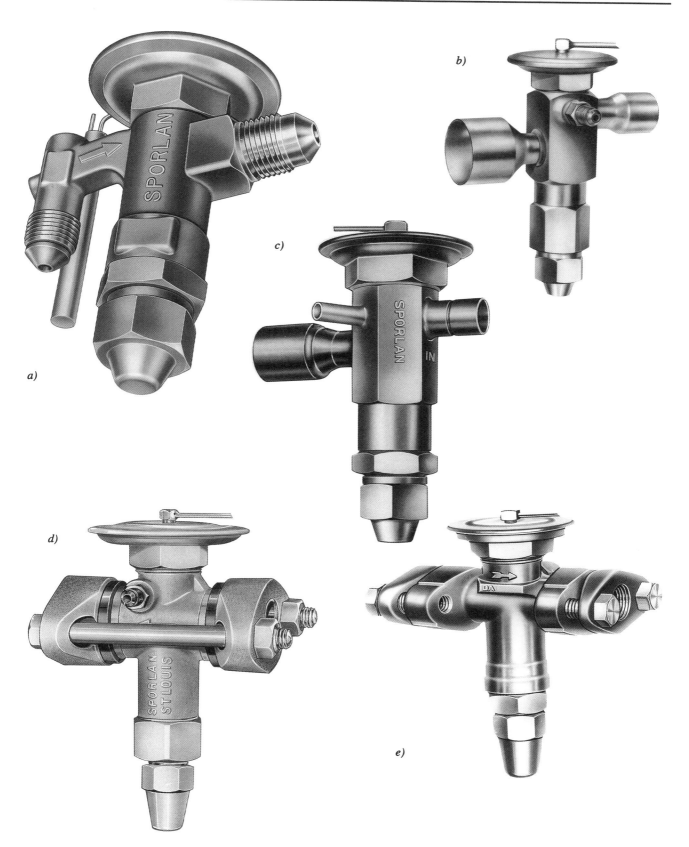

Figure 4-10. Examples of different TXVs (Courtesy, Sporlan Valve Company)

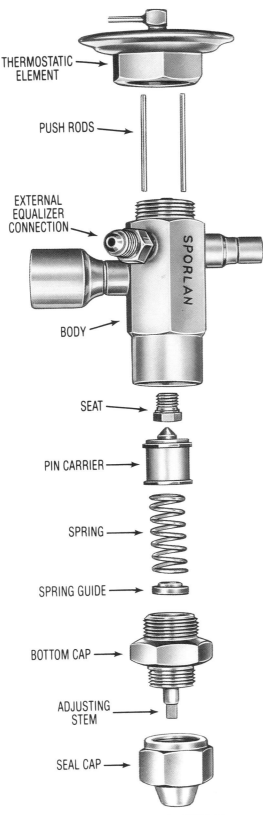

THERMOSTATIC ELEMENT

PUSH RODS

EXTERNAL EQUALIZER CONNECTION

SPORLAN

BODY

SEAT

PIN CARRIER

SPRING

SPRING GUIDE

BOTTOM CAP

ADJUSTING STEM

SEAL CAP

Figure 4-11. Exploded diagram of a TXV (Courtesy, Sporlan Valve Company)

The TXV does not control evaporator pressure, cycle the compressor, control running time, or control box temperature. The main function of the TXV is to meter the right amount of refrigerant into the evaporator coil under all load conditions. By doing this, the valve controls a set amount of evaporator superheat to maintain active evaporators and keep refrigeration and air conditioning systems running safely and efficiently.

One of the main differences between a TXV system and a capillary tube system is the TXV has the ability to control the right amount of evaporator superheat under varying evaporator heat loads. A capillary tube cannot do this and will run very high evaporator superheat at a high evaporator load situation. The TXV will fill the evaporator coil with refrigerant at all loads but will not cause overflow into the suction line if adjusted properly. The action of this dynamic valve is from the interaction of three separate pressures acting on the valve: *remote bulb pressure* (opening force), *spring pressure* (closing force), and *evaporator pressure* (closing force), Figure 4-13.

Remote Bulb Pressure

In Figure 4-13, notice that the pressure (P_1) from the volatile fluid in the remote bulb acts on the top side of the flexible bellows through the connecting capillary tube, causing the valve to open as pressure increases. This changing pressure in the remote bulb is caused by the evaporator outlet temperature becoming hotter or colder from varying heat loads on the evaporator, which momentarily affect evaporator superheat. Higher superheat causes more fluid to vaporize in the remote bulb, producing higher pressures. Less superheat causes more vapor to condense in the remote bulb, decreasing its pressure and closing the valve. As soon as the TXV realizes that the evaporator superheat is changing it either opens or closes, putting more or less refrigerant into the evaporator from changing remote bulb pressures. This maintains a set amount of evaporator superheat.

The remote bulb is clamped firmly to the suction line for good thermal contact. The remote

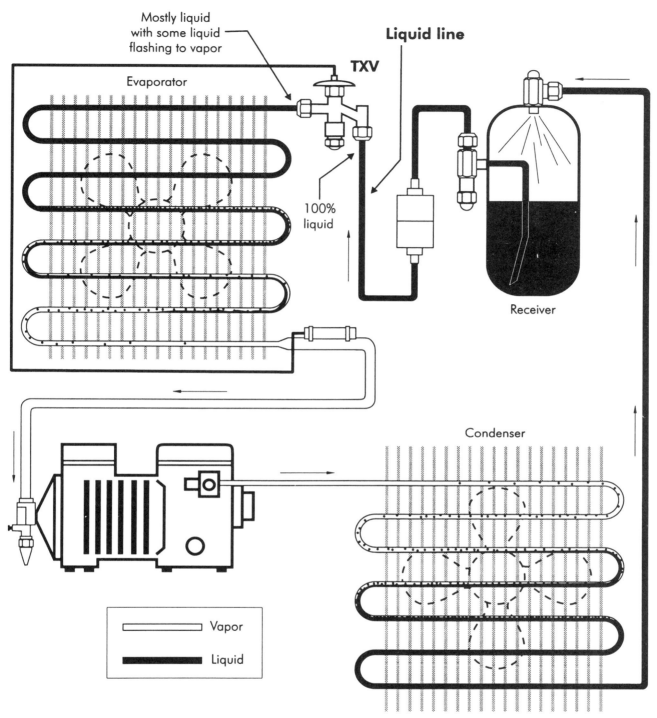

Mostly liquid
with some liquid
flashing to vapor

Liquid line

TXV

Evaporator

100%
liquid

Receiver

Condenser

Vapor

Liquid

Figure 4-12. TXV location

bulb is the TXV's feedback mechanism that gives the valve an indication of current evaporator superheat. Remote bulbs come in a variety of charges for special applications. Consult with the TXV manufacturer for specific information. Figure 4-14 illustrates the do's and don'ts for mounting a remote bulb onto the evaporator outlet.

$$P_1 = P_2 + P_3$$

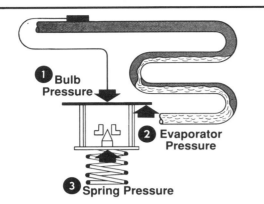

① Bulb Pressure

② Evaporator Pressure

③ Spring Pressure

Figure 4-13. Forces acting on a TXV (Courtesy, Sporlan Valve Company)

Evaporator Pressure and Spring Pressure

As heat loads on the evaporator change, the rate of vaporization of the refrigerant in the evaporator also changes. Higher heat loads cause higher rates of vaporization of the refrigerant in the evaporator, which increases evaporator pressure. Lower heat loads on the evaporator decrease the rate of vaporization of the refrigerant and decrease evaporator pressure.

Notice in Figure 4-13 that evaporator pressures act on the underside of the flexible bellows, causing the valve to close as the evaporator pressure increases. Through a drilled passageway called the internal equalizer, evaporator pressure from the coil entrance acts on the underside of the bellows. In Figure 4-13, evaporator pressure is designated as P_2. This evaporator pressure opposes the remote bulb pressure.

The spring pressure acts in the same direction on the bellows as the evaporator pressure. It is a closing force and is designated as P_3 in Figure 4-13. When the spring pressure (P_3) and the evaporator pressure (P_2) equal the remote bulb pressure (P_1), or $P_1 = P_2 + P_3$, the valve is said to be in dynamic balance and will neither open nor close until the evaporator heat load changes. If the suction pressure rises or falls, the valve tends to open or close. If the remote bulb pressures change from changing superheat, the valve will also open or close.

Spring pressure in most valves is set for a predetermined amount of evaporator superheat (usually between 7° and 10°F) by the manufacturer. This spring is often referred to as the

Bulb installation

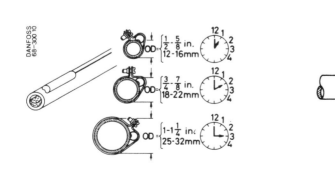

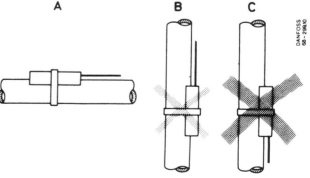

A. Recommended
B. Applicable but not recommended
C. Not applicable
   (charge will drop from inside bulb)

Figure 4-14. Do's and don'ts for connecting remote bulbs to evaporator outlet pipes (Courtesy, Danfoss Automatic Controls, Division of Danfoss, Inc.)

superheat spring, since it is the only way a technician can change the amount of superheat in a TXV system. Increasing the spring tension by turning the spring adjustment clockwise tightens the spring. This, in turn, gives the TXV more closing force and causes the superheat setting to increase, giving the system more operating evaporator superheat. There are some TXVs on the market that come with a set amount of superheat that cannot be changed by a technician. These TXVs are usually used in applications in which superheat settings are critical for proper performance, such as in some ice makers.

Consider a TXV whose superheat spring is set to control 10°F of evaporator superheat for a walk-in cooler. The valve is in a static equilibrium with forces ($P_1 = P_2 + P_3$). As long as there is no change in the evaporator load, meaning no change in evaporator superheat, the valve will remain static with a constant refrigerant flow rate controlling 10°F of evaporator superheat. The degree of evaporator superheat will be 10°F, which is exactly needed to offset the pressure exerted by the superheat spring. The evaporator pressure will also remain constant.

Now assume that someone opens the door to load the cooler. The added heat load on the evaporator will instantaneously cause the 100% saturated vapor point in the evaporator to climb. This happens because of the added rate of vaporization of the refrigerant, which causes the evaporator superheat to rise over the set 10°F. The remote bulb will then experience a higher evaporator superheat and temperature, and the pressure in the remote bulb will exceed the combined evaporator and spring pressure. This will cause the valve to open and let more refrigerant into the evaporator, which will have a tendency to reduce the amount of superheat until the required 10°F is again established. The evaporator will try to fill with refrigerant to gain maximum efficiency and capacity. The valve will again be in a state of static equilibrium controlling 10°F of evaporator superheat but at a higher evaporator pressure.

Additional Pressure

An additional pressure affecting TXV performance arises from the actual pressure drop across the valve. This pressure is a product of the pressure drop across the valve and the ratio of the port area to the effective area of the diaphragm:

Pressure = (Pressure drop) (Port area/effective diaphragm area)

This pressure is an opening force, and the liquid flow through the valve tends to move the TXV in an opening direction. It is sometimes referred to as liquid pressure, but this nomenclature is often misleading. This pressure only becomes significant when the pressure drop across the valve becomes large and/or the ratio of port area to effective diaphragm area becomes large. Because of this pressure, the balanced port TXV was introduced, which will be discussed later in this chapter.

Color Codes

Many times manufacturers of TXVs will color code the top of the valve. This helps the service technician quickly distinguish what refrigerant is in the system. The type of refrigerant is often stamped on the valve top along with a color-coding decal. Figure 4-15 is an example of one manufacturer's color-coding system.

Refrigerant Color Code Used on decals	R-11 — Blue R-12 — Yellow R-13 — Blue R-13B1 — Blue R-22 — Green R-113 — Blue	R-114 — Blue R-134a — Blue R-500 — Orange R-502 — Purple R-503 — Blue R-717 — White

Figure 4-15. Color-coding chart for TXVs (Courtesy, Sporlan Valve Company)

Special Application TXVs

There are special application TXVs available that should be considered when selecting the correct valve for an application. These special application TXVs include the *externally equalized valve*, the *maximum operating pressure valve*, and the *balanced port valve*.

Externally Equalized Valves

The externally equalized valve senses evaporator pressure at the evaporator outlet, not the evaporator inlet, Figure 4-16. This compensates for any friction pressure drops throughout the length of the evaporator. By sensing evaporator outlet pressure, there is less pressure on the bottom of the TXV bellows than if evaporator inlet pressure were sensed. This lower pressure causes a smaller closing force, so the valve remains more open and fills out more of the evaporator by compensating for the pressure drop through the evaporator. Evaporator outlet pressure is sensed through a small tube at the tailpipe of the evaporator leading underneath the TXV bellows, Figure 4-17. Externally equalized TXVs were shown in Figure 4-10 (b), (c), and (d).

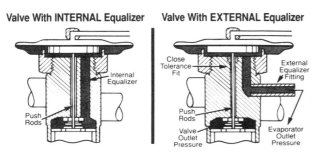

Figure 4-16. Cutaway view of an externally equalized TXV (Courtesy, Sporlan Valve Company)

An externally equalized TXV is used under the following conditions:

- When evaporator pressure drops cause a 2°F drop in temperature in the evaporator. This is for evaporator temperature applications above 0°F.

- When evaporator pressure drops cause a 1°F drop in temperature in the evaporator. This is for evaporator temperature applications below 0°F.

- Anytime distributors feed an evaporator.

Maximum Operating Pressure (MOP) Valves

MOP valves limit evaporator pressure to a maximum but not a minimum. MOP valves prevent high amp draws and overloads of compressors on high heat loads or pulldowns. They also prevent overfeeding at start-ups. This type of TXV has a limited amount of liquid in its remote bulb and will run out of liquid after reaching a certain temperature or maximum operating pressure, Figure 4-18. This prevents the further opening of the TXV and protects the compressor from high evaporator pressures during high loads or pulldowns, because the

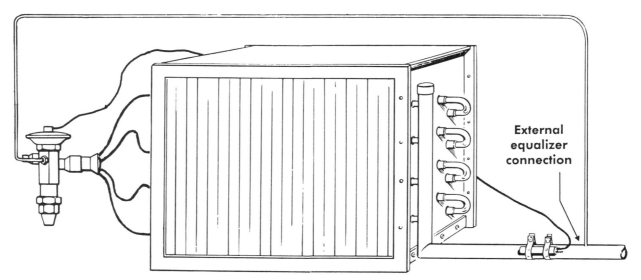

Figure 4-17. External equalizer connection for a TXV (Courtesy, Sporlan Valve Company)

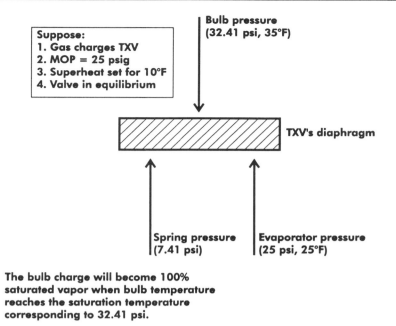

Suppose:
1. Gas charges TXV
2. MOP = 25 psig
3. Superheat set for 10°F
4. Valve in equilibrium

Bulb pressure
(32.41 psi, 35°F)

TXV's diaphragm

Spring pressure
(7.41 psi)

Evaporator pressure
(25 psi, 25°F)

The bulb charge will become 100%
saturated vapor when bulb temperature
reaches the saturation temperature
corresponding to 32.41 psi.

Figure 4-18. Forces on a maximum operating pressure (MOP) TXV

evaporator pressure will never exceed the MOP setting of the valve. In some valves, turning the superheat spring will set a new MOP set point. The MOP TXV may have a mechanical means of preventing a maximum pressure from occurring. The MOP of the valve is usually stamped on the valve in an obvious location.

Balanced Port TXVs

Balanced port TXVs have balanced pressures across the valve to negate the effects of varying head pressures or pressure drops across the valves. The liquid pressure is actually canceled out because it acts on equal areas but in opposite directions. The design uses a double seating piston operated by a single pushrod. The two port construction divides the refrigerant flow into opposite directions, providing a balanced pressure differential across the piston, Figure 4-19. Balanced port TXVs have a large port compared to conventional TXVs, and liquid pressure can affect a large ported valve if not balanced. This pressure only becomes significant when the pressure drop across the valve becomes large and/or the ratio of port area to effective diaphragm area becomes large. Because of the balanced port design, the valve can operate satisfactorily down to approximately 25% of its rated capacity.

The balanced port valve operates with the valve pin controlling very close to its seat. This provides a very stable control of superheat at minimum valve stroke movements, which allows a large port to handle both large and small loads. This type of design helps refrigeration equipment pull down faster. The larger port size can handle some liquid line flash gas if it exists.

Balanced port TXVs should be used if any of the following conditions exist:

- Liquid line flash gas

- Large varying head pressures

- Large varying pressure drops across the TXV affecting capacity of the valve

- Widely varying evaporator loads

- Very low liquid line temperatures

Selection Procedures

When selecting a TXV, always follow the manufacturer's selection procedures and nomenclature. The procedures used in selecting a TXV in this section are provided by Sporlan Valve

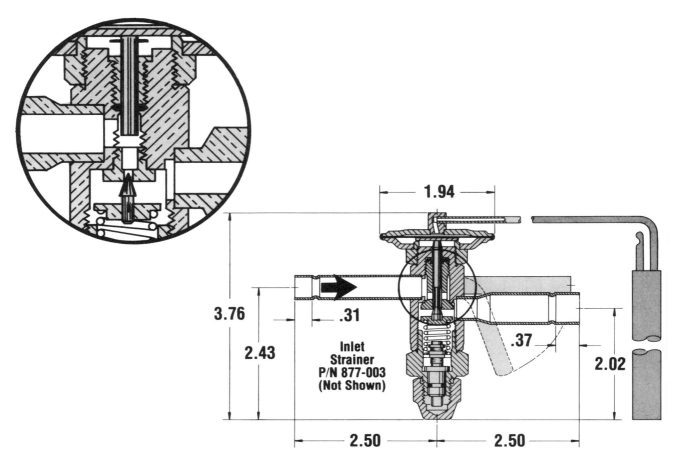

Figure 4-19. Cutaway of balanced port TXV (Courtesy, Sporlan Valve Company)

Company. Similar procedures are used when selecting TXVs from other manufacturers:

1. Determine the pressure drop across the valve by subtracting the evaporating pressure from the condensing pressure. The condensing pressure should be the minimum operating condensing pressure of the system, which usually occurs at lower loads and lower ambients. From this value, subtract all other pressure losses to obtain the net pressure drop across the valve. Pressure drop across the valve is an important measurement, because that pressure is the driving force for the refrigerant to flow across the valve. Without a pressure drop, there will not be any flow. As the pressure drop increases, more refrigerant will flow across the valve. Consider all sources of pressure drop, including:

    a. friction losses through refrigeration lines including the evaporator, suction line, and condenser.

    b. pressure drop across liquid line and suction line accessories such as solenoid valves, strainers, filters, and filter driers (see manufacturer catalogs).

c. static pressure losses or gains due to the vertical lift or drop of the liquid line, Figure 4-20.

d. pressure drop across a refrigerant distributor (if used).

REFRIG-ERANT	VERTICAL LIFT — FEET				
	20	40	60	80	100
	Static Pressure Loss — psi				
12	11	22	33	44	55
22	10	20	30	40	50
500	10	19	29	39	49
502	10	21	31	41	52
717 (Ammonia)	5	10	15	20	25

Figure 4-20. Static pressure losses vs vertical lift for different refrigerants (Courtesy, Sporlan Valve Company)

2. Determine the liquid temperature of the refrigerant entering the valve. The closer the liquid temperature is to the evaporating temperature, the less flash gas that will occur at the entrance of the valve. The less flash gas, the more net refrigeration effect. The TXV capacity table in Table 4-1 is based on a liquid temperature of 100°F for R-22. If the liquid temperature is higher or lower than 100°F, correction factors must be applied, which are also listed at the bottom of Table 4-1. Correction factors must be multiplied by the appropriate capacity in the selection table to either increase or decrease capacity (see example in bottom right hand corner of Table 4-1). Notice that if the liquid temperature falls below 100°F, a higher correction factor is applied. Higher correction factors increase the capacity of the TXV, because the liquid is colder and closer to the evaporating temperature, which causes less flash gas at the evaporator entrance and more net refrigeration effect.

3. Select the valve from the capacity table. **Select a valve based on the design evaporating temperature and the available pressure drop across the**

valve from **Table 4-1**. If possible, the valve capacity should equal or slightly exceed the design rating of the system, Figure 4-21. Be sure to multiply the appropriate liquid temperature correction factor to the valve capacity rating shown in the tables. Once the desired valve capacity is located, determine the nominal capacity of the valve from the second column of the tables. On multiple evaporator systems, select each valve on the basis of individual evaporator capacity.

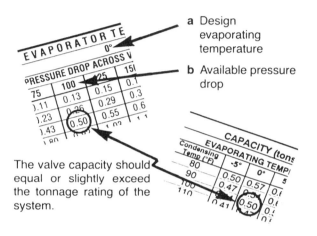

a Design evaporating temperature

b Available pressure drop

The valve capacity should equal or slightly exceed the tonnage rating of the system.

Figure 4-21. Valve capacity should equal or slightly exceed design rating (Courtesy, Sporlan Valve Company)

In Table 4-1, notice that for a constant pressure drop across the valve, as the evaporator temperature increases, the capacity of the TXV increases. This is caused by a lower compression ratio from a higher evaporator temperature and pressure, which causes a higher volumetric efficiency. In turn, the mass flow rate of refrigerant increases through the entire system including the TXV. Less flash gas is also experienced at the evaporator entrance, because the liquid in the liquid line is closer to the higher evaporator temperature. This causes a higher capacity and more net refrigeration effect.

4. Determine if an external equalizer is required. The amount of pressure drop between the valve outlet and remote bulb location (evaporator pressure drop)

R-22

THERMOSTATIC EXPANSION VALVE CAPACITIES for REFRIGERANT-22
TONS OF REFRIGERATION

AIR CONDITIONING, HEAT PUMP and COMMERCIAL REFRIGERATION APPLICATIONS

VGA, VCP100 AND VC THERMOSTATIC CHARGES

EVAPORATOR TEMPERATURE (°F) — PRESSURE DROP ACROSS VALVE (PSI)

VALVE TYPE	NOMINAL CAPACITY	40° 75	40° 100	40° 125	40° 150	40° 175	40° 200	20° 75	20° 100	20° 125	20° 150	20° 175	20° 200	0° 75	0° 100	0° 125	0° 150	0° 175	0° 200
G-EG	1/5	0.17	0.20	0.22	0.24	0.26	0.28	0.17	0.19	0.22	0.24	0.26	0.28	0.15	0.17	0.19	0.21	0.23	0.24
NI-F	1/4	0.22	0.25	0.28	0.31	0.33	0.35	0.21	0.24	0.27	0.30	0.32	0.34	0.20	0.23	0.25	0.28	0.30	0.32
G-EG	1/3	0.30	0.35	0.39	0.43	0.46	0.49	0.30	0.34	0.38	0.42	0.45	0.48	0.26	0.30	0.33	0.37	0.39	0.42
NI-F-G-EG	1/2	0.39	0.45	0.50	0.55	0.60	0.64	0.38	0.44	0.49	0.54	0.58	0.62	0.33	0.38	0.43	0.47	0.51	0.54
G-EG	3/4	0.65	0.75	0.84	0.92	0.99	1.06	0.63	0.73	0.82	0.90	0.97	1.03	0.55	0.64	0.71	0.78	0.85	0.90
NI-F-G-EG	1	0.87	1.00	1.12	1.22	1.32	1.41	0.84	0.97	1.09	1.19	1.29	1.38	0.74	0.85	0.95	1.04	1.13	1.21
F-G-EG	1 1/2	1.39	1.60	1.79	1.96	2.12	2.26	1.35	1.56	1.74	1.91	2.06	2.21	1.18	1.36	1.52	1.67	1.80	1.93
F(Ext)-G & EG(Ext)-S	2	1.73	2.00	2.24	2.45	2.65	2.83	1.69	1.95	2.18	2.39	2.58	2.76	1.48	1.70	1.91	2.09	2.26	2.41
F(Int)-G(Int)	2 1/2	2.17	2.50	2.80	3.06	3.31	3.54	2.11	2.44	2.72	2.98	3.22	3.45	1.85	2.13	2.38	2.61	2.82	3.01
F(Ext)-G & EG(Ext)-S	3	2.77	3.20	3.58	3.92	4.23	4.53	2.70	3.12	3.49	3.82	4.13	4.41	2.36	2.73	3.05	3.34	3.61	3.86
C-S	4	3.90	4.50	5.03	5.51	5.95	6.36	3.80	4.39	4.90	5.37	5.80	6.20	3.32	3.84	4.29	4.70	5.07	5.42
C-S	5	4.50	5.20	5.81	6.37	6.88	7.35	4.39	5.07	5.67	6.21	6.70	7.17	3.84	4.43	4.96	5.43	5.86	6.27
C&S(Ext)	8	6.93	8.00	8.94	9.80	10.6	11.3	6.75	7.80	8.72	9.55	10.3	11.0	5.45	6.29	7.03	7.70	8.32	8.89
S(Ext)	10	8.66	10.0	11.2	12.2	13.2	14.1	8.44	9.75	10.9	11.9	12.9	13.8	6.81	7.86	8.79	9.63	10.4	11.1
H	2 1/2	2.17	2.50	2.80	3.06	3.31	3.54	2.07	2.39	2.67	2.92	3.16	3.38	1.85	2.13	2.38	2.61	2.82	3.01
H	5 1/2	4.85	5.60	6.26	6.86	7.41	7.92	4.63	5.35	5.98	6.55	7.08	7.56	4.13	4.77	5.34	5.85	6.32	6.75
H	7	6.06	7.00	7.83	8.57	9.26	9.90	5.79	6.69	7.48	8.19	8.85	9.46	5.17	5.97	6.67	7.31	7.89	8.44
H	11	9.09	10.5	11.7	12.9	13.9	14.8	8.69	10.0	11.2	12.3	13.3	14.2	7.75	8.95	10.0	11.0	11.8	12.7
H	16	13.2	15.2	17.0	18.6	20.1	21.5	12.6	14.5	16.2	17.8	19.2	20.5	11.2	13.0	14.5	15.9	17.1	18.3
H	20	19.2	22.2	24.8	27.2	29.4	31.4	18.4	21.2	23.7	26.0	28.1	30.0	16.4	18.9	21.2	23.2	25.0	26.8
M	21	18.6	21.5	24.0	26.3	28.4	30.4	18.1	21.0	23.4	25.7	27.7	29.6	17.5	20.2	22.5	24.7	26.7	28.5
M	26	22.9	26.5	29.6	32.5	35.1	37.5	22.4	25.8	28.9	31.6	34.2	36.5	21.5	24.8	27.6	31.9	32.9	35.1
M	34	29.4	34.0	38.0	41.6	45.0	48.1	28.7	33.1	37.1	40.6	43.8	46.9	27.6	31.9	35.6	39.0	42.2	45.1
M	42	36.4	42.0	47.0	51.4	55.6	59.4	35.5	40.9	45.8	50.1	54.2	57.9	34.1	39.4	44.0	48.2	52.1	55.7

BALANCED PORT THERMOSTATIC EXPANSION VALVES

VALVE TYPE	NOMINAL CAPACITY	40° 75	40° 100	40° 125	40° 150	40° 175	40° 200	20° 75	20° 100	20° 125	20° 150	20° 175	20° 200	0° 75	0° 100	0° 125	0° 150	0° 175	0° 200
BF-EBF	AA	0.65	0.75	0.84	0.92	0.99	1.06	0.63	0.73	0.82	0.90	0.97	1.03	0.55	0.64	0.71	0.78	0.85	0.90
BF-EBF	A	1.39	1.60	1.79	1.96	2.12	2.26	1.35	1.56	1.74	1.91	2.06	2.21	1.18	1.36	1.52	1.67	1.80	1.93
BF-EBF	B	2.42	2.80	3.13	3.43	3.70	3.96	2.36	2.73	3.05	3.34	3.61	3.86	2.07	2.39	2.67	2.92	3.16	3.38
BF-EBF	C	4.50	5.20	5.81	6.37	6.88	7.35	4.39	5.07	5.67	6.21	6.70	7.17	3.84	4.43	4.96	5.43	5.86	6.27
EBS	11	10.0	11.5	12.9	14.1	15.2	16.3	9.22	10.6	11.9	13.0	14.1	15.1	7.64	8.82	9.86	10.8	11.7	12.5
O	15	13.0	15.0	16.8	18.4	19.8	21.2	12.0	13.9	15.5	17.0	18.4	19.6	10.1	11.7	13.0	14.3	15.4	16.5
O	20	19.2	22.2	24.8	27.2	29.4	31.4	17.8	20.6	23.0	25.2	27.2	29.1	14.9	17.2	19.3	21.1	22.8	24.4
O	30	26.4	30.5	34.1	37.4	40.3	43.1	24.5	28.2	31.6	34.6	37.4	39.9	20.5	23.6	26.5	29.0	31.3	33.5
O	40	34.9	40.3	45.1	49.4	53.3	57.0	33.7	38.9	43.5	47.6	51.4	55.0	24.8	28.6	32.0	35.1	37.9	40.5
O	55	47.6	55.0	61.5	67.4	72.8	77.8	46.0	53.1	59.3	65.0	70.2	75.1	33.8	39.1	43.7	47.9	51.7	55.3
O	70	63.2	73.0	81.6	89.4	96.6	103	61.0	70.4	78.8	86.3	93.2	99.6	44.9	51.9	58.0	63.5	68.6	73.3
V	52	45.0	52.0	58.1	63.7	68.8	73.5	43.5	50.2	56.1	61.5	66.4	71.0	41.8	48.3	54.0	59.1	63.9	68.3
V	70	63.2	73.0	81.6	89.4	96.6	103	61.0	70.4	78.8	86.3	93.2	99.6	58.7	67.8	75.8	83.0	89.6	95.8
V	100	86.6	100	112	122	132	141	83.6	96.5	108	118	128	136	80.4	92.8	104	114	123	131
W	135	124	143	160	175	189	202	119	138	154	169	183	195	115	133	148	163	176	188
W	180	156	180	201	220	238	255												

REFRIGERANT LIQUID TEMPERATURE CORRECTION FACTORS

Refrigerant Liquid Temperature °F	0°	10°	20°	30°	40°	50°	60°	70°	80°	90°	100°	110°	120°	130°	140°
Correction Factor	1.56	1.51	1.45	1.40	1.34	1.29	1.23	1.17	1.12	1.06	1.00	0.94	0.88	0.82	0.76

These factors include corrections for liquid refrigerant density and net refrigerating effect and are based on an average evaporator temperature of 0 F. However they may be used for any evaporator temperature from −40 F to 40 F since the variation in the actual factors across this range is insignificant.

EXAMPLE: Actual capacity of nominal 5 ton valve at 0 F evaporator, 175 psi pressure drop and 120 F liquid temperature = 5.86 tons × 0.88 = 5.16 tons.

Table 4-1. R-22 TXV capacity with liquid subcooled correction factors (Courtesy, Sporlan Valve Company)

will determine if an external equalizer is required (see section on externally equalized TXVs earlier in this chapter).

5. Select the body type according to the style connections desired. These include flare, solder, and flange connections. Brass, bronze, and gray cast iron are some other options. Different connection angles are also available (see Figure 4-10).

6. Select the selective thermostatic charge according to the design evaporating temperature from Table 4-2.

Example 4-1
Select a TXV for an R-22 air conditioning system with the following criteria:

- Design evaporator temperature = 40°F

- Design condensing temperature = 105°F

- Refrigerant liquid temperature = 90°F

- Design system capacity = 2 tons

- Condensing pressure = 211 psig

- Evaporating pressure = 69 psig

- Liquid line, suction line, and accessory losses = 7 psi

- Distributor and tube losses = 35 psi

- Refrigerant liquid correction factor = 1.06 (90°F liquid at TXV)

Solution 4-1
It is important to note that since there is a distributor used on the evaporator, an external equalizer valve must be used due to the large pressure drop created by the distributor.

First determine the available pressure drop across the TXV, which is the condensing pressure minus the evaporating pressure:

211 psig - 69 psig = 142 psig

From that total, subtract the liquid line, suction line, and accessory losses and the distributor and tube losses:

142 psig - 7 psi - 35 psi = 100 psi
(100 psi is the actual pressure drop across the TXV)

Using Figure 4-22, the valve capacity is as follows:

At a 100 psi pressure drop across the valve and a 40°F design evaporating temperature, the rated capacity of 2 tons exactly meets the design system capacity of 2 tons. However, we must take into consideration the 90°F liquid correction factor, which will increase the capacity rating of the 2 ton rated TXV to 2.12 tons (see below).

(2.00 rated tons)(1.06 liquid correction factor) = 2.12 tons

REFRIGERANT	AIR CONDITIONING OR HEAT PUMP	COMMERCIAL REFRIGERANT +50°F to −10°F	LOW TEMPERATURE REFRIGERANT 0°F to −40°F	EXTREME LOW TEMPERATURE REFRIGERANT −40°F to −100°F
12	FCP60	FC	FZ FZP	—
22	VCP100 and VGA	VC	VZ VZP40	VX
134a	JCP60	JC	—	—
502	RCP115	RC	RZ RZP	RX
717	REFER TO THE SECTION ON AMMONIA REFRIGERATION IN BULLETIN 10-9			

Table 4-2. Selective charges (Courtesy, Sporlan Valve Company)

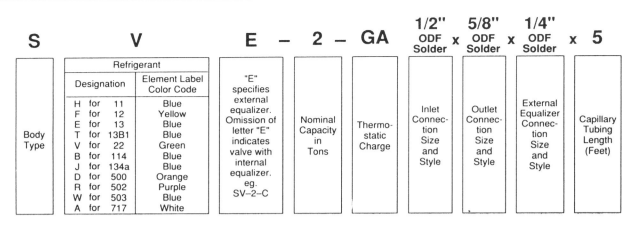

S	V		E – 2 – GA			1/2" ODF Solder	x	5/8" ODF Solder	x	1/4" ODF Solder	x 5

Body Type	Refrigerant		"E" specifies external equalizer. Omission of letter "E" indicates valve with internal equalizer. eg. SV–2–C	Nominal Capacity in Tons	Thermo- static Charge	Inlet Connec- tion Size and Style	Outlet Connec- tion Size and Style	External Equalizer Connec- tion Size and Style	Capillary Tubing Length (Feet)
	Designation	Element Label Color Code							
	H for 11 F for 12 E for 13 T for 13B1 V for 22 B for 114 J for 134a D for 500 R for 502 W for 503 A for 717	Blue Yellow Blue Blue Green Blue Blue Orange Purple Blue White							

Figure 4-22. TXV selection (Courtesy, Sporlan Valve Company)

Remember, we want to select a valve with a rated valve capacity that equals or slightly exceeds the design tonnage rating of the system, which is 2 tons.

The valve that is rated for 2.12 tons, after figuring in the liquid correction factor, does slightly exceed the design tonnage of 2.0 tons. The choice would then be a 2.12 ton rated capacity valve, which would be a 2.00 ton nominal capacity valve in Table 4-1.

All that is left to do is to specify a body type; type of refrigerant; external or internal equalizer; thermostatic element charge type; inlet, outlet, and external equalizer connection sizes; and capillary tube length for the thermostatic remote bulb (see Figure 4-20).

Automatic Expansion Valves (AXVs)

The AXV is located between the liquid line and the evaporator. Its main function is to maintain a constant pressure in the evaporator under all evaporator loading conditions without regard to evaporator superheat; in fact, there is no remote bulb or feedback mechanism to measure evaporator superheat. The AXV either starves or fills out the evaporator coil with refrigerant in response to different heat loads on the evaporator. Because of the AXV's constant pressure characteristic, it cannot be used with a low pressure motor control.

The AXV is made of a valve casing, pressure bellows or diaphragm, needle and seat, spring, adjustment stem or screw, and a liquid inlet strainer. The two forces that throttle the AXV open and closed are the spring and the evaporator pressure, Figure 4-23. Constant evaporator pressure is maintained by the interaction of the spring and evaporator pressures. The evaporator pressure is a closing force, and the spring pressure is an opening force. During the compressor on cycle, the valve will hold the evaporator pressure in equilibrium with the spring pressure. The valve stem or adjustment screw should be adjusted for the desired evaporator pressure. As the compressor cycles off, the valve will close due to increasing evaporator pressure from residual refrigerant in the evaporator boiling off, and refrigerant flow to the evaporator will cease.

AXV efficiency is considered poor when compared to other metering devices. This is because in order to keep a constant pressure in the evaporator, the rate of vaporization in the evaporator must be kept constant under all evaporator loads. At heavy evaporator loads, the refrigerant delivered to the evaporator must be throttled back because of the increased rate of refrigerant vaporization in the evaporator, momentarily causing a higher evaporator pressure. This throttling effect causes an inactive evaporator with an abundance of superheat, Figure 4-24; however, a constant evaporator pressure is still maintained.

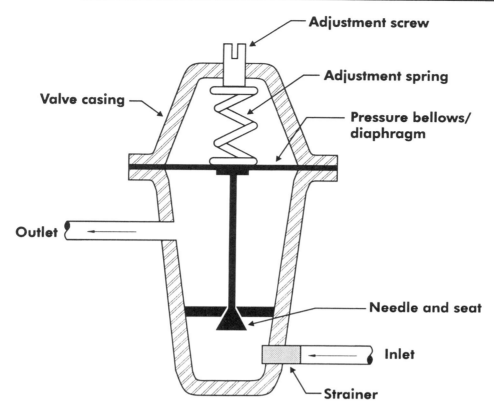

Figure 4-23. *Spring and evaporator pressure on an AXV*

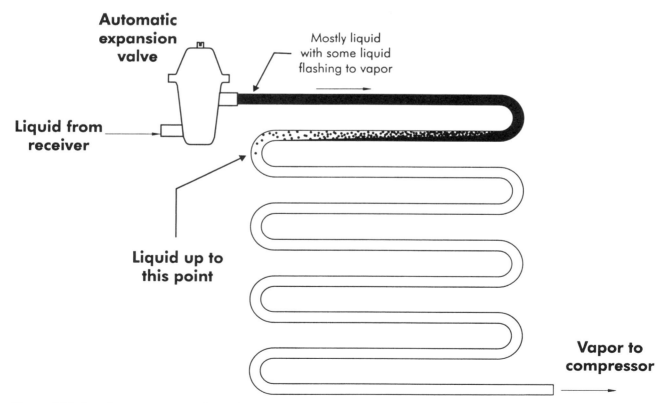

Figure 4-24. *Inactive evaporator with an abundance of superheat*

At low loads, the evaporator pressure momentarily decreases, causing the spring force to throttle the valve more open. This maintains the valve's constant pressure characteristics. At a low evaporator load, more of the evaporator surface must be wetted with refrigerant in order to maintain a constant rate of vaporization, Figure 4-25. This lowers the evaporator superheat. If the load is permitted to be lowered too far without shutting the compressor off, the valve could open too far and flood the compressor in order to try and maintain a constant evaporator pressure. The thermostat should be adjusted to shut the compressor off before flooding of the entire evaporator occurs. The AXV should be adjusted to maintain an evaporator pressure that corresponds to the lowest evaporator temperature desired through the entire running cycle of the compressor.

AXVs are usually found on smaller equipment with constant evaporator loads, including ice cream freezers and makers and drinking fountains. AXVs are often used to prevent products or water from freezing.

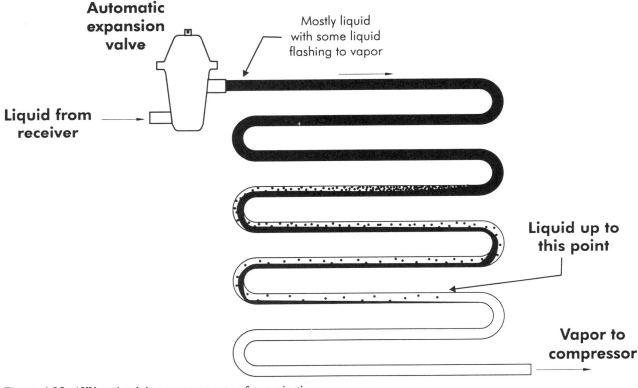

Figure 4-25. AXV maintaining constant rate of vaporization

CHAPTER FIVE

System Charge

One of the most puzzling questions a technician will ask is how much refrigerant should be put in a refrigeration system when charging. There is really no straightforward answer, because all refrigeration systems differ in the amount of charge they hold. However, there are guidelines, charts, and techniques all technicians should follow when charging a system. The techniques discussed in this chapter do not apply to automotive or any transportation air conditioning system charging.

CHARGING TXV/RECEIVER/SIGHTGLASS REFRIGERATION SYSTEMS

Most large commercial refrigeration systems employ TXVs as metering devices. Because of the throttling action of the TXV metering device to satisfy refrigeration loads, a liquid receiver must be employed to act as a liquid refrigerant reservoir. Liquid receivers also act as storage vessels to store the condenser charge variations between summer and winter operations. When the TXV is throttled down at lower loads, the receiver will contain more liquid. The TXV throttles open more at higher heat loads and draws liquid refrigerant from the receiver. The remote bulb sends more of an opening pressure to the TXV diaphragm, which opens the valve more. This is why it is of utmost importance to put the refrigeration equipment under a high load when charging with refriger-

ant. High loads ensure that the TXV metering device is fully open and delivering the maximum amount of refrigerant to the evaporator. At these higher loads, the receiver will be at its lowest level. High loads can be achieved by placing a false heat load on the evaporator.

At higher loads, a sightglass located in the liquid line will bubble if undercharged. The sightglass in the liquid line may not bubble when a system is undercharged if the TXV is throttled partly closed, the receiver has some liquid, and the system is at a lower heat load. Unless severely undercharged, a sightglass will only bubble under higher loads. The service technician must make sure the sightglass does not bubble under high and low loads in order for the system to be properly charged.

Service technicians often confuse a bubbling sightglass with a low flow rate sightglass. If undercharged and at a high load, the receiver will be at its lowest level and the sightglass will have entrained refrigerant gas bubbles in it, Figure 5-1. If the system is charged correctly but experiences a very low heat load on the evaporator, a low flow rate sightglass, not a bubbling sightglass, may be seen. Low flow rate sightglasses do not have refrigerant gas bubbles entrained in the flowing liquid. A low flow rate sightglass only partially fills the volume of the sightglass and does not indicate an undercharge, Figure 5-2. The low flow rate of refrigerant through the system is caused from the lower loads on the evaporator. Lower loads mean higher compression ratios and lower volumetric efficiencies. The density of the return gas to the compressor is also decreased at lower loads and lower suction pressures, causing low mass flow rates of refrigerant. Horizontal liquid lines with sightglasses encounter this phenomenon much more often than vertical liquid lines with sightglasses. Gravity will force the liquid in the liquid line to settle at the lowest point, and this can entirely fill a sightglass volume even at low flow rates.

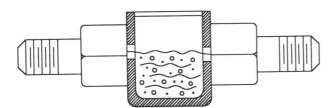

Figure 5-1. Bubbling sightglass (Courtesy, Todd Rose, Ferris State University)

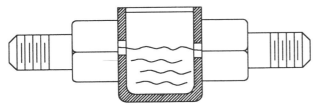

Figure 5-2. Low flow rate sightglass (Courtesy, Todd Rose, Ferris State University)

In these cases, the technician must make sure the liquid line is full of liquid before and after the sightglass. This can be done with an electronic sightglass, Figure 5-3. An electronic sightglass determines whether there is vapor entrained in the liquid. This assists the technician in charging the system and ensures 100% liquid at the metering device. Electronic sightglasses are especially helpful when the refrigeration system has no sightglass. The electronic sightglass works when its sensors are clamped about 1-1/2" apart on the liquid line. The technician then waits for beeping sounds or flashing lights to indicate the presence of vapor and liquid in the liquid line. Electronic sightglasses can also be used on the evaporator outlet or start of the suction line to let the technician know if there is liquid entrained with the vapors. Electronic sightglasses are also used

extensively in automotive and commercial/residential air conditioning system charging.

Charging Rules

The following rules should be used when charging TXV/receiver/sightglass systems:

1. Always charge a TXV/receiver/sightglass system under a high load. Remember, systems with 50 lb or more of refrigerant that leak significantly must be leak checked and repaired (see Chapter Ten). Once the leak is found, the system refrigerant recovered, and the leak repaired, the system is ready to be evacuated. If the vacuum is pulled on a refrigeration system with a deep vacuum pump, once the desired vacuum is reached and the pump is isolated from the system, charge liquid refrigerant into the high side of the system until high and low side pressures equalize and refrigerant stops flowing. It is preferable to charge refrigerant into the receiver if valving allows. Putting liquid refrigerant into the low side of the system can damage the compressor at start-up and

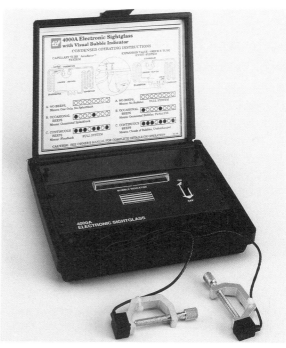

Figure 5-3. Electronic sightglass (Courtesy, TIF Instruments, Inc.)

dilute the oil in the crankcase with refrigerant, causing oil flash and scored bearings.

2. Once liquid has been charged into the high side of the system and system pressures have equalized, let the system set idle for about 10 minutes. This helps vaporize any liquid the TXV has released into the evaporator while charging. After a 10 minute wait, start the system and monitor the amp draw. Keep the doors open and the system under a high load to allow the TXV to throttle open fully. This will draw a maximum amount of liquid from the receiver. Let the system run for a while at this high heat load to reach an equilibrium. If the sightglass bubbles, charge refrigerant vapor into the low side of the system until the sightglass stops bubbling. Refrigerant vapor can be charged into the low side of any running system as long as the technician knows that the undercharge is not caused by a leak. Remember, all leaks should be fixed, and systems with 50 lb and more of refrigerant cannot be left to leak or severe fines will be paid (see Chapter Ten). *Note: If the refrigerant is a newer refrigerant that has a temperature glide associated with it, liquid refrigerant must be throttled through the low side to avoid fractionation. The proper technique is to employ a sightglass and valve, which allow liquid to be removed from the cylinder while manually adjusting the valve, causing enough restriction to "flash" the liquid before it enters the suction valve. This flashing will be visible through the sightglass, Figure 5-4. When properly charged, the sightglass should not bubble under high or low loads.*

3. Take evaporator superheat, compressor superheat, and condenser subcooling readings. Evaporator superheat reading guidelines are listed in Table 5-1. If evaporator superheat settings are not within the guidelines, adjust the TXV

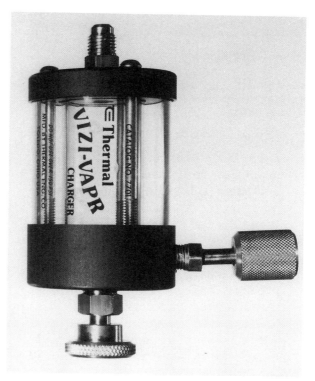

Figure 5-4. Charging device that allows visibility of both liquid and vapor while charging (Courtesy, Thermal Engineering Company)

superheat spring. Always give the TXV enough time to react to the change before taking the next reading.

Make sure condenser subcooling is at least 6°F. This is just a starting figure, because actual subcooling amounts will vary depending on the type of refrigerant used and how much pressure drop the subcooled liquid will see from the condenser outlet to the inlet of the TXV. This ensures that the condenser is producing enough liquid to deliver a solid column of 100% liquid to the TXV.

Compressor superheats should always be between 20° and 30°F. This ensures that the compressor will not see any liquid refrigerant at lower loads. This superheat range will also prevent compressor overheating and high amp draws.

Application	Air Conditioning and Heat Pump	Commercial Refrigeration	Low Temperature Refrigeration
Evaporator temperature	40° to 50°F	0° to 40°F	-40° to 0°F
Suggested superheat setting	8° to 12°F	6° to 8°F	4° to 6°F

Table 5-1. Guidelines for evaporator superheat settings

4. Under high heat load, take evaporator superheat, compressor superheat, and condenser subcooling readings and compare them to the suggested guidelines. Record both the condensing and evaporating pressure. Monitor the amp draw of the compressor with an ammeter. Take a voltage reading at the compressor terminals. Check it with the manufacturer amp curve for the pressure readings and voltage taken (see previous section on performance curves), and make sure it is within the range (±5%) for the pressures and voltage recorded.

5. Record the condensing and evaporating pressure of the system. Compare the amp reading of the compressor to the manufacturer amp curves for the pressures recorded. Make sure the amp draw is within range (±5%) on the curves for the two pressures and voltage recorded.

After applying these basic rules, a TXV/receiver/sightglass refrigeration system can be charged with the proper amount of refrigerant.

CHARGING CAPILLARY TUBE OR FIXED ORIFICE REFRIGERATION SYSTEMS

Capillary tube systems vary somewhat from TXV systems in their charging procedures. Most capillary tube systems do not have sightglasses in their liquid lines. Capillary tube systems have fixed orifices and do not throttle open and closed like TXV or AXV metering devices, so a receiver is usually not needed to hold any liquid coming from the condenser. This is one reason why a capillary tube system is a critically charged system.

Critically charged systems are systems that specify an exact amount of refrigerant charge. These charges are usually specified to the quarter ounce. The manufacturer nameplate, usually located on the condensing unit, evaporator faceplate, or unit frame, will specify the critical charge. Manufacturers critically charge capillary tube systems in order to achieve peak system efficiency and performance. Capillary tube systems are also critically charged to prevent floodback of liquid refrigerant under low heat loads or high ambients exposed to the condenser.

If the system has been running for some time and an undercharge is suddenly suspected, a leak probably exists. It is recommended procedure to leak check the system. For systems that have over 50 lb of refrigerant for commercial and industrial refrigeration, leakage rates of more than 35% annually are considered substantial leaks and must be repaired. If the equipment is a chiller or comfort cooling air conditioner containing over 50 lb of refrigerant, a 15% annual leakage rate is considered substantial and must be repaired. Once the refrigerant is recovered and the leak is repaired, a deep vacuum must be pulled on the system. After successful evacuation, the nameplate charge must be measured and charged into the system. Most capillary tube systems are small systems, and it is good practice to recover, leak check, evacuate, and then critically charge the system. These procedures take very little time if systems are small. This eliminates guesswork as to whether the system has the right amount of charge.

Charging from a Vacuum

Once the desired vacuum is reached and the vacuum pump is isolated from the system, add the specified nameplate amount of liquid refrigerant to the high side of the refrigeration system through a gauge manifold. Adding this refrigerant can be done with a programmable electronic charging device, Figure 5-5, or a critical charging cylinder, Figure 5-6, designed for capillary tube systems. Often, a combination critical charging cylinder, vacuum pump, and manifold set can be purchased, Figure 5-7. Once the critical amount of liquid refrigerant is charged into the system, let the liquid charge bleed from the high side of the system through the capillary tube. This will cause a good portion of the liquid to vaporize and end up in the evaporator as a vapor. Evidence of this can be seen by letting the system idle for about 10 minutes after liquid charging to the high side. Within a minute, the low side pressure will gradually rise, demonstrating that the capillary tube is not plugged and that refrigerant is traveling through it. Figure 5-8 shows how to charge a system with an electronic, programmable charging device.

Figure 5-6. Critical charging cylinder (Courtesy, Thermal Engineering Company)

Once the low side pressure rises to equal the high side pressure, the system can be started. It is important to allow the high and low side pressures to equalize, because this allows the compressor motor to start more easily, Figure 5-9. Capillary tube systems are often not equipped with start capacitors (hard start kits), and system pressures must equalize before attempting to start the compressor. This will prevent locked rotor situations and often tripped compressor overloads.

To illustrate the need for a critically charged system, consider a 3-ton system that calls for 5 lb of refrigerant to be critically charged. A 10% undercharge is only 8 oz of refrigerant, but this undercharge will reduce system efficiency by 20%. A 23% undercharge will reduce system efficiency by 50%. The correct charge is critical, so do not to use cheap scales, long charging hoses, or equipment that has not been kept up. *Note: It is of utmost importance to weigh in the exact charge into a capillary tube or fixed orifice system.*

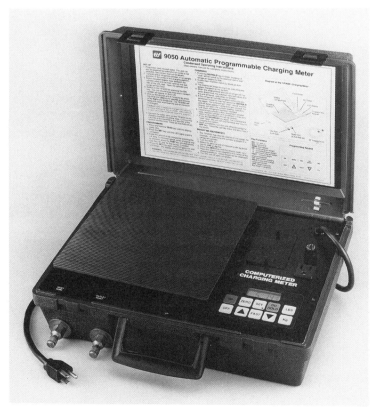

Figure 5-5. Programmable electronic charging device (Courtesy, TIF Instruments, Inc.)

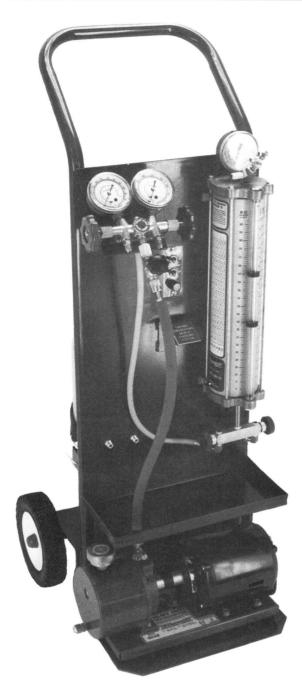

Figure 5-7. Portable combination charging cylinder, vacuum pump, and manifold gauge set (Courtesy, Thermal Engineering Company)

Compressor motors employing capillary tube systems are usually low starting torque motors that cannot start with large pressure differences between high and low side pressures. Not waiting for the system to equalize will cause the compressor to lock rotor and open its overload

protection device. The overload protector will continue to open and reset until the pressures are just about equal, then the low starting torque motor will start.

Once the system starts, take a compressor amp reading and measure voltage at the compressor terminals. Make sure the amp reading is the compressor only and not the total amps of the system. Compare the amp reading to the manufacturer compressor curve or system curves to make sure the system is charged and running as efficiently as possible (see Chapter Three). System performance curves are preferred in this case, because compressor curves reveal only compressor performance, not system performance. If the system is operating correctly or potentially correctly, then compressor curves can be used.

Always remember to correct for voltage when using compressor curves. If undercharged, add refrigerant vapor to the low side of the refrigeration system while it is running. *Note: If the refrigerant is a newer refrigerant with a temperature glide, liquid must be throttled in the low side of the system with the compressor running to avoid fractionation. The proper technique is to employ a sightglass or a charging device, which allows liquid to be removed from the system while manually adjusting a valve. This causes enough restriction to "flash" the liquid before it enters the suction valve.*

An overcharge of refrigerant will cause elevated head and suction pressures. The amp draw will also be much higher than the curves indicate it should be. An undercharge of refrigerant will cause the head pressure and suction pressure to go down. The amp draw will also be much lower than the curves indicate if undercharged. A filter drier change is always recommended before recharging a system. Often, the replacement filter drier will be larger than the original drier. The larger drier will hold more liquid refrigerant, and the system may be a bit undercharged even if the nameplate charge is used. This is why it is important to use performance curves after charging.

Figure 5-8. Charging a system using a programmable charging scale manufactured by TIF Instruments, Inc. (Photo by Bill Bitzinger, Office of University Communication Services, Ferris State University)

CHARGING CAPILLARY TUBE OR FIXED ORIFICE AIR CONDITIONING SYSTEMS

Charging an air conditioning comfort system is much different than charging a refrigeration system. Air conditioning systems rely on certain volumetric amounts of air flow referred to as cubic feet per minute (cfm) across their evaporator coils. Air conditioning processes are both sensible and latent processes, which take both heat and moisture from the air. Because of this, both wet bulb (wb) and dry bulb (db) temperatures, along with air flows (cfm) and compressor superheats are needed to charge these systems.

The most accurate method of charging a capillary or fixed orifice system is to weigh the charge into an empty system; however, time does not always allow recovering and evacuat-

Figure 5-9. Waiting for system pressures to equalize (Photo by Bill Bitzinger, Office of University Communication Services, Ferris State University)

ing before charging. The next most efficient and accurate way of charging a capillary tube system is to use a superheat charging table or curve, Figure 5-10. Charging by the compressor superheat method best balances the system's ability to absorb heat with the available heat load on the evaporator. Superheat charging can be used on capillary or fixed orifice metering devices when charging in ambients above 60°F. Compressor superheat in these types of systems will vary depending on the air conditions flowing over the condenser and evaporator coils. The amount of superheat at the compressor must be adjusted to meet these varying conditions. The superheat charging method shows compressor superheat for various entering air conditions at the evaporator and condenser. Remember that compressor superheat will vary as conditions vary, and no one superheat value can be used for all conditions. If this charging method is performed correctly, the system will work safely and efficiently at all conditions as long as the ambient is over 60°F at the condenser.

To use the superheat charging method perform the following steps:

1. Make sure the evaporator has the rated cfm of air flow across its coil (400 cfm/ton ±10% is a good rule). Velometers and/or manometers with proper equations will assist in calculating system cfm. Check to see if blower assemblies and coils are free of dirt and that the duct system is sized properly. Remember, the unit manufacturer controls condenser air volume by design, and the installer and designer control evaporator air volume. Fan speed must be adjusted to overcome design and installation variables to obtain 400 cfm/ton ±10%.

CONDENSER AIR INLET TEMP.-°F-D.B.	EVAPORATOR AIR INLET TEMP.-°F-W.B.											
	54	56	58	60	62	64	66	68	70	72	74	76
60	13	17	18	20	24	26	28	30				
65	11	13	17	17	18	22	25	28	30			
70	8	11	12	14	16	18	22	25	28	30		
75		7	10	12	14	16	18	23	26	28	30	
80			6	8	12	14	16	18	23	27	28	30
85				6	8	12	14	17	20	25	27	28
90					6	9	12	15	18	22	25	28
95						7	11	13	16	20	23	27
100							8	11	14	18	20	25
105							6	8	12	15	19	24
110								7	11	14	18	23
115									8	13	16	21

Figure 5-10. Superheat charging table (Courtesy, Addison Products Company)

To measure cfm across the evaporator coil, use the following equation (TD is the temperature difference across the unit coil or heat exchanger):

$$cfm = \frac{Btu}{(1.08)(TD)}$$

For fossil fuel furnaces in the heating mode, the formula becomes:

$$cfm = \frac{(Btuh\ input)(Combustion\ efficiency)}{(1.08)(TD)}$$

For single-phase electric heating elements, the formula becomes:

$$cfm = \frac{((Average\ volts)(Average\ amps))(3.414\ Btu/watt)}{(1.08)(TD)}$$

If the heating element is three phase, the formula becomes:

$$cfm = \frac{((Average\ volts)(Average\ amps))(3.414\ Btu/watt)(1.73)}{(1.08)(TD)}$$

Note: When measuring the average current, it must include the current of the heating element and fan motor. This is because the fan motor's heat also enters the air stream. In addition, temperature sensors must be placed out of the direct sight line of the heating elements so as not to include any radiant heating effects. This is usually done after the first elbow in the ductwork.

If a technician measures the wattage input to an electric heater coil or gas flow to a gas-fired furnace in the ductwork, the cfm can be calculated using the above formulas.

Subtract the static pressure (Sp) from the total pressure (Tp) in the ductwork, Figure 5-11, with an inexpensive U-tube water manometer to obtain air velocity pressure (Vp). The actual air velocity in feet per minute (fpm) can then be calculated using the following equation:

$$Velocity\ (fpm) = (\sqrt{Vp})(4005)$$

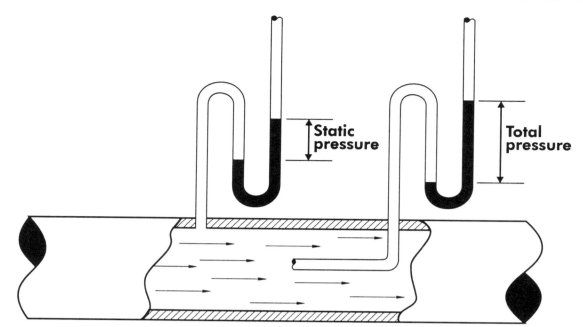

Figure 5-11. Measuring static and total pressure

Some instruments measure both total and static pressures at the same time and are connected to a delicate meter in such a way that the subtraction is done automatically, and the actual velocity of the air is read out in feet per minute (fpm), Figures 5-12 and 5-13. These instruments usually require the operator to traverse the duct with a long pitot tube at certain traverse points, Figure 5-14. An average velocity is then calculated.

Because the velocity pressure is a square root function, they cannot be first averaged together and then converted to velocity. This is because it is not mathematically possible to average a square root; a reading must be converted to velocity in feet per minute (fpm), then averaged together. The following table shows how to determine an average (Courtesy, Refrigeration Service Engineers Society):

Pitot tube reading	Calculation		Ft/min
.04	$4005\sqrt{.04}$	=	800
.05	$4005\sqrt{.05}$	=	895
.06	$4005\sqrt{.06}$	=	980
.07	$4005\sqrt{.07}$	=	1060
.05	$4005\sqrt{.05}$	=	895
.06	$4005\sqrt{.06}$	=	980
			5610

The average is as follows:

$$\frac{5610 \text{ fpm}}{6} = 935 \text{ fpm}$$

Once the air velocity in fpm is known, use the following equation to obtain unit cfm (area is the cross-sectional area of the air duct in square feet):

cfm = (Velocity in fpm) (Area in square feet)

Area for rectangular or square ducts = (length) (width)

Area for round or circular ducts = $(3.14)r^2$, where r = the radius of the round or circular duct

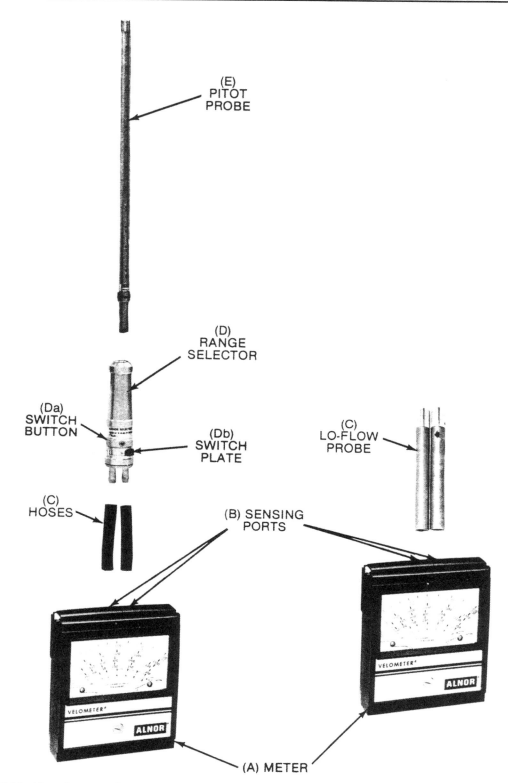

Figure 5-12. Air velocity reading instruments (Courtesy, Alnor Instrument Company)

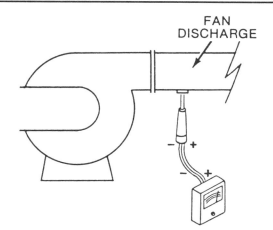

FAN DISCHARGE

Figure 5-13. Air velocity reading instrument in discharge duct (Courtesy, Alnor Instrument Company)

Static pressures (Sp) may also be measured across a working air conditioning coil (wet coil), and cfm may be obtained directly from manufacturer tables based on cfm for coil air-side pressure drops, Figure 5-15. Always refer to manufacturer literature for specific cfm based on static pressure drops across coils.

2. Measure the compressor superheat at the compressor, Figure 5-16.

3. With the unit running and stabilized, measure condenser inlet air dry bulb (db) temperature and evaporator inlet air wet bulb (wb) temperature.

4. Using Figure 5-10, find the proper compressor superheat at the intersection of db and wb temperatures.

5. If the compressor superheat measured in Step #2 is too high, refrigerant must be added. If the compressor superheat is too low, refrigerant must be recovered from the system. *Caution: Wait at least 15 minutes after adding or removing refrigerant from the system before attempting to re-measure compressor superheat.*

Example 5-1

Determine if the charge is correct for an R-22 capillary tube or fixed orifice system with the following information:

- Condenser air inlet = 85°F

- Evaporator air inlet (wb temperature) = 66°F and 400 cfm/ton (±10%) flow rate

- Suction pressure measured at compressor = 60 psig (34°F), Figure 5-17

- Compressor inlet temperature = 56°F

Solution 5-1

From the information given, it is possible to determine that the compressor superheat is 22°F (56° - 34°F). Looking at Figure 5-10, the correct superheat at the intersection of the condenser air inlet (85°F) and the evaporator air inlet (66°F) is 14°F. This system is running a compressor superheat of 22°F. The superheat is too high, so the system is undercharged. Refrigerant vapor must be added to the low side of the system while it is running. After charging is finished, let the system run for at least 15 minutes and check the compressor superheat again. *Note: If the refrigerant is a newer refrigerant that has a temperature glide associated with it, liquid refrigerant must be throttled through the low side to avoid fractionation. The proper technique is to employ a sightglass and valve, which allow liquid to be removed from the cylinder while manually adjusting the valve, causing enough restriction to "flash" the liquid before it enters the suction valve. This flashing will be visible through the sightglass, Figure 5-4. When properly charged, the sightglass should not bubble under high or low loads.*

As the technician adds refrigerant, the suction pressure should increase and the compressor inlet temperature should decrease or become colder because of less superheat. These two temperatures will come closer together as refrigerant is added. *Caution: If, during charging, the indoor wet bulb (evaporator air inlet) temperature changes more than 2°F or the outdoor ambient temperature changes more than 3°F, check the superheat table for a new superheat and charge to this new superheat.*

Measuring Duct Velocity

The velocity of an air stream in a duct is not unifom throughout the cross-section; air near the walls moves more slowly due to friction. Elbows, transitions and obstructions also cause variations in the velocity at any one cross-section.

To obtain the average velocity in ducts of 4″ diameter or larger, drill a ½″ diameter or larger hole in the duct and take a series of duct velocity readings (commonly referred to as a traverse), with the Pitot Probe, at points of equal area across the duct. A formal pattern of sensing points is recommended and these points are referred to as traverse point readings. Shown are recommended velocity reading point locations for traversing round and square (or rectangular) ducts.

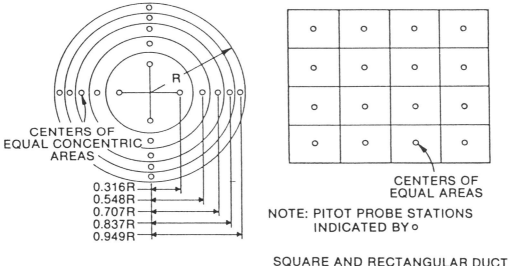

0.316R
0.548R
0.707R
0.837R
0.949R

CENTERS OF EQUAL CONCENTRIC AREAS

R

ROUND DUCT TRAVERSE POINTS

CENTERS OF EQUAL AREAS

NOTE: PITOT PROBE STATIONS INDICATED BY ○

SQUARE AND RECTANGULAR DUCT TRAVERSE POINTS

In round ducts, take velocity readings at the center of equal concentric areas; take at least 20 readings along two diameters. In square or rectangular ducts, take a minimum of 16 and a maximum of 64 readings at centers of equal areas. Calculate the average of all readings.

For maximum accuracy, observe the following precautions:

1. Perform the traverse in a section of the duct where the air stream is as uniform as practical. This is generally a location of eight or more duct diameters of straight duct upstream from the traverse location.

2. Do not take the traverse near a duct elbow, transition or obstruction.

3. Make a complete, careful and accurate traverse and record the results on a worksheet.

Figure 5-14. Method for measuring air velocity in an air duct using a pitot tube traverse (Courtesy, Alnor Instrument Company)

Rated air flow		Cooling capacity Btu/hr
CFM	H₂0*	
820	.15	24,400
800	.15	24,600
1050	.20	30,000
800	.10	28,200
1050	.21	30,800
1200	.30	36,000
1200	.25	38,000
1200	.20	35,800
1325	.20	42,000
1230	.15	40,000

* Static pressure loss for add-on coils (wet coil) and available static pressure for duct system on blower coils.

Figure 5-15. Static pressure drop vs cfm for evaporator coil

Superheat Charging Curves

Manufacturers may vary the style of superheat charging curves they offer; however, the same underlying principle holds for all tables and curves. Figure 5-18 is an example of a superheat charging curve instead of a table. The curve is based on 400 cfm/ton air flow at 50% relative humidity across the evaporator coil. The steps to charge a system according to this curve are as follows:

1. Measure indoor db temperature. This is the return air at the air handler. *Note: Use wb temperature if the relative humidity is above 70% or below 20%.*

2. Measure outdoor db temperature at outdoor unit. This is air temperature at the condenser.

3. Measure suction pressure at the compressor, and convert to a temperature using a pressure/temperature chart.

4. Measure compressor temperature on suction line near compressor.

5. Calculate the amount of compressor superheat.

6. Find the intersection where the outdoor temperature and indoor temperature meet and read compressor superheat.

If the compressor superheat of the system is more than 5°F above what the chart reads, add refrigerant vapor into the low side of the operating system until the superheat is within 5°F of the chart. If the compressor superheat of the system is more than 5°F below what the chart reads, recover refrigerant until the superheat is within 5°F of the chart. *Note: Always let the system run at least 15 minutes after adding or recovering refrigerant from the system before recalculating compressor superheat. If using a newer alternate refrigerant that has a temperature glide, liquid refrigerant must be throttled into the low side of the system while the system is running to avoid fractionation.*

Example 5-2
Determine if the charge is correct for an R-22 capillary tube or fixed orifice system with the following information:

- Indoor dry bulb temperature = 80°F

- Outdoor dry bulb temperature = 90°F

- Suction pressure at compressor = 60 psig (34°F)

- Compressor inlet temperature on suction line = 54°F

Solution 5-2
From the information given, it is possible to determine that the compressor superheat is 20°F (54° - 34°F). Looking at Figure 5-18, the correct superheat at the intersection of 90°F outdoor temperature with 80°F indoor temperature is about 17°F. This system is running a

The diagram below illustrates the proper Manifold Gauge and Temperature Analyzer connections for charging a split cooling system.

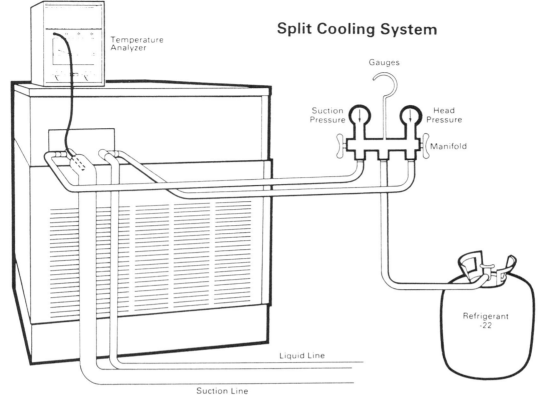

Split Cooling System

Figure 5-16. Measuring superheat at the compressor (Courtesy, Trane Company)

compressor superheat of 20°F. This is within 5°F of the superheat chart, meaning the system is fully charged and no refrigerant needs to be added to the system.

Superheat Curve Theory

The theory behind superheat tables and curves is simple. Notice in Figure 5-18 that when moving to the right on the bottom axis, the outdoor temperature rises. For a constant indoor db or wb temperature (lines that slant downward from left to right), as outdoor ambient increases the operating compressor, superheat decreases. The reason for this is that there is more head pressure pushing the subcooled liquid out of the condenser bottom through the liquid line and the capillary tube. This forces more refrigerant into the evaporator and gives less superheat, which is why some systems flood and slug liquid at hot outdoor ambients when

SUCTION PRESSURE PSIG	SATURATED SUCTION TEMP. °F.
54.9	30
57.5	32
60.1	34
62.8	36
65.6	38
68.5	40
71.5	42
74.5	44
77.6	46
80.8	48
84.0	50
87.4	52
90.8	54
94.3	56
97.9	58
101.6	60

Figure 5-17. Pressure/temperature chart for R-22

Charging Capillary Tube — Cooling Mode Only

Current Superheat Method

1. Measure indoor dry bulb temperature. (Return air at air handler).

2. Measure outdoor dry bulb temperature. (Measure at outdoor unit).

3. Measure suction pressure at pressure tap on outdoor unit.

4. Measure suction temperature near pressure tap on outdoor unit.

5. Determine degrees superheat from temperature pressure chart or low side manifold gage.

6. *Find the intersection where the outdoor temperature and indoor temperature meet and read degrees superheat. If unit superheat is more than 5° above chart value, add R-22 until within 5°. If unit superheat is more than 5° below chart value, remove R-22 until within 5°.

7. If superheat is below the 5° limit line, DO NOT ADD R-22.

Chart based on 400 CFM/Ton indoor air flow and 50% relative humidity.

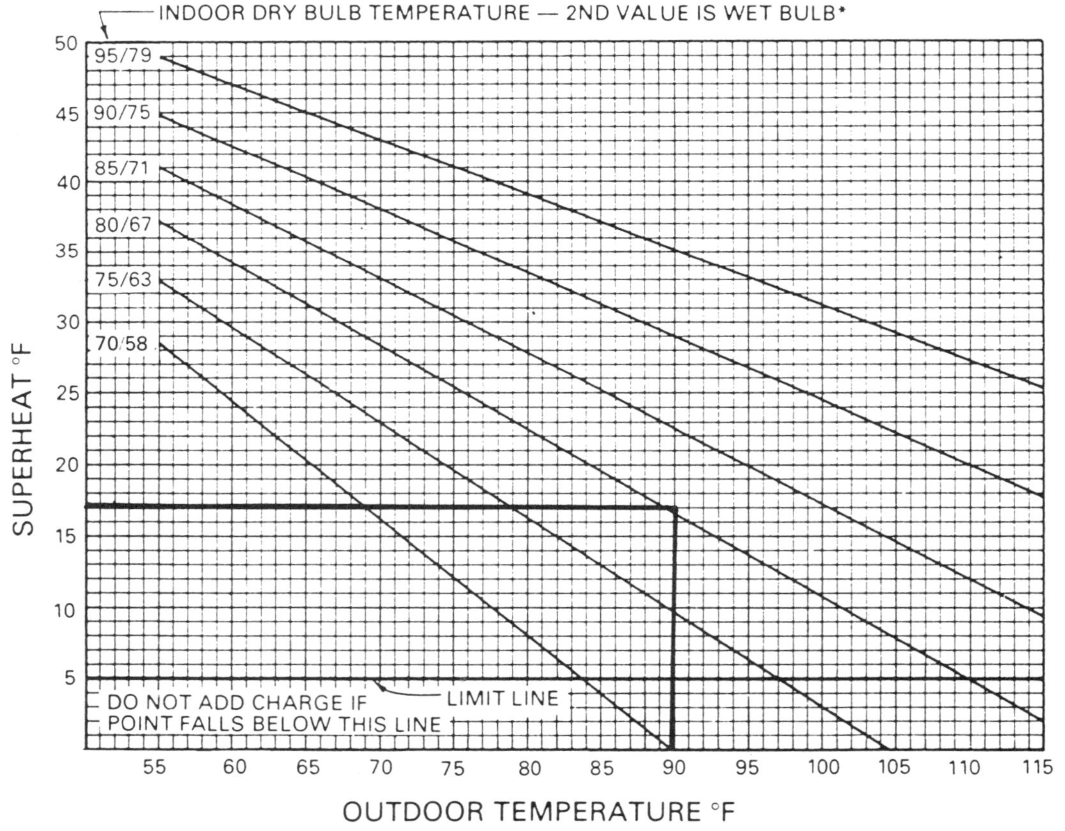

Figure 5-18. Superheat charging curve (Courtesy, Trane Company)

they are overcharged. The superheat curve prevents this from occurring if followed properly.

Referring to Figure 5-18, consider a constant indoor db temperature across the evaporator coil of 75°F. If outdoor db temperature is increased from 70° to 105°F, the operating compressor superheat will fall from 23° to 0°F. This is caused by hotter outdoor ambients, resulting in higher head pressures pushing more liquid through the capillary tube and into the evaporator. It is normal for the system to run 23°F of compressor superheat when the outdoor ambient is 70°F. Adding additional refrigerant to this system would not be wise, because if the outdoor ambient climbs to 95°F later in the day, the system compressor will slug or flood. *Note: If the relative humidity is above 70% or below 20%, use wb temperatures instead of db temperatures across the evaporator coil for Figure 5-18 to compensate for the varying latent (moisture) loads.*

If the outdoor temperature stays constant and the indoor db or wb temperature increases, the operating superheat will increase. This loading of the indoor coil with sensible heat, latent heat, or both, will cause a more rapid vaporization of refrigerant in the evaporator, causing high compressor superheats. This is a normal occurrence. Many technicians will add refrigerant in this case and overcharge the system. It is completely normal for a capillary or fixed orifice metering device system to run high superheat at high evaporator loads.

Again referring to Figure 5-18, consider if the outdoor ambient stays constant at 95°F. In this case, as the indoor db temperature across the evaporator coil rises from 75° to 95°F, the operating compressor superheat will rise from 6° to 33°F. At a 95°F indoor air db temperature and a 95°F outdoor air db temperature, the superheat should normally be 33°F according to the chart. This seems like an inefficient system with an inactive evaporator, which is the greatest disadvantage of a fixed orifice metering device. However, this is the only way a fixed orifice system can prevent slugging and flooding of refrigerant with varying indoor and outdoor loads that air conditioning systems ex-

perience. The main advantage of fixed orifice metering devices is their inexpensive cost. If this system had a TXV as a metering device, it would control a constant superheat and keep the evaporator active at high or low heat loads.

Technicians are often hesitant to measure and use wb temperatures when working on air conditioning systems. However, it is very important to get these measurements when in the field. Wet bulb temperature gives an indication of both the latent (moisture) and sensible heat loads on the coil. A simple thermocouple with some moist cotton wrapped around it placed in a hole drilled in the air duct will suffice. There are more sophisticated devices on the market for measuring wb temperatures. A psychrometer is probably one of the most popular devices used. It consists of a sock wrapped around an ordinary db thermometer. Psychrometers that are slung around a swivel connection are referred to as sling psychrometers, Figure 5-19. The technician simply wets the sock with distilled water and places it in the air stream until a temperature stabilizes. With both a db and wb temperature, the technician may obtain the relative humidity (rh) of the air by referring to charts or graphs, such as that found in Table 5-2.

When charging capillary tube and fixed orifice air conditioning systems, consult with the manufacturer of the air conditioning system and use their exact method of charging. Some manufacturers use different curves and tables for different models of their equipment, Figure 5-20. Other manufacturers have eliminated the need for a wb temperature because of custom-made charging curves that represent their laboratory tests on the equipment. Some manufacturers use a slide-rule superheating charging calculator, Figure 5-21. In general, the charging table in Figure 5-10 can be used with all capillary tube and fixed orifice air conditioning systems if wb and db temperatures are used, and the 400 cfm/ton (±10%) of evaporator air is established.

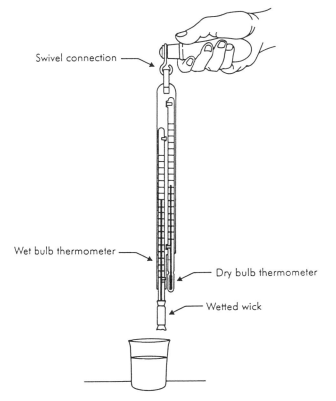

Figure 5-19. Sling psychrometer

Swivel connection
Wet bulb thermometer
Dry bulb thermometer
Wetted wick

Wet Bulb Depression

Dry Bulb Temp. °F	1	2	3	4	5	6	7	8	9	10	11	12	13	14	15	16	17	18	19	20	21	22	23	24	25	26	27	28	29	30
32	90	79	69	60	50	41	31	22	13	4																				
36	91	82	73	65	56	48	39	31	23	14	6																			
40	92	84	76	68	61	53	46	38	31	23	16	9	2																	
44	93	85	78	71	64	57	51	44	37	31	24	18	12	5																
48	93	87	80	73	67	60	54	48	42	36	31	25	19	14	8	3														
52	94	88	81	75	69	63	58	52	46	41	36	30	25	20	15	10	6	0												
56	94	89	82	77	71	66	61	55	50	45	40	35	31	26	21	17	12	8	4											
60	94	90	84	78	73	68	63	58	53	49	44	40	35	31	27	22	18	14	6	2	9									
64	95	90	85	79	75	70	66	61	56	52	48	43	39	35	31	27	23	20	16	12	14									
68	95	90	85	81	76	72	67	63	59	55	51	47	43	39	35	31	28	24	21	17	19									
72	95	91	86	82	78	72	69	65	61	57	53	52	48	45	39	35	32	28	25	22	23									
76	96	91	87	83	78	74	70	67	63	59	55	54	51	47	42	38	35	32	29	26	27									
80	96	92	87	83	79	76	72	68	64	61	57	56	53	50	44	41	38	35	32	29	30	24	21	18	16	13	11	8	6	4
84	96	92	88	84	80	77	73	70	66	63	59	58	55	54	47	44	41	38	35	32	33	27	25	22	20	17	15	12	10	8
88	96	92	88	85	81	78	74	71	67	64	61	59	57	55	49	46	43	41	38	35	35	30	28	25	23	21	18	16	14	12
92	96	93	89	85	82	78	75	72	69	65	62	61	58	57	51	48	45	43	40	38	37	33	30	28	26	24	22	19	17	15
96	96	93	89	86	82	79	76	73	70	67	64	62	59	58	53	50	47	45	42	40	40	35	33	31	29	26	24	22	20	18
100	96	94	90	86	83	80	77	74	71	68	65	63	61	59	54	52	49	47	44	42	41	37	35	33	31	29	27	25	23	21
104	97	95	90	87	84	80	77	74	72	69	66	64	62		56	53	51	48	46	44	43	39	37	35	33	31	29	27	25	24
108	97	96	90	87	84	81	78	75	72	70	67				57	54	52	50	47	45	45	41	39	37	35	33	31	29	28	26

Note: Wet bulb depression is the difference between the dry bulb and wet bulb temperatures.

Table 5-2. Psychrometric table: Percent relative humidity from dry bulb temperature and wet bulb depression (Fahrenheit temperature scale)

SUPERHEAT METHOD — CAPILLARY TUBE SYSTEM

This method of charging is effective **only** when the indoor conditions are within 2F of desired indoor comfort and the suction line pressure and temperature are stabilized.

CAUTION: USE LIQUID LINE HEAD PRESSURE METHOD IF INDOOR CONDITIONS ARE ABOVE NORMAL TEMPERATURE AND HUMIDITY.

1. Read and record outdoor ambient air dry bulb (DB) temperature entering condenser unit.
2. Read and record suction line pressure and temperature at the service valve or service port at compressor.
3. Enter proper table below as **determined by the model number** at the intersection of suction pressure and outdoor ambient temperature. Suction line temperature should coincide with the table reading.
4. If suction line temperature is not the same, adjust refrigerant charge.
   Adding R-22 will raise suction pressure and lower suction line temperature.
   Removing R-22 will lower suction pressure and raise suction line temperature.
   CAUTION — If adding R-22 raises both suction pressure and temperature, the unit is overcharged.
5. Fifteen to thirty minutes after adjusting the charge, Steps 1 through 3 are to be repeated after suction line pressure and temperature have stabilized.
6. Should the intersection of the suction pressure and outdoor ambient temperature fall in the open areas of the tables, the following are likely causes:

Left of Numbers	Right of Numbers
A. Low indoor airflow	A. Gross overcharge
B. Restricted refrigerant line	B. Defective compressor
C. Low charge	C. Indoor conditions above desired comfort

MODEL SERIES (-)ACC-, (-)ACW- STANDARD EFFICIENCY

OUTDOOR AMBIENT °F	52	54	56	58	60	62	64	66	68	70	72	74	76	78	80	82	84
ABOVE 100							43	44	45	46	48	49	50	52	53		
100						43	44	46	47	48	50	51	52	54	55		
95					45	47	48	50	51	52	54	55	56	57	58		
90					50	52	53	55	56	57	59	60	61	63			
85				53	54	56	57	59	60	61	63	64					
80			56	58	59	61	62	64	65	66	68						
75		59	60	62	63	65	66	68	69	70							
70	61	63	64	66	67	69	70	72	73								

MODEL SERIES (-)AFC-, (-)AFW- HIGH EFFICIENCY

OUTDOOR AMBIENT °F	52	54	56	58	60	62	64	66	68	70	72	74	76	78	80	82	84
ABOVE 100							42	43	44	45	46	48	49	50	52	53	
100						45	47	48	50	51	52	54	55	56	58	59	
95					50	52	53	55	56	57	59	60	61	63			
90				55	57	58	60	61	62	64	65	66					
85			58	60	61	63	64	66	67	68	70						
80			63	65	66	67	68	70	71	72							
75		66	68	69	71	72	74	75									
70		71	73	74	76												

MODEL SERIES (-)AGD-, (-)AHD- SUPER HIGH EFFICIENCY

OUTDOOR AMBIENT °F	52	54	56	58	60	62	64	66	68	70	72	74	76	78	80	82	84
ABOVE 100							42	43	44	45	46	48	49	50	52	53	54
100						43	45	46	48	49	50	52	53	54	56	58	
95					50	52	53	55	56	57	59	60	61	63	64		
90				56	57	59	60	62	63	64	66	67	68				
85				63	64	66	67	69	70	72	74						
80			68	70	71	73	74	76	77	78							
75		70	72	74	75	77	79										
70	72	74	76	77	78												

(REF. 92-20477-84-01)

92-20551-05-01

Figure 5-20. Superheat method for capillary tube system (Courtesy, Rheem Air Conditioning Division)

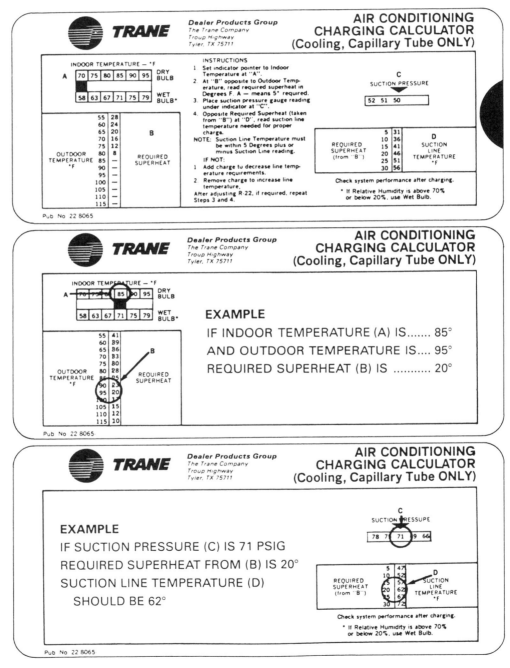

Figure 5-21. Slide-rule calculator (Courtesy, Trane Company)

CHAPTER SIX

Diagnosing Air Conditioning Systems

Air conditioning system diagnosis is not easy. A service technician must be a trained professional to diagnose a system efficiently and correctly, because it is no longer good practice to rely on rules of thumb for coil temperatures or pressures.

Air conditioning system problems mainly fall into two categories: air flow problems and refrigerant cycle problems. Both of these types of problems will be discussed in this chapter.

AIR FLOW

When there is an air side problem in an air conditioning system, it is usually because there is either too much air or too little air. Assuming that the entire air conditioning (a/c) unit was originally set up and operated properly, too much air probably will not be a problem area. This is because air handling systems do not just suddenly increase their air volume flow (cfm) without some outside help.

If the driven shieve on the blower motor becomes loose, the driven belt will ride lower on the driven shieve, decreasing fan speed instead of increasing it. It is also possible to rule out fan speed being too high, because most a/c units call for higher fan speeds to move the higher densities of colder air associated with air conditioning. This is why the majority of air flow problems involve not enough air flow rather than too much air flow. Remember, however, that air conditioning systems are not always set up properly in the first place. The duct system may not have been designed properly, resulting in oversized, undersized, or leaky ducts.

To determine if there is an air flow or refrigerant flow problem, first record the air temperatures into and out of the evaporator coil and determine if they are higher or lower than they should be. An air flow that is too low will give greater temperature differences across the coil than too much air flow. This greater temperature difference is from the air being in contact with the coil longer, thus decreasing its temperature coming out of the coil. By comparing the measured temperature difference to the manufacturer's required temperature differences, a technician can establish whether there is an air flow problem or a refrigerant flow problem.

To determine the required temperature difference across the coil, a technician must obtain the wb and db temperatures of the air entering the coil. A psychrometer, which was discussed in Chapter Five, is the only instrument needed for these measurements. In fact, a thermistor or thermocouple with a wet piece of cotton wrapped around it can give the wb temperature accurately enough for air conditioning work. Once the wb and db temperatures of the entering air are measured, the relative humidity of the entering air can be obtained from a chart or table (see Table 5-2 in Chapter Five). In Figure 6-1, for a constant air entering or return air db temperature, the temperature difference across the coil increases with decreasing relative humidities. The reason for this larger temperature difference with lower relative humidities is the decreased moisture (latent) load that the a/c coil has to condense. If the coil does not have to condense as much moisture out of the air, it can perform more sensible cooling because the coil temperature will be lower. Sensible cooling is exactly what is being measured when obtaining the temperature difference across a cooling coil with a db temperature.

If the temperature difference is greater than the required temperature difference across the coil, there is an air flow problem. The problem is not enough air flow, causing the air to stay in contact with the coil much too long and giving a greater temperature difference across the coil. If the temperature difference across the coil is less than the required temperature difference, the problem would be a refrigerant flow problem. This is because a/c systems hardly ever increase in air flow without some kind of human intervention. Some causes of decreased air flow in an a/c system include the following:

• Dirty air filters

• Faulty duct design

• Loose fan pulleys

• Slipping fan belts

• Blower motor running slow (burning out)

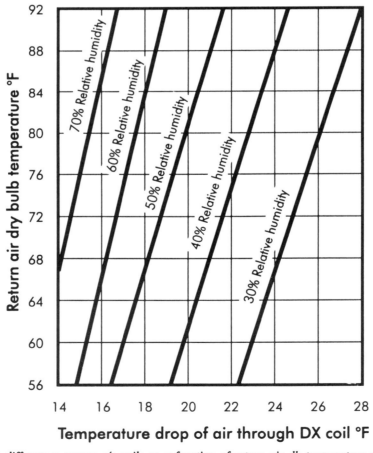

Figure 6-1. Temperature differences across a/c coils as a function of return air db temperature and relative humidity

- Restriction in duct system

- Dirty evaporator coil

- Dirty or missing fan blades

- Direct drive blower with wrong speed tap on the motor

REFRIGERANT CYCLE

As stated earlier, if there is a low temperature difference across the evaporator coil, the problem is refrigerant flow. The low temperature difference also indicates a capacity drop, meaning that the heat handling capabilities of the system have failed. Assuming the a/c system has been running for some time under these conditions, the service problem cannot be electrical.

The larger the coil surface area of the evaporator, the closer the coil temperature will be to the entering air temperature. This forces the coil temperature to a higher temperature, which means the a/c unit will run higher evaporating (suction) pressures and temperatures. With this increase in vapor pressure and coil temperature, the a/c unit will experience higher efficiencies from the higher pressure refrigerant gases entering the reciprocating compressor with each revolution of its crankshaft. The compression ratio will decrease from the higher evaporator temperatures and pressures, causing the mass flow rate of refrigerant vapors through the compressor to increase. These are the reasons why manufacturers produce larger, more efficient a/c coils. The larger and more efficient the coil, the smaller the coil temperature from the entering air temperature. This causes higher evaporator pressures and more efficient a/c units. This larger and more efficient coil does increase the manufacturing cost, but increased unit efficiency should offset the higher costs. The term used to describe an a/c unit's efficiency is the *Energy Efficiency Rating (EER)*.

Remember that coil surface area may vary from one manufacturer to another. One manufacturer may choose large coil areas for high suction pressures and smaller compressors, while another may use smaller surface area coils, which cause lower suction pressures and use larger compressors. As long as the a/c unit meets the required Btu requirements and EERs, it is a trade-off. Coil surface area is often a function of geographic region. Large surface coils running higher suction pressures may not have low enough coil temperatures to remove enough moisture (latent heat). Their apparatus dew points will be too high to condense the right amount of moisture from the air passing through the coil, causing high humidity, mold, and human discomfort.

Condensers

Other than having different pressures and temperatures, the condenser and evaporator are very similar. Condensers are usually a bit larger than the evaporator in order to handle the evaporator load, suction line superheat heat gain, heat of compression, and motor heat loads of the compressor. Like the evaporator, the surface area of the condenser affects the design temperature difference between the condensing temperature and the ambient temperature. The larger the condenser, the more expensive it is to manufacture, but the EER will be much higher. Condensing temperatures and pressures vary with coil surface area size and EER, just like evaporator temperatures and pressures.

As stated in a previous chapter, subcooling is defined as the difference between the measured liquid temperature and the saturation temperature at a given pressure. Condenser subcooling can be measured at the condenser outlet with a thermometer or thermocouple and a pressure gauge. Simply subtract the condenser outlet temperature from the saturation temperature at the condenser outlet to obtain the amount of liquid subcooling in the condenser. The saturation pressure must be measured at the condenser outlet and converted to a temperature. Always take the pressure at the same point the temperature is taken to alleviate any pressure drop error through the condenser.

A forced air condenser should have from 6° to 10°F of liquid subcooling if charged properly. However, the amount of condenser subcooling depends on the static and friction line pressure losses in the liquid line and will vary from system to system. The 6° to 10°F of liquid subcooling is assuming no liquid amplification pump is pressurizing the liquid out of the condenser. Condenser subcooling can be an indicator of the refrigerant charge in the system. For receiverless systems, less refrigerant charge means less subcooling. The rated EER has little or no affect on condenser subcooling.

Another factor that affects condenser subcooling is the air entering the condenser. As the condenser air entering temperature increases, the liquid subcooling decreases. This is because higher condensing (head) pressures force more of the subcooled liquid through the metering device to the evaporator. As a result, evaporator superheat is less due to increased flow rate through its coil.

Liquid Line Restrictions

Liquid line restrictions may be caused by a number of factors, including the following:

- Restricted filter drier

- Restricted TXV screen

- Kinked liquid line

- Kinked or bent U bend on lower condenser coil

- Restricted solder joint in the liquid line

- Oil logged capillary tube

A restricted liquid line starves the evaporator of refrigerant, which causes low pressures in the evaporator. If the evaporator is starved of refrigerant, the compressor and condenser will also be starved, because the evaporator will not absorb much heat for the condenser to reject. Most of the refrigerant will be in the condenser, which will not necessarily cause high head pressures because of the reduced heat load on the evaporator. Because most of the refrigerant

charge is in the condenser, liquid subcooling in the condenser will increase. This is very different from an undercharge of refrigerant, which results in low condenser subcooling. If the system has a receiver, most of the refrigerant will be in the receiver, causing lower than normal head pressures.

Symptoms of a restricted liquid line include the following:

- Local cool spot after a severe restriction from expansion of refrigerant at the local pressure drop.

- Low ampere draw at the compressor from reduced refrigerant flow.

- Bubbles in the sightglass if the restriction is before the sightglass. (Sightglasses are optional on some systems.)

- High superheats from a starved evaporator.

- Low to normal head pressures in a system with a receiver due to liquid backed up in the condenser, but no heat load for condenser to reject.

Suction Line Restrictions

The suction line is a much more sensitive refrigerant line than the liquid line, because vapor instead of liquid flows through it. A restricted suction line will cause low suction pressures and a starved compressor and condenser. A starved compressor will lead to low compressor amp draw because of its lightened load. The condensing pressure will also be low because of its light load. Since suction line restrictions starve compressors of refrigerant, the entire mass flow rate of refrigerant will decrease through the system, causing high superheats from inactive evaporators. Restricted and/or dirty suction filters are the major cause of suction line restrictions.

If there is a suction line restriction, liquid subcooling in the condenser will be normal to a bit high, because a lot of refrigerant will be in the condenser coil but will not be circulated very fast. The condenser subcooling may be normal to a bit high if there is a receiver in the system. Condenser subcooling lets the technician know there is refrigerant in the system and that an undercharge of refrigerant can be ruled out.

Undercharge

Undercharged systems mean less mass flow rate throughout the entire system. Low suction and discharge pressures with high superheat in the evaporator are all indications of an undercharge. Severely undercharged systems will run very low condenser subcooling, because there is no refrigerant to subcool. If the subcooling drops to zero, the hot gas in the condenser will start to leave the condenser with some liquid, and bubbles will form in the sightglass if the system is fortunate enough to have one. Compressor amp draw will be low because of the decreased refrigerant flow. Service technicians may be confused as to whether the problem is an undercharge of refrigerant or a liquid line restriction. Remember, a liquid line restriction will give the system a lot of subcooling in the condenser and an undercharge will not. Otherwise, symptoms are very similar.

Overcharge

A system with an overcharge of refrigerant will have higher than normal condensing temperatures because of liquid backing up in the condenser causing high subcooling and robbing the condenser of useful condensing area. The elevated head pressure will cause the volumetric efficiency of the compressor to decrease because of higher pressures of re-expanding volume vapors in the clearance pocket of the compressor. The amp draw of the compressor will increase from the higher head pressure, creating higher compression ratios, and the entire system will have reduced capacities. If the system has a TXV metering device, the TXV will still try to maintain its superheat and the evaporator pressure will be normal to slightly high, depending on the amount of overcharge. The higher evaporator pressure will be caused from the decreased mass flow rate from the

higher compression ratio, and the evaporator will have a hard time keeping up with the higher heat load of the warm entering air temperature. The TXV will have a tendency to overfeed on its opening strokes due to the high head pressures.

If the system uses a capillary tube metering device, the same symptoms occur except for evaporator superheat. Remember, one reason a capillary tube system is critically charged is to prevent flooding of the compressor on low evaporator loads. The higher head pressures of an overcharged capillary tube system will have a tendency to overfeed the evaporator, resulting in decreased superheat. If the system is over 10% overcharged, liquid may enter the suction line and reach the suction valves or crankcase, resulting in compressor damage and failure.

Compressor Inefficiencies

Compressors are responsible for circulating refrigerant throughout a system; therefore, inefficient compressors will decrease the heat transfer ability of an air conditioning system. Leaky valves or worn piston rings are two of the major problems that lead to compressor inefficiencies. A symptom for an inefficient compressor is high suction pressure along with low discharge pressure. The evaporator will not be able to handle the load due to decreased refrigerant flow, and the conditioned space temperature will start to rise. This rise in return air temperature will overload the evaporator, causing high suction pressures and higher than normal superheats.

Piston ring blow-by and reed valve leakage can also cause high suction pressures. The condenser will see a reduced load from the decreased mass flow rate of refrigerant being circulated through it. The reduced condenser load will cause a low condensing pressure. The compressor amp draw will be lowered because of a lower work load from the low mass flow rate of refrigerant. Subcooling in the condenser should be a bit low from the reduced heat load on the condenser.

Symptoms of an inefficient compressor include the following:

- High suction pressures

- Low head pressures

- Low compressor amp draw

- High superheat

- High return air temperature

- Condenser subcooling (low to normal)

Noncondensibles

Air and water vapor are probably the best known noncondensibles in a refrigeration or air conditioning system. Noncondensibles usually enter a system through poor service practices and/or leaks. A technician forgetting to purge hoses can let air and water vapor into a system. The air and water vapor will pass through the evaporator and compressor because the compressor is a vapor pump. Once the air reaches the condenser, it will remain at its top and not condense. The subcooled liquid seal at the condenser bottom will prevent the air from passing out of the condenser. This air and water vapor will take up valuable condenser surface area and cause high head pressures. Subcooling will be high because the high head pressures cause a greater temperature difference between the liquid temperature in the condenser and the ambient. This will allow for more subcooling because of the greater temperature difference. Noncondensibles in a system and an overcharge of refrigerant have very similar symptoms when a TXV metering device is used. Often, the only way to distinguish between noncondensibles in a system and an overcharge is to shut the unit off for a while and let the condenser fan run. This will cool the system down quickly. Once the system is cooled down, if the pressures on the high side of the system still remain relatively high while the unit is off, noncondensibles are in the system. There will be no pressure/temperature relationship with the refrigerant and the ambient if noncondensibles are in the condenser. With an overcharge of

refrigerant, the high side pressure should decrease rapidly with the system off and the condenser fan running.

Symptoms of noncondensibles in the system include the following:

- High head (condensing) pressures

- High subcooling

- High compression ratios

- High discharge temperatures

Low Condenser Air Entering Temperature

Low condenser air entering temperature will cause a low head pressure from the excessive heat transfer between this cool ambient and the condenser coil. Low head pressures may reduce flow through metering devices, which have capacity ratings dependent on the pressure differences across them. A 30 psi pressure difference is usually the minimum across TXVs. This reduced refrigerant flow causes a starved evaporator, which will cause low suction pressures and high superheats. However, this may be offset by increased subcooling at lower ambients.

This entire drop in capacity may decrease the heat removal ability of an air conditioner if it is not designed for it. If not designed properly, liquid will start to back up in the condenser. Because of a low heat transfer rate caused from the lower condenser temperature, the liquid temperature in the bottom of the condenser will be low, causing liquid subcooling in the condenser to increase. Less refrigerant circulated also means less work for the entire system, so the ampere draw of the compressor will be lowered.

If the system is set up for reduced condenser air entering temperature, the head pressure can be designed to float or change with the changing ambient temperature. This will give lower head pressures and increased efficiencies. A properly matched TXV to handle reduced pressure drops across its orifice may have to be incorporated into the design (see Chapter Two). Symptoms of reduced condenser air entering temperature include the following:

- Low entering air temperature

- Low suction pressure (if not designed for low head pressure at TXV)

- Low discharge (condensing) pressure

- High superheat (if not designed for low head pressure at TXV)

- Low amp draw

- High subcooling

High Condenser Air Entering Temperature

Higher outdoor ambients will cause head pressures to elevate in order to complete the heat rejection task. The temperature difference (TD) between condensing temperature and ambient will go down, and the refrigerant gas will not condense until the head pressure rises. The condenser cannot reject as much heat at this lower TD and will accumulate the heat. The accumulated heat forces the condensing temperature to elevate to a TD where the heat can be rejected. Remember, the temperature difference is the driving potential for heat transfer. However, this heat rejection happens at a higher condensing temperature, forcing the system to have higher compression ratios and lower efficiencies.

High head pressures cause the compression ratio to increase, causing low volumetric efficiencies from higher pressure vapors re-expanding in the clearance volume of the piston cylinder on each downstroke. As volumetric efficiencies decrease, mass flow rates decrease and the compressor is less efficient. High head pressures will also elevate liquid temperatures entering the metering device, which will increase evaporator flash gas and decrease the net refrigeration effect. Because of these inefficiencies, the suction pressure may be a bit higher.

The system will have a hard time maintaining temperature and humidity of the conditioned space. Evaporator superheat will vary depending on the type of metering device.

Capillary Tubes
Flow rates through a capillary tube metering device or any fixed orifice metering device depend on the pressure difference across the metering device. Higher head pressures will increase the flow rate through the metering device, pushing the subcooled liquid at the condenser bottom through the metering device at a faster rate. Because of this, condenser subcooling will decrease. Evaporator superheat will also decrease because of a flooded evaporator coil with a lot of flash gas at its entrance.

TXVs
TXV systems will try to maintain evaporator superheat even though the pressure drop across the valve may be out of its control range at the higher ambient temperatures. Here, the condenser subcooling may be normal to a bit high.

CHAPTER SEVEN

Systematic Troubleshooting

Troubleshooting modern refrigeration and air conditioning equipment requires a thorough knowledge of systems and their functions. In order to troubleshoot well, a technician must use the proper instrumentation and have certain organizational skills. Organizational skills help the technician solve even the toughest problems more quickly.

SERVICE CHECKLIST

One way to organize information is to use a service checklist similar to that shown in Table 7-1.

The information collected in Table 7-1a will give the technician the information necessary to calculate the values in Table 7-1b.

TROUBLESHOOTING

The service checklist shown in Table 7-1 will be used throughout this chapter in order to troubleshoot systems that are not working properly. In all examples that follow, assume the system components were properly sized originally and are still in the system. All systems will be low temperature TXV/receiver systems using R-134a as the refrigerant. The desired box temperature is 0°F. Remember that establishing a valid standard for a "normal" system in the

Item to be measured	Measured value
Compressor discharge temperature	°F
Condenser outlet temperature	°F
Evaporator outlet temperature	°F
Compressor inlet temperature	°F
Ambient temperature	°F
Box temperature	°F
Compressor volts	V
Compressor amps	A
Low side (evaporator) pressure	psig
High side (condensing) pressure	psig

(a)

Item to be calculated	Calculated value
Condenser split	°F
Condenser subcooling	°F
Evaporator superheat	°F
Compressor superheat	°F

(b)

Table 7-1. Service checklist

field of refrigeration and air conditioning can sometimes be difficult. Especially when it has often been said that there is no exact science when troubleshooting in the hvac/r field. Many systems, especially field assembled refrigeration and air conditioning systems, have fewer standards for "normal" and require greater judgment on the part of the service technician. Please keep this in mind throughout this chapter.

Undercharge of Refrigerant

Table 7-2 gives the information for an undercharged system.

Item to be measured	Measured value
Compressor discharge temperature	170°F
Condenser outlet temperature	78°F
Evaporator outlet temperature	10°F
Compressor inlet temperature	50°F
Ambient temperature	70°F
Box temperature	20°F
Compressor volts	230 V
Compressor amps	Low
Low side (evaporator) pressure	3.94 in. Hg (-20°F)
High side (condensing) pressure	86.4 psig

(a)

Item to be calculated	Calculated value
Condenser split	10°F
Condenser subcooling	2°F
Evaporator superheat	30°F
Compressor superheat	70°F

(b)

Table 7-2. Service checklist for undercharged system

The symptoms for this undercharged system are as follows and are explained in detail following the list:

- Medium to high compressor discharge temperatures
- High evaporator superheat
- High compressor superheat
- Low condenser subcooling
- Low compressor amps
- Low evaporator pressures
- Low condensing temperatures

Compressor discharge

This temperature is a bit high compared to normal system operations. The 170°F discharge temperature is caused by the evaporator and compressor running high superheat along with high compression ratios. When undercharged, do not expect the TXV to control superheat. The TXV may have vapor and liquid at its entrance. The evaporator will be starved of refrigerant and running high superheat. The compressor will see increasingly high superheat with the compression stroke adding even more.

Compression ratios will also be elevated, giving the system a higher than normal heat of compression. Compression ratios will be high from low evaporator pressures, which will give the system very low volumetric efficiencies and cause unwanted inefficiencies with low flow rates. The compressor will have to compress much lower pressure vapors coming from the suction line to the condensing pressure. This requires a greater compression range and a higher compression ratio. This greater compression range from the lower evaporator pressure to the condensing pressure is what causes compression work and generates some heat of compression. The increased heat may be seen by the medium to high compressor discharge temperature. However, because of the lower flow rates from the lower volumetric efficiencies, the compressor will see a somewhat low load, which keeps the discharge temperature from getting too hot. Higher compression ratios and higher superheat are the causes of somewhat high discharge temperatures.

The absolute limit for any discharge temperature measured about 3 inches from the compressor on the discharge line is 225°F. The back of the discharge valve is usually about 50° to 75°F hotter than the discharge line. This would make the back of the discharge valve about 275° to 300°F. This could vaporize oil around the cylinders and cause excessive wear. At 350°F, oil will break down, and the compressor will overheat. Compressor overheating is one of the most serious field problems. Try to keep discharge temperatures below 225°F for longer compressor life.

High evaporator superheat

Since the evaporator is starved of refrigerant from the undercharge, there will be high evaporator superheat, which will lead to high compressor (total) superheat. The receiver will not be getting enough liquid refrigerant from the condenser because of the shortage of refrigerant in the system. This will starve the liquid line and may even bubble a sightglass if the condition is severe enough. The TXV will not see normal pressures and may even try to pass liquid and vapor from the starved liquid line. The TXV will also be starved and cannot be expected to control superheat.

High compressor superheat

Again, since the liquid line, TXV, and evaporator will be starved of refrigerant from the undercharge, so will the compressor. This can be seen in the high compressor superheat reading.

Low condenser subcooling

Because the compressor will have very hot vapors from the high superheat readings, the gases entering the compressor will expand and have a low density. The compression ratio will be high from the low suction pressure, causing low volumetric efficiencies. The compressor will not pump as much refrigerant, and all components in the system will be starved of refrigerant. The 100% saturated liquid point in the condenser will be very low, causing low condenser subcooling. The condenser will not receive enough refrigerant vapor to condense it to a liquid and feed the receiver. Condenser subcooling is a good indicator of how much refrigerant charge is in the system. Low condenser subcooling may mean a low charge, while a high condenser subcooling may mean an overcharge.

This is not true for capillary tube systems, because they have no receiver. A capillary tube system can run high subcooling simply from a restriction in the capillary tube or liquid line. The excess refrigerant will accumulate in the condenser, causing high subcooling and high head pressures. If a TXV/receiver system is restricted in the liquid line, some refrigerant will accumulate in the condenser but most will collect in the receiver. This will cause low subcooling and low head pressure.

Low compressor amps

High superheat will cause compressor inlet vapors from the suction line to expand, which decreases their density. Low density vapors entering the compressor will mean low refrigerant flow through the compressor. This will cause a low amp draw, because the compressor does not have to work as hard compressing the low density vapors. Low refrigerant flow will also cause refrigerant-cooled compressors to overheat.

Low evaporator pressure

Low evaporator pressure is caused by the compressor being starved of refrigerant. The compressor tries to draw refrigerant into its cylinders, but there is not enough refrigerant to satisfy it. The entire low side of the system will experience low pressure.

Low condensing pressure/temperature

Because the evaporator and compressor are being starved of refrigerant, the condenser will also be starved. Starving the condenser will reduce the heat load on the condenser, because it cannot reject heat when it is short of refrigerant. Without the ability to accept or reject heat, the condenser will be at a lower temperature. This lower temperature will cause a lower pressure in the condenser because of the pressure/temperature relationship at saturation.

The temperature difference between the condensing temperature and the ambient is called the condenser *delta tee* or *split*. The service industry usually uses the term *condenser split*, so that is the term that will be used in this book. The equation for the condenser split is as follows:

$$\text{Condensing temperature} - \text{Ambient temperature} = \text{Condenser split}$$

As the condenser receives less heat from the starved compressor, the condenser split will decrease. No matter what the ambient is, the condenser split, or difference between the condensing temperature and the ambient, will remain the same if the load remains the same on the evaporator. Condenser split will change if the load on the evaporator changes. Some common condenser splits for refrigeration applications are listed in Table 7-3. Box temperatures will tell the technician the evaporator load; low box temperature means a low load, and a high box temperature means a high load.

Low evaporator loads	10° to 15°F
Medium evaporator loads	15° to 25°F
High evaporator loads	25° to 35°F

Note: These common condenser splits are assuming the head pressure is not being floated with the ambient.

Table 7-3. Common condenser splits for refrigeration applications only

Overcharge of Refrigerant

The system checklist in Table 7-4 shows a system with an overcharge of refrigerant.

Item to be measured	Measured value
Compressor discharge temperature	240°F
Condenser outlet temperature	90°F
Evaporator outlet temperature	15°F
Compressor inlet temperature	25°F
Ambient temperature	70°F
Box temperature	10°F
Compressor volts	230 V
Compressor amps	High
Low side (evaporator) pressure	8.8 psig (5°F)
High side (condensing) pressure	172 psig (120°F)

(a)

Item to be calculated	Calculated value
Condenser split	50°F
Condenser subcooling	30°F
Evaporator superheat	10°F
Compressor superheat	20°F

(b)

Table 7-4. Service checklist for overcharged system

The symptoms for this overcharged system are as follows and are explained in detail following the list:

- High discharge temperature
- High condenser subcooling
- High condensing pressures
- Higher condenser splits
- Normal to high evaporator pressures

- Normal evaporator superheat

- High compression ratios

High discharge temperature

With an overcharged system, the high discharge temperature of 240°F is caused from the high compression ratio. Liquid backed up in the condenser will flood some of the condenser surface area, causing high head pressures. All of the heat being absorbed in the evaporator and the suction line, along with motor heat and high heat of compression from the high compression ratio, must be rejected into a smaller condenser because of the backed up liquid.

High condenser subcooling

Because of the overcharge of refrigerant in the system, the condenser will have too much liquid backed up at its bottom, causing high subcooling. Remember, any liquid in the condenser lower than the condensing temperature is considered subcooling. This can be measured at the condenser outlet with a thermometer or thermocouple. Subtract the condensing out temperature from the condensing temperature to obtain the amount of liquid subcooling. A forced air condenser should have at least 8°F of liquid subcooling; however, subcooling amounts depend on system piping configurations, liquid line static, and friction pressure drops. Condenser subcooling is an excellent indicator of the system refrigerant charge; the lower the refrigerant charge, the lower the subcooling, and the higher the charge, the higher the subcooling.

High condensing pressures

Subcooled liquid backed up in the condenser will cause a reduced condensing surface area and raise condensing pressures. Once the condensing pressures are raised, there is more of a temperature difference between the ambient and condensing temperature, causing greater heat flow to compensate for the reduced condensing surface area. The system will still reject heat, but at a higher condensing pressure and temperature.

Higher condenser splits

Because of the higher condensing pressures, thus higher condensing temperatures, there will be a greater temperature difference (split) between the ambient and condensing temperature.

Normal to high evaporator pressures

The TXV will try to maintain its evaporator superheat, and the evaporator pressure will be normal to slightly high depending on the amount of overcharge. If the overcharge is excessive, the evaporator's higher pressure will be caused by the decreased mass flow rate through the compressor from high compression ratios, which cause low volumetric efficiencies. The evaporator will have a harder time keeping up with the higher heat loads from the warmer entering air temperature. The TXV may have a tendency to overfeed on its opening stroke due to the high head pressures.

Normal evaporator superheat

The TXV will try to maintain superheat even at an excessive overcharge. As mentioned earlier, the TXV may overfeed slightly during its opening strokes, but it should stabilize if still in its operating pressure ranges.

High compression ratios

The condenser flooded with liquid during the overcharge will run high condensing pressures. This will cause high compression ratios and volumetric efficiencies, resulting in low refrigerant flow rates.

Overcharged Capillary Tube Systems

If the system uses a capillary tube metering device instead of a TXV, the same symptoms will occur, with the exception of the evaporator superheat. Remember, capillary tube systems are critically charged to prevent floodback of refrigerant to the compressor during low evaporator loads. The higher head pressures of an overcharged system will have a tendency to overfeed the evaporator, thus decreasing the superheat. If the system is overcharged by more than 10%, liquid can enter the suction line and reach the suction valves or crankcase, causing compressor damage and failure (see Chapter Five).

Dirty Condenser or Restricted Air Flow Over Condenser

The system checklist in Table 7-5 shows a system with a dirty condenser or restricted air flow over the condenser.

Item to be measured	Measured value
Compressor discharge temperature	250°F
Condenser outlet temperature	110°F
Evaporator outlet temperature	10°F
Compressor inlet temperature	25°F
Ambient temperature	70°F
Box temperature	15°F
Compressor volts	230 V
Compressor amps	High
Low side (evaporator) pressure	6.2 psig (0°F)
High side (condensing) pressure	185.5 psig (125°F)

(a)

Item to be calculated	Calculated value
Condenser split	55°F
Condenser subcooling	15°F
Evaporator superheat	10°F
Compressor superheat	25°F

(b)

Table 7-5. Service checklist for a system with a dirty condenser or restricted air flow over the condenser

The symptoms for this system with a dirty condenser or with restricted air flow over the condenser are as follows and are explained in detail following the list:

- High discharge temperatures

- High condensing pressures

- High condenser splits

- Normal to high condenser subcooling

- Normal to high evaporator pressures

- Normal superheat

- High compression ratios

- High amp draw

High discharge temperatures

With the outside of the condenser coil plugged with dirt, grease, weeds, cottonwood fuzz, or dust, or poor airflow from an inoperable fan, the discharge temperature will be high. The high discharge temperature is caused from the refrigeration system not being able to reject heat in the condenser. The condensing pressure rises and because of the pressure/temperature relationship, the condensing temperature will also rise. The high heat of compression from the high compression ratio will also cause the discharge temperature to be high.

High condensing pressures

Since heat from the evaporator, suction line, motor, and heat of compression is rejected in the condenser, the condenser coil must be kept clean with the proper amount of air flow through it. A dirty condenser or restricted air flow across the condenser cannot reject this heat fast enough, and the condensing temperature and pressure will elevate. Once this temperature is elevated, the condenser split will become greater and heat will be rejected at the needed rate. However, the system will be operating at elevated condensing temperatures and pressures, causing unwanted inefficiencies from high compression ratios.

High condenser splits

As mentioned earlier, when the condensing temperature rises above the ambient, the temperature difference between the ambient and the condensing temperature will become greater. This is the higher condenser split. At the higher splits, heat can be rejected because a greater temperature difference will enhance heat transfer. The system will suffer at these higher condensing pressures and temperatures because of the unwanted low volumetric efficiencies from the higher compression ratios.

Normal to high condenser subcooling

High condensing pressures cause high compression ratios, which in turn, cause low volumetric efficiencies. Low volumetric efficiencies cause low refrigerant flow rates, which will not form much subcooling. However, what subcooling is formed in the condenser will be at an elevated temperature and will reject heat to the ambient faster because of the higher condenser split. Because of this faster heat rejection, the liquid in the condenser will cool faster and have a greater temperature difference when compared to the condensing temperature. This is one of the big differences between an overcharge of refrigerant and a blocked condenser. An overcharge of refrigerant can cause very high condenser subcooling, but a blocked condenser will not.

Normal to high evaporator pressures

Again, the TXV will try to maintain a constant amount of evaporator superheat. Because of the low refrigerant flows caused by low volumetric efficiencies, the evaporator may not keep up with the heat load. This may cause high box temperatures and slightly higher evaporator pressures. The TXV may let too much refrigerant out during the beginning of its opening stroke from the higher head pressures, causing slightly higher evaporator pressures. Otherwise, because the TXV is maintaining superheat and doing its job, there may be normal evaporator pressures depending on the severity of the condenser condition.

Normal superheat

The TXV will maintain the set evaporator superheat unless the condensing pressure exceeds the range of the valve. Each TXV has a pressure range that it can operate within. Read the top of the TXV or consult with the manufacturer for more precise information on TXV temperature and pressure ranges.

High Compression Ratios

Higher condensing pressures cause higher compression ratios, which lead to lower volumetric efficiencies.

High amp draw

The high compression ratio will cause a greater pressure range for the suction vapors to be compressed, requiring more work from the compressor and increasing the amp draw. A look at any amp performance curve will show a higher amp draw at higher head pressures if the suction pressure is constant.

Restricted Liquid Line After the Receiver

The system checklist in Table 7-6 shows a system with a restricted liquid line after the receiver (before the filter drier).

The causes of a restricted liquid line include the following:

- Restricted filter drier

- Restricted TXV screen

- Kinked liquid line

- Restricted liquid line solder joint

Many technicians believe that when any part of the system's high side is restricted or plugged, head pressures will be elevated. *This is not the case, especially in a TXV/receiver system.* A restricted liquid line will starve the evaporator of refrigerant, causing low evaporator pressures. With a starved evaporator, the compressor and condenser will also be starved, causing lower condenser and evaporator pressures.

Item to be measured	Measured value
Compressor discharge temperature	215°F
Condenser outlet temperature	70°F
Evaporator outlet temperature	30°F
Compressor inlet temperature	60°F
Ambient temperature	75°F
Box temperature	30°F
Compressor volts	230 V
Compressor amps	Low
Low side (evaporator) pressure	1.8 psig (-10°F)
High side (condensing) pressure	95 psig (85°F)

(a)

Item to be calculated	Calculated value
Condenser split	10°F
Condenser subcooling	15°F
Evaporator superheat	40°F
Compressor superheat	70°F

(b)

Table 7-6. Service checklist for a system with a restricted liquid line after the receiver (before the filter drier)

The symptoms for this system with a restricted liquid line after the receiver are as follows and are explained in detail following the list:

- Higher than normal discharge temperature

- High superheat

- Low evaporator pressures

- Low condensing pressures

- Normal condenser subcooling

- Low condenser splits

- May be a local cool spot or frost after the restriction

- Bubbles in sightglass if restriction is before the sightglass

- Low amp draw

- Short cycle the low pressure control (LPC)

Higher than normal discharge temperature

High discharge temperatures are caused from high compressor superheat, which results from a starved evaporator. High compression ratios from the low evaporator pressure will cause high heat of compression and high discharge temperatures.

High superheat

Both evaporator and compressor superheat will be high. This is caused from the TXV, evaporator, and compressor being starved of refrigerant from the liquid line restriction.

Most of the refrigerant will be in the receiver, with some in the condenser.

Low evaporator pressures

The low evaporator pressure is caused from the compressor being starved of refrigerant. The compressor will try to draw refrigerant from the evaporator through the suction line, but the liquid line restriction will prevent refrigerant from entering the evaporator. This will cause the compressor to put the evaporator in a low pressure situation.

Low condensing pressures

The evaporator, compressor, and condenser will all be starved of refrigerant. Reduced refrigerant to the evaporator will cause a reduced heat load to be delivered to the condenser. The condenser will not have to elevate its temperature and pressure to reject heat.

Normal condenser subcooling

Since the condenser will be starved of refrigerant, it will not condense much vapor to liquid. Much of the liquid in the condenser will probably remain there and subcool because of the low refrigerant flow caused from the restriction. The receiver will have a reduced flow in and out. Most of the refrigerant will be in the receiver, with some in the condenser. If the receiver is in a hot ambient, subcooling may be lost as refrigerant sits in the receiver. This is why some commercial systems have receiver bypasses for certain situations. Receiver bypasses are a liquid line solenoid valve controlled by a thermostat, which bypasses liquid around the receiver, to the liquid line.

Low condenser splits

Because the condenser is being starved of refrigerant, there is not much heat to reject. This will cause low condenser splits.

Local cold spot or frost after restriction

Liquid refrigerant flashing to vapor might occur at the restriction if the restriction is severe enough. Feeling the liquid line and the filter drier may reveal a local cold spot. A thermistor on the liquid line about 12 inches before the entrance of the TXV should not be colder than the ambient that surrounds it. If it is, there is a sure restriction somewhere upstream.

There are many times when a filter drier or line may be partially plugged and the technician cannot feel a temperature difference across it with his or her hands. Humans can distinguish a temperature difference of more than 10°F with their hands, but only if the temperature differences are a little higher than their body temperature of 98.6°F. A filter drier in an R-12 system with a condensing temperature of 110°F would need a 20 psi pressure drop to exhibit a 10°F temperature difference. R-22 and R-502 systems would need about a 30 psi pressure drop to exhibit a 10°F temperature difference.

Because of this, many filter drier restrictions go unchecked by technicians, because they are difficult to sense by touch. The use of a sightglass after the filter drier to show flashing will assist the technician. The same sightglass will also assist in system charging. A sightglass that indicates moisture will alert the technician if the system is contaminated with moisture.

Bubbles in sightglass

If there is a restriction in the liquid line before the sightglass, bubbles are sure to occur in the sightglass. Many technicians believe that a bubbling sightglass means nothing but an undercharge of refrigerant. *This is not true.* On start-up of some refrigeration systems when there is a large load on the system, bubbling and flashing may occur in the sightglass downstream of the receiver. This bubbling is caused from a pressure drop at the entrance of the outlet tube of the receiver, Figure 7-1. Bubbling may also occur during rapid increases in loads. The TXV may be opened wide during an increase in load, and some flashing could occur even though the receiver has sufficient liquid. Sudden changes in head pressure control systems, which may dump hot gas into the receiver to build up head pressure, often will bubble a sightglass even though there is sufficient liquid in the receiver to form a seal on the receiver dip tube outlet. A sightglass on the receiver will prevent technicians from overcharging in this case but would cost the manufacturer a bit more money initially.

A sightglass on the liquid line before the TXV will also help let the technician know if any liquid flashing is occurring before the TXV. This flashing could be from loss of subcooling or too much static and/or friction pressure drop in the liquid line before it reaches the TXV causing liquid line flash gas (see Chapter Five).

As stated in Chapter Five, there is a big difference between a bubbling sightglass and a low flow rate sightglass. Bubbles entrained in the liquid are a sign of a pressure drop causing liquid flashing, or an undercharge of refrigerant causing vapor and liquid to exit the receiver due to no subcooling. Remember, condenser

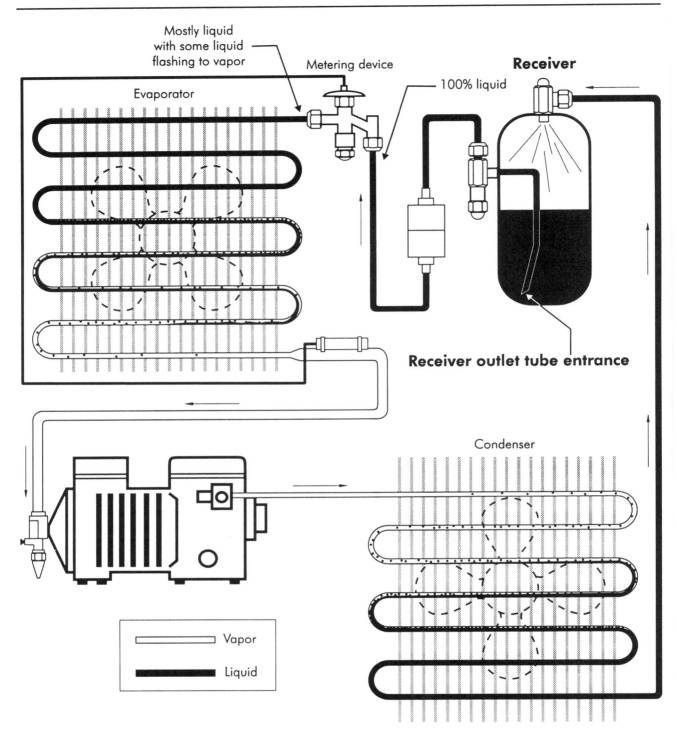

Figure 7-1. Refrigeration system showing entrance of receiver outlet tube

subcooling will be low if an undercharge causes the bubbling of the sightglass. Otherwise, the bubbling sightglass could mean a restricted liquid line, restricted filter drier, loss of receiver or liquid line subcooling from hot ambients, or static and friction losses in the liquid line that are too great.

On the other hand, a low flow rate sightglass is an indication that the system is about ready to cycle off because the box temperature has pulled down to a low enough temperature. It is at these times that the system is at its lowest heat load and the refrigerant flow rate through the system will be the lowest. The sightglass may be only 1/4 to 1/2 full with no entrained bubbles, especially when there are horizontal liquid lines. Do not add refrigerant in this situation or the system will be overcharged. The overcharge will be noticed at the higher heat loads. The low heat load causes the system to be at its lowest suction pressure, so the density of refrigerant vapors entering the compressor will also be lowest. Because of the lowest evaporator pressures, the compression ratio will be high, causing low volumetric efficiencies and low refrigerant flow rates. There is usually plenty of subcooling in the condenser, but the sightglass will only be partially filled. Do not confuse a low flow rate sightglass with a bubbling sightglass that has bubbles entrained in the liquid.

A sightglass after the filter drier is a good method to use when determining if the drier is starting to plug. Refrigerant will flash from the added pressure drop in the restricted drier. As mentioned earlier, a sightglass right before the TXV will tell the technician if liquid flashing is occurring there. Just because the sightglass is bubbling does not necessarily mean an undercharge, so do not automatically add refrigerant. A lot of systems are found with the receiver completely filled with liquid, because the service technician kept charging refrigerant while trying to clear up the sightglass. An electronic sightglass can also be used before the TXV or after the drier to sense refrigerant vapors.

Low amp draw
Because the compressor is being starved of refrigerant from the restriction in the liquid line, it will not have to work as hard in compressing what vapors do pass through it. The low density of the vapors from the low evaporator pressure will require less work from the compressor, requiring a low amp draw.

Short cycle the low pressure control
The LPC will cycle the compressor off and on

from the low evaporator (suction) pressure. Once off, refrigerant will slowly enter the evaporator and cycle the compressor back on. The compressor will cycle on and off until the problem is fixed or the compressor overheats.

Restricted Liquid Line Before the Receiver
The system checklist in Table 7-7 shows a system with a restricted liquid line before the receiver.

When the restriction occurs before the receiver, the evaporator and compressor will be starved. The refrigerant that does enter the condenser will be trapped there, causing high condenser subcooling. This high subcooling will take up valuable condenser surface area, causing the condensing (head) pressure to elevate.

The symptoms for this system with a restricted liquid line before the receiver are as follows and are explained in detail following the list:

- High discharge temperatures

- High condensing (head) pressures

- High condenser subcooling

- High condenser splits

- Low evaporator pressures

- Low amp draw

- Bubbling sightglass

- Short cycling on the low pressure control (LPC)

High discharge temperature
High discharge temperatures are caused from high compression ratios, which result from the elevated condensing pressures. High condensing temperatures are caused from the liquid backed up in the condenser from the restriction between the condenser and the receiver.

Item to be measured	Measured value
Compressor discharge temperature	270°F
Condenser outlet temperature	90°F
Evaporator outlet temperature	20°F
Compressor inlet temperature	60°F
Ambient temperature	75°F
Box temperature	25°F
Compressor volts	230 V
Compressor amps	Low
Low side (evaporator) pressure	0.28 in. Hg (-15°F)
High side (condensing) pressure	215 psig (135°F)

(a)

Item to be calculated	Calculated value
Condenser split	60°F
Condenser subcooling	45°F
Evaporator superheat	35°F
Compressor superheat	75°F

(b)

Table 7-7. Service checklist for a system with a restricted liquid line before the receiver

High condensing (head) pressures
High head pressures are caused by the liquid back-up in the condenser, which causes a reduced condensing surface area. The restriction before the receiver causes most of the liquid to remain in the condenser and slowly move into the receiver. This is much different than if the restriction is after the receiver as in the previous example. A restriction after the receiver will cause most of the refrigerant to be in the receiver. This will cause low head pressures, because the receiver is designed to carry all of the

refrigerant plus a refrigerant vapor head for safety purposes.

High condenser subcooling
With the restriction in the liquid line between the receiver and condenser, liquid will remain mainly in the condenser and cause high condenser subcooling. Do not confuse this with an overcharge of refrigerant. *A system overcharged with refrigerant will not have low evaporator pressures and high superheat as these symptoms do.*

High condenser splits
The elevated condensing pressures and temperatures will cause the temperature difference between the surrounding ambient and the condensing temperature to increase. This temperature difference is the condenser split.

Low evaporator pressures
Because of the restriction in the liquid line, the receiver, TXV, and evaporator will be starved. The compressor will keep pumping, trying to pull vapors from the evaporator. This will cause low pressure in the evaporator. How low the suction pressure will get depends on the severity of the restriction.

Low amp draw
Amp draws will usually be low in this situation because of the compressor being starved of refrigerant. The low density of the small amount of refrigerant vapors coming into the compressor will also cause low amp draws.

Bubbling sightglass
The sightglass will be bubbly in this situation because of the starved receiver and liquid line after the receiver. Depending on the severity of the restriction, the sightglass may look empty, because it will have very little refrigerant running through it.

Short cycling on the LPC
Since the compressor is starved of refrigerant, the low pressure control (LPC) will cycle the

compressor off and on from the low (evaporator) suction pressures.

Air in the System

The system checklist in Table 7-8 shows a refrigeration system that contains air.

When air enters a system, it collects in the top of the condenser and is trapped. Air is a noncondensible and cannot be condensed like refrigerant vapors. The liquid seal (subcooled liquid) at the condenser bottom will prevent air from leaving the condenser. Air causes a reduction in condensing surface area and high condensing (head) pressures.

Air can enter the system through a leak in the low side of the refrigeration system. Refrigerant leaks will eventually lead to an undercharged system. Severely undercharged systems will run vacuums in the low side of a refrigeration system. These vacuums will pull in air from the atmosphere, because the low side pressure is lower than the atmospheric pressure.

Air may enter a refrigeration system because of the following:

- Leaks - tube, gasket, or flanges

- Poor charging procedures

- Poor recovery or recycling procedures

- Forgetting to purge hoses when accessing systems

The symptoms for this system that contains air are as follows and are explained in detail following the list:

- High discharge temperatures

- High condensing (head) pressures

- High condenser subcooling

- High condenser split

- High compression ratios

Item to be measured	Measured value
Compressor discharge temperature	235°F
Condenser outlet temperature	85°F
Evaporator outlet temperature	17°F
Compressor inlet temperature	40°F
Ambient temperature	75°F
Box temperature	15°F
Compressor volts	230 V
Compressor amps	High
Low side (evaporator) pressure	8.8 psig (5°F)
High side (condensing) pressure	185.5 psig (125°F)

(a)

Item to be calculated	Calculated value
Condenser split	50°F
Condenser subcooling	40°F
Evaporator superheat	12°F
Compressor superheat	35°F

(b)

Table 7-8. Service checklist for a system containing air

- Normal to slightly high evaporator (suction) pressures

- Normal superheat

- High amp draws

High discharge temperatures

High discharge temperatures are caused from high compression ratios. High heat of compression is associated with high compression ratios, which are associated with the high condensing (head) pressure. The compressor must compress suction vapors through a greater pressure range, generating more heat.

High condensing (head) pressures
High head or condensing pressures are generated from the air taking up condensing surface area. The air will stay at the top of the condenser and not condense, which will leave a smaller condenser to desuperheat, condense, and subcool.

High condenser subcooling
The elevated condensing temperatures and pressures make the subcooled liquid in the bottom of the condenser hotter. There will be more of a temperature difference between the subcooled liquid and the ambient, into which heat is rejected. This will increase the rate of heat transfer from the subcooled liquid, because a temperature difference is the driving potential for heat transfer to take place. The higher subcooling does not necessarily mean there is physically more liquid at the condenser bottom. It just means that there is more cooling of the same amount of liquid to make the temperature difference greater. Remember, condenser subcooling is a temperature difference between the liquid temperature at the condenser outlet and the condensing temperature.

High condenser split
Since the air settles at the top of the condenser causing elevated condensing pressures and temperatures, the temperature difference between the surrounding ambient and the condensing temperature will be high. This temperature is defined as the condenser split.

High compression ratios
The higher condensing (head) pressures will cause the compression ratio to increase, causing low volumetric efficiencies and loss of capacity.

Normal to slightly high evaporator (suction) pressures
The TXV will control superheat as long as the pressure range of the valve is not exceeded. It takes a very high head pressure to exceed the pressure range of most TXVs. The TXV may overfeed slightly on its opening strokes because of the greater pressure difference across its orifice. This may give the evaporator a slightly higher than normal suction pressure.

If the air in the condenser is extreme, the compression ratio will skyrocket, causing very low volumetric efficiencies. This will cause a low capacity, and the box temperature may rise. This added heat in the box may cause the evaporator pressure to increase because of the added heat load.

If the head pressure increases on a capillary tube system, the capillary tube will overfeed the evaporator with refrigerant and flood or slug the compressor.

Normal superheat
As mentioned earlier, the TXV will try to maintain evaporator superheat as long as the valve pressure range is not exceeded. The opening strokes of the TXV may momentarily overfeed the evaporator, but it will start to control shortly thereafter.

High amp draws
The high compression ratio will cause a greater pressure range for the suction vapors to be compressed, requiring more work from the compressor and increasing the amp draw. A look at any amp performance curve will show a higher amp draw at higher head pressures if the suction pressure is constant.

Restricted Metering Device
The system checklist in Table 7-9 shows a system with a restricted metering device.

A restricted metering device acts similarly to a liquid line restriction that occurs after the receiver. A restricted TXV will cause the evaporator, compressor, and condenser to be starved of refrigerant, causing low suction pressures, high superheat, low amp draws, and low head pressures.

The causes for a restricted metering device include the following:

- Plugged inlet screen

- Foreign material in orifice

- Oil logged from flooding

- Wax build-up in valve from wrong oil in system

- Sludge in moving parts from a compressor burnout

- Manufacturer defect in valve

The symptoms for this system with a restricted metering device are as follows and are explained in detail following the list:

- Somewhat high discharge temperatures

- Low condensing (head) pressures

- Low condenser splits

- Normal to slightly high condenser subcooling

- Low evaporator (suction) pressure

- High superheat

- Low amp draw

- Short cycling on the low pressure control (LPC)

High discharge temperatures
Somewhat high discharge temperatures will be caused from the higher superheat due to the evaporator being starved of refrigerant. The compressor will experience a lot of sensible heat from the evaporator and suction line, along with its heat of compression and motor heat. The

Item to be measured	Measured value
Compressor discharge temperature	200°F
Condenser outlet temperature	70°F
Evaporator outlet temperature	30°F
Compressor inlet temperature	65°F
Ambient temperature	70°F
Box temperature	30°F
Compressor volts	230 V
Compressor amps	Low
Low side (evaporator) pressure	1.8 psig (-10°F)
High side (condensing) pressure	104.2 psig (85°F)

(a)

Item to be calculated	Calculated value
Condenser split	15°F
Condenser subcooling	15°F
Evaporator superheat	40°F
Compressor superheat	75°F

(b)

Table 7-9. Service checklist for a restricted metering device

compressor will probably overheat from the lack of refrigerant cooling if it is a refrigerant-cooled compressor.

Low condensing (head) pressures
The condenser will also be starved of refrigerant, and there will be little heat to reject to the ambient surrounding the condenser. This will allow the condenser to operate at a lower temperature and pressure.

Lower condenser splits

Since the condenser will be starved of refrigerant, it can operate at a lower temperature and pressure. This is because it does not need a large temperature difference between the ambient and the condensing temperature to reject the small amount of heat it is getting from the evaporator, suction line, and compressor. This temperature difference is the condenser split. If there were large amounts of heat to reject in the condenser, the condenser would accumulate heat until the condenser split was high enough to reject this large amount of heat. High heat loads on the condenser mean large condenser splits. Low heat loads on the condenser mean low condenser splits.

Normal to slightly high condenser subcooling

Most of the refrigerant will be in the receiver and some will be in the condenser, resulting in normal to slightly high condenser subcooling. The refrigerant flow rate will be low through the system from the restriction, which will cause what refrigerant is in the condenser to remain there longer and subcool more.

Low evaporator pressure

The compressor will be starving for refrigerant and will pull itself into a low pressure situation. It is the amount and rate of refrigerant vaporizing in the evaporator that keeps the pressure up. A small amount of refrigerant vaporizing will cause a lower pressure.

High superheat

High superheat is caused by the evaporator and compressor being starved of refrigerant. With the TXV restricted, the evaporator will become inactive and run high superheat, causing high compressor superheat. The 100% saturated vapor point in the evaporator will climb up the evaporator coil, causing high superheat.

Low amp draw

High compressor superheat and low suction pressures will cause low density vapors to enter the compressor. In addition, the compressor will be partly starved from the TXV being restricted. These factors will put a very light load on the compressor, causing the amp draw to be low.

Short cycling on the low pressure control (LPC)

The compressor may short cycle on the LPC depending on the severity of the restriction in the TXV. The low suction pressures may cycle the compressor off prematurely. After a short period of time, the evaporator pressure will slowly rise due to small amounts of refrigerant and the heat load. This will cycle the compressor back on. This short cycling may keep occurring until the compressor overheats. Short cycling is hard on controls, capacitors, and motor windings.

Restricted Air Flow Over Evaporator

The system checklist in Table 7-10 shows a system with restricted air flow over the evaporator.

Whenever the evaporator coil sees reduced air flow across its face, there is a reduced heat load on the coil. No air flow will cause the refrigerant in the coil to remain a liquid and not vaporize. This liquid refrigerant will travel past the evaporator coil and eventually reach the compressor, causing compressor damage from flooding and/or slugging.

Many technicians will change out a compressor because of broken internal parts, but they will not find the actual cause of the problem. Broken internal parts may result from a faulty time clock or an open defrost heater not letting the system defrost. This frosts or ices the evaporator coil, resulting in flooding or slugging of the compressor. If the technician does not run a system checklist and run the system through its modes after changing the compressor, the new compressor is sure to fail for the same reasons. Compressors installed by service technicians fail at a rate of six to seven times that of the original equipment.

Item to be measured	Measured value
Compressor discharge temperature	88°F
Condenser outlet temperature	82°F
Evaporator outlet temperature	-8°F
Compressor inlet temperature	-5°F
Ambient temperature	75°F
Box temperature	25°F
Compressor volts	230 V
Compressor amps	Slightly high
Low side (evaporator) pressure	2.2 psig (-9°F)
High side (condensing) pressure	104 psig (90°F)

(a)

Item to be calculated	Calculated value
Condenser split	15°F
Condenser subcooling	8°F
Evaporator superheat	1°F
Compressor superheat	4°F

(b)

Table 7-10. Service checklist for a system with restricted air flow over the evaporator

Compressor manufacturers ask technicians to examine broken compressors for the cause of failure. Opening a semi-hermetic compressor and examining its internal parts does not void the warranty as long as all of the parts are returned with the old compressor. The technician should make a list of the causes that could be blamed for the failure and eliminate them one by one once the system is up and running. As mentioned earlier, an open defrost heater may be the cause. If the system is not put through the defrost mode or systematically checked with an ohmmeter and voltmeter, the real problem of an open defrost heater will never be found, and the replacement compressor will soon fail. Causes and symptoms should be listed and system checklists made when systematically troubleshooting systems.

The causes for a system with restricted air flow over the evaporator include the following:

- Frosted evaporator coil from bad defrost heater

- Frosted evaporator coil from high humidity

- Frosted evaporator coil from evaporator fan out

- Frosted evaporator coil from defrost component malfunctions

- Frosted evaporator from no load on the evaporator coil

- Dirty evaporator coil

- Defrost intervals set too far apart

The symptoms for this system with restricted air flow over the evaporator are as follows and are explained in detail following the list:

- Low discharge temperatures

- Low condensing (head) pressures

- Low condenser splits

- Low to normal evaporator (suction) pressures

- Low superheat

- Cold compressor crankcase

- High to normal amp draw

Low discharge temperatures
Since the superheat is low and the evaporator and compressor could be flooding, the com-

pression stroke could contain liquid entrained with vapor (wet compression). The heat of compression should vaporize any liquid. This will take heat away from the cylinder and leave a colder discharge temperature. If discharge temperatures are cooler than the condensing temperature, liquid is being vaporized by the compression stroke. This is called wet compression. If wet compression is severe enough, head bolts may be stripped and discharge valves ruined from hydraulic pressures. These pressures build up from trying to compress liquid refrigerant.

Low condensing (head) pressures

The restricted air flow over the evaporator coil will not allow the refrigerant in the evaporator to experience a heat load, so it will not be completely vaporized. With no heat load to be rejected in the condenser, the condensing pressure and temperature will not have to elevate to reject heat to the ambient. Low condensing pressures will result.

Low condenser splits

Low condensing pressures and temperatures will result in a low condensing split, because the condenser will not have to elevate its temperature to reject the small heat load.

Low to normal evaporator (suction) pressures

Because of the reduced heat load in the evaporator coil, the refrigerant vaporization rate and amount will be reduced. This will give lower vapor pressures in the low side of the system.

Low superheat

Because the heat load on the evaporator coil will be reduced, not much refrigerant will be vaporizing. The 100% saturated vapor point in the evaporator will end up past the end of the evaporator, and the TXV will usually lose control. Compressors may slug and/or flood in these situations.

Cold compressor crankcase

The compressor (total) superheat will be low, resulting in the compressor flooding or slugging at some point during the on cycle. There will be liquid refrigerant in the compressor crankcase boiling off, flashing the oil, and causing compressor damage. It is the boiling of refrigerant in the crankcase that will make the crankcase cold to the touch. The crankcase may even sweat or frost if conditions are right.

High to normal amp draw

Droplets of liquid refrigerant will be entrained with the suction vapors, causing the density of the refrigerant coming from the suction line to be high. Some refrigerant may even be in liquid form. This will require more work from the compressor, and the amp draw may be slightly high depending on the severity of the flooding or slugging.

Oil-Logged Evaporator

The system checklist in Table 7-11 shows a system with an oil-logged evaporator.

Oil may gather in the evaporator because it is the coldest component with the largest tubes, thus the slowest refrigerant flow rate. Oil in the evaporator will coat the inner wall of the coil and reduce heat transfer through the walls, causing a loss of capacity and poor performance. The compressor will be robbed of some of its crankcase oil and run with a lower than normal oil level. This may score or ruin mechanical parts in the compressor.

The causes for a system with an oil-logged evaporator include the following:

- Flooding compressor circulating oil into the system

- Too much oil in the system

- System not piped correctly (no oil traps or piping too large)

- Liquid migration to compressor crankcase during off cycle

Item to be measured	Measured value
Compressor discharge temperature	190°F
Condenser outlet temperature	78°F
Evaporator outlet temperature	-15°F
Compressor inlet temperature	0°F
Ambient temperature	75°F
Box temperature	10°F
Compressor volts	230 V
Compressor amps	Slightly high
Low side (evaporator) pressure	3.94 in. Hg (-20°F)
High side (condensing) pressure	104 psig (90°F)

(a)

Item to be calculated	Calculated value
Condenser split	15°F
Condenser subcooling	12°F
Evaporator superheat	5°F
Compressor superheat	20°F

(b)

Table 7-11. Service checklist for a system with an oil-logged evaporator

- TXV out of adjustment (too little superheat)

- Not enough defrost cycles for low temperature machines

- Wrong type or viscosity of oil

The symptoms for this system with an oil-logged evaporator are as follows and are explained in detail following the list:

- Noisy compressor

- Low oil level in sightglass on compressor crankcase

- TXV having a hard time controlling superheat (hunting)

- Low evaporator and compressor superheat

- Warmer than normal box temperatures with loss of capacity

Noisy compressor

The compressor may be noisy because of a lack of oil. Other metallic sounds may also be heard because of a lack of lubrication.

Low oil level in compressor crankcase sightglass

Most of the oil will be in the evaporator, condenser, receiver, and connecting lines so the crankcase will be low on oil. Other components, excluding the compressor, may have too much oil, which will cause a low oil level in the compressor crankcase sightglass.

A compressor crankcase that is flooded with refrigerant will often pump oil. This is because the crankcase will foam due to liquid refrigerant flashing in it, and small oil droplets entrained in the oil will be pumped through the compressor. This will oil log many components in the system. The velocity of the refrigerant traveling through the lines and P-traps will try to return the oil from the system to the crankcase. However, oil will continue to get into the system if the flooding situation is not remedied.

TXV having a hard time controlling superheat (hunting)

The TXV will have too much oil passing through it and will start to hunt. The evaporator tailpipe will be oil logged and the inside of the tubes coated with oil. The remote bulb of the TXV at the evaporator outlet will have a hard time sensing a true evaporator outlet temperature because of the reduced heat transfer through the line. A constant superheat will not be maintained, because the TXV remote bulb may sense a warmer than normal temperature from the oil insulating the inside of the line. This may make the TXV run a low superheat and flood or slug the compressor. The sightglass in the liquid line will often be discolored with a yellow or brown tint from a refrigerant and oil mixture flowing through it.

Low evaporator and compressor superheat

The TXV may be running low superheat, which will cause lower compressor superheat.

Warmer than normal box temperatures with capacity losses

The reduced heat transfer in both the condenser and evaporator from the excess oil coating the inner tubing will decrease capacity. The compressor will run longer trying to maintain a desired box temperature.

Inefficient Compressor from Bad Valves

The system checklist in Table 7-12 shows a system with an inefficient compressor due to bad valves.

A system that has bad valves will have a high evaporator (suction) pressure along with a low condensing (head) pressure. There is no other situation that will give a system both low head and high suction at the same time other than worn piston rings causing blow-by of gases around the rings. Anytime the gauge reads low head with high suction pressures, there must be a valve problem, worn rings, or cylinder damage.

The causes for a system with an inefficient compressor due to bad valves or worn rings include the following:

- Slugging of refrigerant and/or oil

- Migration problems

- Flooding problems

- Overheating the compressor

- Acids and/or sludge in the system deteriorating parts

- TXV set wrong (too little superheat)

The symptoms for this system with an inefficient compressor due to bad valves or worn rings are as follows and are explained in detail following the list:

- Higher than normal discharge temperatures

- Low condensing (head) pressures and temperatures

- Normal to high condenser subcooling

- Normal to high superheat

- High evaporator (suction) pressures

- Low amp draw

Higher than normal discharge temperatures

A discharge valve that is not seating properly because it is damaged will cause low head pressure. Refrigerant vapor will be forced out of the cylinder and into the discharge line during the upstroke of the compressor. On the downstroke, this same refrigerant that is in the discharge line will be drawn back into the cylinder because of the discharge valve not seating properly. This short cycling of refrigerant will cause the discharge gases to heat over and over again, causing higher than normal discharge temperatures. However, if the valve problem has progressed to where there is not much re-

Item to be measured	Measured value
Compressor discharge temperature	220°F
Condenser outlet temperature	75°F
Evaporator outlet temperature	25°F
Compressor inlet temperature	55°F
Ambient temperature	75°F
Box temperature	20°F
Compressor volts	230 V
Compressor amps	Low
Low side (evaporator) pressure	11.6 psig (10°F)
High side (condensing) pressure	95 psig (85°F)

(a)

Item to be calculated	Calculated value
Condenser split	10°F
Condenser subcooling	10°F
Evaporator superheat	15°F
Compressor superheat	45°F

(b)

Table 7-12. Service checklist for a system with an inefficient compressor due to bad valves

frigerant flow rate through the system, there will be a lower discharge temperature from the low flow rate.

Low condensing (head) pressures and temperatures

Bad valves and/or worn piston rings will cause some of the discharge gases to be short cycled in and out of the compressor cylinder, causing a low refrigerant flow rate to the condenser.

This will result in a reduced heat load on the condenser and reduced condensing (head) pressures and temperatures.

Normal to high condenser subcooling

There will be reduced refrigerant flow through the condenser and the entire system. Most of the refrigerant will be in the condenser and receiver, which may give the condenser slightly higher subcooling.

Normal to high superheat

Because of the reduced refrigerant flow through the system, the TXV may not be receiving the refrigerant flow rate it needs, resulting in high superheat. However, the superheat may be normal if the valve or worn ring problem is not severe.

High evaporator (suction) pressures

Refrigerant vapor will be drawn from the suction line into the compressor cylinder during the downstroke of the compressor. However, during the upstroke, this same refrigerant may leak back into the suction line because of the suction valve not seating properly. The result is high suction pressure. Worn rings will also short cycle refrigerant causing high suction pressures.

Low amp draw

Low amp draw is caused by reduced refrigerant flow rate through the compressor. During the compression stroke, some of the refrigerant will leak through the suction valve and back into the suction line, reducing the refrigerant flow. During the suction stroke, some of the refrigerant will leak through the discharge valve (because it is not seating properly) and return into the compressor cylinder. In both situations, there will be reduced refrigerant flow rate, causing a lower amp draw. The low head pressure that the compressor must pump against will also reduce the amp draw.

CHAPTER EIGHT

Alternative Refrigerants, Refrigerant Blends, and Oils

CHEMISTRY

Methane and ethane are the two molecules from which most refrigerants are derived, Figure 8-1. These two molecules contain nothing but carbon (C) and hydrogen (H) and are referred to as pure hydrocarbons. Pure hydrocarbons, which are flammable, were widely used as refrigerants before the ozone depletion and global warming scares. Ammonia is one refrigerant that is neither methane or ethane based. It contains only nitrogen and hydrogen (NH_3) and is not an ozone depleting refrigerant.

Whenever hydrogen atoms are removed from either the methane or ethane molecule and are replaced with either chlorine or fluorine, the new molecule is said to be either chlorinated, fluorinated, or both. Chlorine and fluorine are referred to as *halocarbons*. When chlorine or fluorine replace some or all of the hydrogen atoms in a pure hydrocarbon molecule, the molecule is said to be *halogenated*. Figure 8-2 shows some of the halogenated refrigerants, including R-11, R-12, R-22, and R-134a.

Chlorofluorocarbons (CFCs)

CFCs are methane or ethane based refrigerants that have had all of their hydrogen atoms replaced with chlorine or fluorine. Once this occurs, the molecule is said to be a fully halogenated hydrocarbon because no hydrogen atoms remain. R-11, R-12, R-13, R-113, and R-114 are all popular CFC refrigerants, Figure 8-3. CFCs are scheduled to be phased out by 1996, because they are very harmful to the earth's

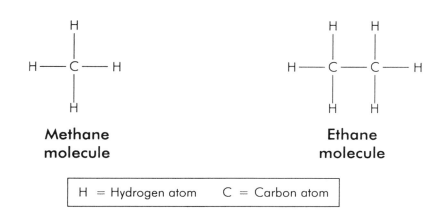

Methane molecule

Ethane molecule

H = Hydrogen atom   C = Carbon atom

Figure 8-1. Methane and ethane molecules

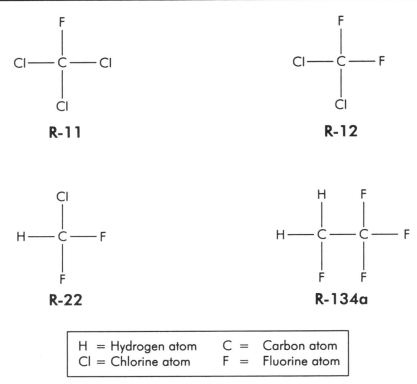

Figure 8-2. Halogenated hydrocarbon molecules

protective ozone layer. Because of their stable chemical configuration, CFC molecules have a very long life when exposed to the atmosphere. Once exposed to the atmosphere, CFC molecules react with ozone molecules in the stratosphere and cause its destruction. CFCs also contribute to global warming. It became illegal to intentionally vent CFC refrigerants to the atmosphere on July 1, 1992.

Hydrochlorofluorocarbons (HCFCs)

HCFCs are methane or ethane based refrigerants that have lost some of their hydrogen molecules to chlorine and fluorine but still have some of their hydrogen atoms. These molecules are *partially halogenated*. R-22, R-123, and R-124 are examples of HCFCs, Figure 8-4. HCFCs are less stable than fully halogenated molecules and will break down in the atmosphere faster than CFCs. HCFCs are less dangerous to the earth's protective ozone layer than CFCs; however, HCFCs do have some global warming potential. HCFCs are scheduled to be phased out by the year 2030; however, a ridgid phase-out schedule will take effect much sooner (see

Chapter Ten). It became illegal to intentionally vent HCFC refrigerants to the atmosphere on July 1, 1992.

Hydrofluorocarbons (HFCs)

When a methane or ethane based refrigerant is halogenated by the fluorine atom alone, it is referred to as a *hydrofluorocarbon* (HFC). These molecules contain no chlorine atoms and do not deplete the earth's protective ozone layer; however, HFCs do have some global warming potential. Some popular HFC refrigerants are R-32, R-134a, R-143a, R-152a, R-125, Figure 8-5. It will become illegal to intentionally vent HFC refrigerants to the atmosphere on November 15, 1995.

REFRIGERANT BLENDS OR MIXTURES

Global warming and ozone depletion scares will force an end to the manufacturing of CFCs by the end of 1995. HCFCs will be outlawed by the year 2030. As stated earlier, the intentional

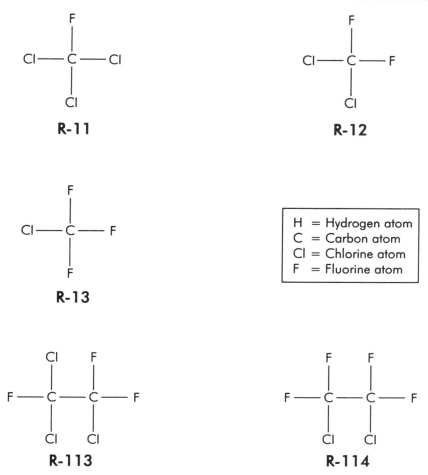

Figure 8-3. Popular chlorofluorocarbon (CFC) refrigerant molecules

venting of CFCs and HCFCs became illegal on July 1, 1992, and the intentional venting of alternative refrigerants, including HFCs, will become illegal November 15, 1995. These environmental scares and legislative actions have accelerated research to find new environmentally safe refrigerants to replace environmentally degrading refrigerants. Refrigerant blends have been used extensively in the original equipment manufacturer (OEM) and retrofit markets and most are here to stay.

Azeotropic Mixtures

Refrigerant blends are not new to the hvac/r industry. Two popular azeotropic binary[1] blends that have been in existence for a long time are R-502 and R-500, both CFCs. These azeotropic mixtures consist of R-22/R-115 and R-12/R-152a, respectively. Azeotropic mixtures are mixtures of two or more liquids, which, when mixed in precise proportions, behave like a compound when changing from liquid to gas. At atmospheric pressure, azeotropic mixtures do not change composition when evaporating or condensing, even though they are made up of two or more components. Pure azeotropic mixtures only exhibit such behavior at one temperature for a specific composition at atmospheric pressure. Deviations from this behavior at other pressures are very slight and essentially undetected. The boiling point of the azeotropic mixture will either be above or below the boiling points of the individual liquid components, but there will still be only one boiling point or condensing point for each given pressure. The molecules usually vaporize at the same rate, and the liquid and vapor have the same composition. Pure azeotropic mixtures essentially behave as pure substances when changing state.

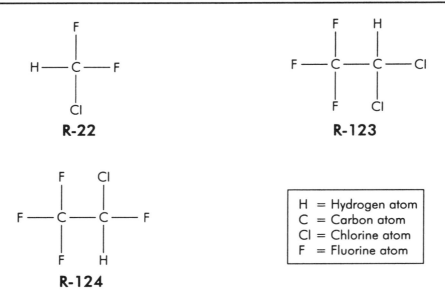

Figure 8-4. Three hydrochlorofluorocarbon (HCFC) refrigerant molecules

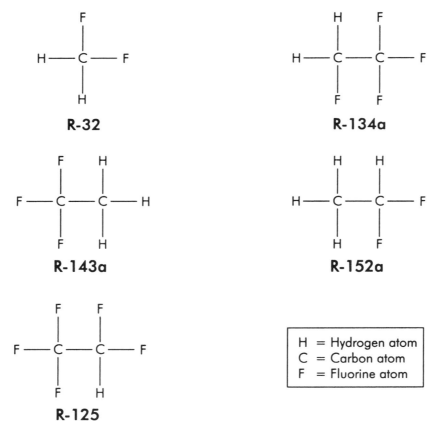

Figure 8-5. Five hydrofluorocarbon (HFC) refrigerant molecules

Many of the blends that are being researched, and some that are on the market today to replace both CFC and HCFC refrigerants, are not pure azeotropic mixtures. They are near-azeotropic mixtures. Because these blends form mixtures and not pure compounds, two or three molecules will be present in any one sample of liquid or vapor. This is where a difference arises in the pressure/temperature relationship of the blend versus pure compounds like R-12, R-22, or R-134a.

Before continuing, it is important to learn the definitions of several important terms:

- A **mixture** is a blend of two or more components that do not have fixed proportions to one another. No matter how thoroughly blended, each component will retain a separate existence, and more than one molecule will be present (e.g., the near-azeotropic blends and saltwater).

- A **compound** is a substance formed by a union of two or more elements in definite proportions by weight. Only one molecule will be present (e.g., R-12, water [H_2O]).

- **Temperature glide** is the range of condensing or evaporating temperatures for one pressure.

Many alternative refrigerants are near-azeotropic mixtures or blends. These blends have a temperature glide when they boil (vaporize) and condense. Temperature glide is when the blend has a range of temperatures as it evaporates and condenses for a given pressure. Temperature glide occurs because the change of state takes place at different pressure/temperature relationships for each component in the blend. As the refrigerant changes state along the length of a heat exchanger, there will be a range of boiling or condensing points for each pressure, Table 8-1.

An examination of Table 8-1 clearly shows that the liquid and vapor phases of this blend have two distinct temperatures for one given pres-

	°F	Pressure per square inch, absolute (psia)
V	75.08	100.0
L	67.79	100.0
V	76.30	102.0
L	69.04	102.0
V	77.51	104.0
L	70.27	104.0
V	78.69	106.0
L	71.48	106.0
V	79.86	108.0
L	72.67	108.0
V	81.02	110.0
L	73.85	110.0

Table 8-1. Pressure/temperature relationships of near-azeotropic refrigerant blend R-22/R-152a/R-124

sure. For example, at 100 psia the liquid temperature is 67.79°F and the vapor temperature is 75.08°F. Temperature glide may range from 2° to 12°F depending on specific blends and system conditions. Pure compounds like R-12 boil at a constant temperature for each pressure, Table 8-2. True azeotropic blends like R-502 and R-500 also boil at a constant temperature for each given pressure.

Most of the blends exhibit low temperature glides, and system performance will not be affected. The percentage composition of the liq-

	°F	Pressure per square inch, absolute (psia)
V	80.83	100.0
L	80.83	100.0
V	82.16	102.0
L	82.16	102.0
V	83.47	104.0
L	83.47	104.0
V	84.76	106.0
L	84.76	106.0
V	86.04	108.0
L	86.04	108.0
V	87.29	110.0
L	87.29	110.0

Table 8-2. Pressure/temperature relationships of R-12

uid and vapor phases of a near-azeotropic mixture or blend will be nearly identical, due to the very similar vapor pressure values of each component. System design conditions must be evaluated when retrofitting with a near-azeotropic blend. A somewhat similar, but not exact condition exists when a pure compound or true azeotropic blend experiences pressure drop through the length of a heat exchanger.

Major advantages of refrigerant blends include the following:

- Blends have comparable capacity and efficiency when compared to CFCs and HCFCs.

- Blends can use oils that are already on the market (certain synthetic alkylbenzenes and esters).

- Blends are compatible with most materials of construction in today's systems.

Fractionation

Fractionation occurs when one or more refrigerant of the same blend leaks at a faster rate than other refrigerants in the same blend. This different leakage rate is caused by the slightly different vapor pressures of each constituent in the near-azeotropic mixture.

Fractionation was initially considered a problem, because it was thought the original refrigerant composition of blend constituents may change over time from leaks and recharges. Depending on the blend, fractionation may also segregate the blend to a flammable mixture if one or two constituents in the blend is flammable. When leaked, refrigerant blend fractionation may also result in faster capacity losses than single component pure compounds like R-12, R-502, or R-22. However, further research proved that most blends were near-azeotropic enough for fractionation to be managed.

Figure 8-6 shows a pure compound (R-12) and a near-azeotropic, HCFC-based refrigerant blend consisting of R-22/R-152a/R-124 (53%/13%/34%). This HCFC-based blend is currently on the market and is an interim replacement for R-

12 in medium and high temperature applications. The graph shows pressure changes resulting from five vapor leaks from a confined container of R-12 (straight line) and the refrigerant blend (jagged lines). Each leak was a 50% weight loss followed by a recharge of the original refrigerant or blend to its initial weight. Notice that R-12 did not lose any vapor pressure when leaked and recharged. This is because it is a pure compound with a constant boiling point for any certain pressure. The blend, however, did lose some vapor pressure after each leak sequence. This is an example of blend fractionation. Once recharged with the original blend to its initial weight, the vapor pressure did recover somewhat, but not quite to its original vapor pressure. After five consecutive leaks of 50% and subsequent recharging, the blend lost less than 10% of its original vapor pressure. This is considered satisfactory and will only slightly impact performance. It is necessary to consider the severity of the leak/recharge test when trying to put an actual system in perspective.

Leaks of this magnitude should never occur, because an environmentally conscious service technician should leak check hvac/r systems before adding any refrigerant. It is illegal to top off systems that leak significant amounts of CFC- or HCFC-based refrigerants without first repairing the leak (see Chapter Ten). The cost of refrigerant, fines, service callbacks, and company reputation should also deter technicians from intentionally topping off leaky systems.

To keep fractionation at a minimum, the charging of a refrigeration system incorporating a near-azeotropic blend should be done with liquid refrigerant whenever possible. It is important to remove liquid only from the charging cylinder to ensure that the proper blend percentage or composition enters the refrigeration system. Refrigerant cylinders containing near-azeotropic refrigerant blends have dip tubes that run to the bottom of the cylinder to allow liquid to be removed from the cylinder when upright. Once the liquid is removed from the cylinder, the refrigerant can be charged as vapor as long as all of the refrigerant vapor is put in the system. When adding liquid refrigerant to a system

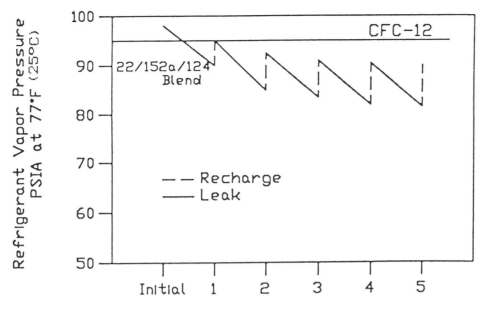

Figure 8-6. Refrigerant vapor pressure changes as a function of leak and recharge numbers (Courtesy, DuPont Company)

that is running, the liquid must be *throttled* (vaporized) into the system to avoid any damage to the compressor. A throttling valve must be used to make sure liquid does not slug the compressor when charging in this fashion.

Table 8-3 shows a few popular HCFC- and HFC-based refrigerant blends.

Example 8-1
Evaluate the condenser subcooling and evaporator superheat for the system shown in Figure 8-7, which uses the blend MP39 (R-401A). Assume an evaporating pressure of 5 psig (20 psia) and a condensing pressure of 105 psig (120 psia). The evaporator outlet temperature is 16°F, and the condenser outlet temperature is 70°F.

Solution 8-1
Referring to Table 8-4 for the MP39 (R-401A) blend at 20 psia, there are two temperatures involved for the one pressure of 20 psia. The temperatures are -2°F for the saturated vapor phase and -13°F for the saturated liquid phase. Pure compounds like R-12 only have one temperature for both liquid and vapor phases at one given pressure. Since evaporator superheat

is calculated starting at the 100% saturated vapor point in the evaporator, -2°F vapor is used for the superheat calculation instead of the -13°F liquid temperature:

16°F - (-2°F) = 18°F evaporator superheat

The temperature glide in the evaporator is 11°F, which is the difference between the liquid and vapor temperatures of -13° and -2°F, respectively. The evaporator vaporizes the refrigerant at a range of temperatures from -13° to -2°F. This is much different than a pure compound like R-12 at 20 psia, which has both liquid and vapor temperatures of -8°F, Table 8-5. With R-12, it does not matter if liquid or vapor temperature at 20 psia is chosen for a superheat calculation, because they have the same temperature of -8°F. However, refrigerant blends with associated temperature glides require choosing the right temperature for a given pressure or else the result will be a superheat calculation error and possibly ruined or inefficient equipment.

Blend	Replaces	Temperature Application	Make-up	Base
MP39 (R-401A)	R-12	High and medium	R-22 R-152a R-124	HCFC
MP66 (R-401B)	R-12	Low temperature and transportation refrigeration	R-22 R-152a R-124	HCFC
AC9000 (R-407C)	R-22	Air conditioning	R-32 R-125 R-134a	HFC
HP80 (R-402A)	R-502	Low and medium	R-22 R-125 R-290	HCFC
HP81 (R-402B)	R-502	Medium and low	R-22 R-290 R-125	HCFC
FX-10 (R-408A)	R-502	Medium and low	R-22 R-125 R-143a	HCFC
KLEA 66 (R-407C)	R-22, R-502	Low, medium, and high	R-32 R-125 R-134a	HFC
KLEA 60 (R-407A)	R-502	Medium and low	R-32 R-125 R-143a	HFC
HP62 (R-404A)	R-502	Medium and low	R-143a R-125 R-134a	HFC
AZ-20 (R-410A)	R-22	High temperature air conditioning	R-32 R-125 (Pure azeotrope)	HFC
AZ-50 (R-507)	R-502	Low and medium commercial refrigeration	R-125 R-143a (Pure azeotrope)	HFC
FX-56 (R-409A)	R-12	Medium and high	R-22 R-124 R-142b	HCFC
KLEA 61 (R-407B)	R-502	Low and medium	R-32 R-125 R-134a	HFC

Note: Blends consisting of two refrigerants are called binary blends; those with three refrigerants are ternary blends.

Table 8-3. HCFC- and HFC-based refrigerant blends

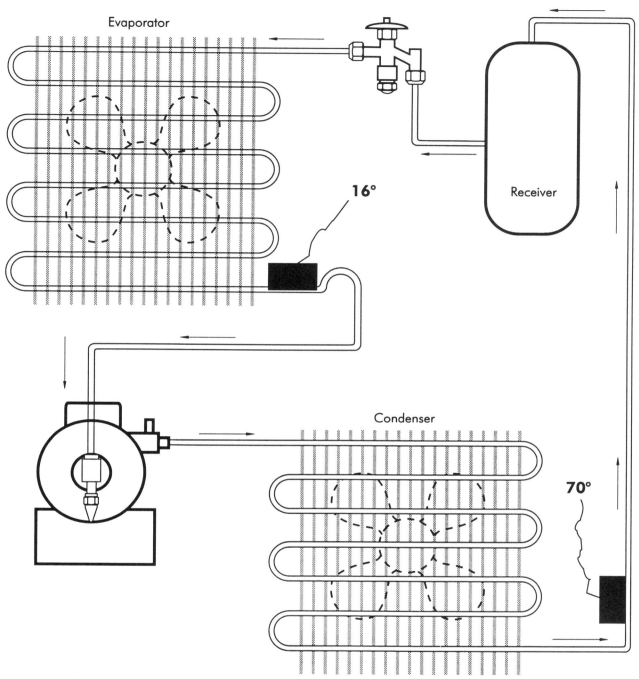

Figure 8-7. Refrigeration system with thermistors on evaporator and condenser outlets

To determine the condenser subcooling for this system, refer to Table 8-6 for the pressure/temperature relationship at 120 psia. Once again, there are two temperatures and two phases associated with a pressure of 120 psia. The phases are liquid and vapor, and their temperatures are 83° and 92°F, respectively. Note that the vapor temperatures for a given pressure are always warmer than the liquid temperatures in both the evaporator and condenser. Again, there is a temperature glide or range of temperatures that the condenser is experiencing as its saturated vapor is condensing. Since condenser subcooling is calculated beginning at the 100% saturated

Saturated vapor phase		Saturated liquid phase
Pressure (psia)	Temperature (°F)	Pressure (psia)
12.91	-20.00	17.10
13.25	-19.00	17.51
13.59	-18.00	17.93
13.95	-17.00	18.36
14.30	-16.00	18.80
14.67	-15.00	19.25
15.04	-14.00	19.70
15.42	-13.00	20.17
15.81	-12.00	20.64
16.21	-11.00	21.12
16.61	-10.00	21.61
17.02	-9.00	22.10
17.44	-8.00	22.61
17.87	-7.00	23.13
18.30	-6.00	23.65
18.74	-5.00	24.18
19.20	-4.00	24.73
19.66	-3.00	25.28
20.13	-2.00	25.84
20.60	-1.00	26.41
21.09	0.00	27.00
21.59	1.00	27.59
22.09	2.00	28.19
22.60	3.00	28.80
23.13	4.00	29.42
23.66	5.00	30.05
24.20	6.00	30.70
24.75	7.00	31.34
25.31	8.00	32.00
25.88	9.00	32.68
26.46	10.00	33.36
27.05	11.00	34.05
27.65	12.00	34.75
28.26	13.00	35.47
28.88	14.00	36.19
29.51	15.00	36.93
30.15	16.00	37.68
30.81	17.00	38.44
31.47	18.00	39.21
32.14	19.00	39.99

Table 8-4. Pressure/temperature relationships for MP39 blend (R-401A)

liquid point in the condenser, the temperature of 83°F saturated liquid is used to calculate condenser subcooling:

83°F - 70°F = 13°F condenser subcooling

The temperature glide in the condenser is 9°F, which is the difference between the saturated liquid and vapor temperatures of 83° and 92°F, respectively. The condenser has a condensing temperature range from 83° to 92°F. This is much different than R-12, which has the same vapor and liquid temperature of approximately 93.5°F at 120 psia, refer to Table 8-5. Again, with R-12, it would not matter if liquid or vapor temperature is chosen for a superheat calculation, because they have the same temperature. However, refrigerant blends with associated temperature glides require choosing the saturated liquid temperature when calculating subcooling amounts.

Because refrigerant blends contain more than one molecule and are not pure compounds like R-12, R-22, or R-134a, they will evaporate and condense at a range of temperatures (temperature glide). System design conditions must be evaluated when retrofitting with blends. Most blends also require a synthetic alkylbenzene or ester lubricant when retrofitted. Because of the high percentage of R-22 in some blends, the condenser may experience higher saturated condensing temperatures and pressures when in operation. R-22 is also noted for its higher than normal heat of compression, so higher than normal discharge temperatures can be expected. Consult the OEM and refrigerant manufacturer on specific retrofitting procedures when retrofitting with near-azeotropic refrigerant blends (see Appendix).

ALTERNATIVE REFRIGERANTS

Widespread research is being conducted to develop a drop-in replacement for R-12, R-502, R-22, and many other refrigerants. Refrigerant blends are mathematically modeled with computers to give maximum system efficiency and performance. Even vapor pressures can be adjusted by varying the percentage of each constituent in the blend. In fact, many refrigerant manufacturers use the same blend but vary the percentage composition of the constituents for use in different evaporating temperature applications. Percentage composition changes are also

Temperature (°F)	Pressure		Temperature (°F)	Pressure	
	Psia	Psig		Psia	Psig
-40	9.3076	10.9709*	70	84.888	70.192
-39	9.5530	10.4712*	71	86.216	71.520
-38	9.8035	9.9611*	72	87.559	72.863
-37	10.059	9.441*	73	88.918	74.222
-36	10.320	8.909*	74	90.292	75.596
-35	10.586	8.367	75	91.682	76.986
-34	10.858	7.814*	76	93.087	78.391
-33	11.135	7.250*	77	94.509	79.813
-32	11.417	6.675*	78	95.946	81.250
-31	11.706	6.088*	79	97.400	82.704
-30	11.999	5.490*	80	98.870	84.174
-29	12.299	4.880*	81	100.36	85.66
-28	12.604	4.259*	82	101.86	87.16
-27	12.916	3.625*	83	103.38	88.68
-26	13.233	2.979*	84	104.92	90.22
-25	13.556	2.320*	85	106.47	91.77
-24	13.886	1.649*	86	108.04	93.34
-23	14.222	0.966*	87	109.63	94.93
-22	14.564	0.270*	88	111.23	96.53
-21	14.912	0.216	89	112.85	98.15
-20	15.267	0.571	90	114.49	99.79
-19	15.628	0.932	91	116.15	101.45
-18	15.996	1.300	92	117.82	103.12
-17	16.371	1.675	93	119.51	104.81
-16	16.753	2.057	94	121.22	106.52
-15	17.141	2.445	95	122.95	108.25
-14	17.536	2.840	96	124.70	110.00
-13	17.939	3.243	97	126.46	111.76
-12	18.348	3.652	98	128.24	113.54
-11	18.765	4.069	99	130.04	115.34
-10	19.189	4.493	100	131.86	117.16
-9	19.621	4.925	101	133.70	119.00
-8	20.059	5.363	102	135.56	120.86
-7	20.506	5.810	103	137.44	122.74
-6	20.960	6.264	104	139.33	124.63
-5	21.422	6.726	105	141.25	126.55
-4	21.891	7.195	106	143.18	128.48
-3	22.369	7.673	107	145.13	130.43
-2	22.854	8.158	108	147.11	132.41
-1	23.348	8.652	109	149.10	134.30
0	23.849	9.153	110	151.11	136.41
1	24.359	9.663	111	153.14	138.44
2	24.878	10.182	112	155.19	140.49
3	25.404	10.708	113	157.27	142.57
4	25.939	11.243	114	159.36	144.66

*Inches of mercury below one atmosphere

Table 8-5. Pressure/temperature relationships for a pure compound (R-12), continued on next page

Temperature (°F)	Pressure		Temperature (°F)	Pressure	
	Psia	Psig		Psia	Psig
5	26.483	11.787	115	161.47	146.77
6	27.036	12.340	116	163.61	148.91
7	27.597	12.901	117	165.76	151.06
8	28.167	13.471	118	167.94	153.24
9	28.747	14.051	119	170.13	155.43
10	29.335	14.639	120	172.35	157.65
11	29.932	15.236	121	174.59	159.89
12	30.539	15.843	122	176.85	162.15
13	31.155	16.459	123	179.13	164.43
14	31.780	17.084	124	181.43	166.73
15	32.415	17.719	125	183.76	169.06

Table 8-5. Continued from previous page

used to lower compression ratios and discharge temperatures for maximum operating performances and efficiencies.

Refrigerant blends can be HCFC based, HFC based, or a combination of both. The HCFC-based blends are only interim CFC replacements because of their chlorine content. Because HCFCs constitute a major percentage of some blends, they have lower ozone depletion and global warming potentials than most CFC refrigerants they are replacing. The HFC-based blends will be long-term replacements for certain CFCs and HCFCs until researchers can find pure compounds to replace them.

R-134a

R-134a, which is an HFC, is the alternate refrigerant of choice to replace R-12 in many medium and high temperature stationary refrigeration and air conditioning applications, as well as automotive air conditioning. R-134a has very similar pressure/temperature and capacity relationships when compared to R-12, and R-134a also suffers small capacity losses when used as a low temperature refrigerant, Figure 8-8. R-134a contains no chlorine in its molecule and has an ozone depletion potential of zero and a global warming potential of 0.27.

Polarity differences between commonly used organic mineral oils and HFC refrigerants make R-134a insoluble, which means it is incompatible with mineral oils used in many refrigeration and air conditioning applications today. R-134a systems must employ synthetic polyol ester (POE) or polyalkylene glycol (PAG) lubricants.

R-134a is also replacing R-12 and R-500 in many centrifugal chiller applications. Ester-based lubricants are also required in these applications, because R-134a is incompatible with organic mineral oils. In many centrifugal chiller applications, R-134a has resulted in improved efficiencies but not without some reductions in capacity.

R-22

Many chemical companies and professional organizations are currently researching possible R-22 alternative replacements. One promising replacement is a mixture of R-152a and R-32. The major drawback in this binary blend is that both refrigerants are flammable. Another promising replacement for R-22 is a binary blend of R-32 and R-134a. Again, the flammability of blends containing R-32 needs further research.

Some of the pure compounds and refrigerant mixtures being tested are as follows (the constituent percentages are in parentheses):

• Propane (R-290), which is a pure hydrocarbon

• Ammonia (R-717), which contains nothing but nitrogen and hydrogen

Saturated vapor phase		Saturated liquid phase
Pressure (psia)	Temperature (°F)	Pressure (psia)
70.76	60.00	83.83
72.01	61.00	85.22
73.27	62.00	86.63
74.56	63.00	88.06
75.86	64.00	89.50
77.17	65.00	90.96
78.51	66.00	92.44
79.86	67.00	93.94
81.23	68.00	95.46
82.61	69.00	96.99
84.02	70.00	98.55
85.44	71.00	100.12
86.88	72.00	101.71
88.34	73.00	103.32
89.82	74.00	104.94
91.32	75.00	106.59
92.83	76.00	108.26
94.37	77.00	109.94
95.92	78.00	111.65
97.49	79.00	113.37
99.09	80.00	115.12
100.70	81.00	116.88
102.33	82.00	118.67
103.98	83.00	120.47
105.65	84.00	122.29
107.34	85.00	124.14
109.06	86.00	126.01
110.79	87.00	127.89
112.54	88.00	129.80
114.32	89.00	131.73
116.11	90.00	133.68
117.93	91.00	135.65
119.77	92.00	137.64
121.63	93.00	139.66
123.51	94.00	141.69
125.41	95.00	143.75
127.34	96.00	145.83
129.29	97.00	147.93
131.26	98.00	150.06
133.25	99.00	152.20

Table 8-6. Pressure/temperature relationships for MP39 blend (R-401A)

- R-32/R-125 (60/40), an azeotropic mixture

- R-32/R-134a (30/70)

- R-134a (100)

- R-32/R-125/R-134a (10/70/20)

- R-32/R-125/R-134a (30/10/60)

- R-32/R-125/R-134a/R-290 (20/55/20/5)

- R-32/R-227ea (35/65)

- R-32/R-134a (25/75)

All refrigerants and blends listed to replace R-22 are void of chlorine and have a zero ozone depletion potential. Propane (HC-290) is not a strong stand-alone candidate because of its extreme flammability.

AlliedSignal's Genetron AZ-20 (R-410A) is an HFC-based, nonflammable, non-ozone depleting azeotropic blend of R-32 and R-125. It has been primarily designed to replace R-22 in residential air conditioning applications. KLEA 66 (R-407C) by ICI America is a binary blend of R-32 and R-134a. It is targeted for breweries and supermarkets, as well as meat, dairy, and refrigerant transport industries. Dupont's SUVA AC9000 (R-407C), consisting of R-32/R-125/R-134a, is an HFC-based, near-azeotropic refrigerant blend to replace R-22 in air conditioning and heat pump applications. SUVA AC9000 is on the market today.

R-502

R-502 has an interim HCFC-based ternary blend replacement consisting of R-22/R-125/R-290 (propane). Its composition is 60/38/2, respectively. Since propane is a pure hydrocarbon, it can assist in oil solubility and simultaneously contribute to the refrigeration effect. The propane is in such small proportions that the blend is not flammable. This blend provides a 90% improvement over R-502 in ozone depletion and global warming potentials and is presently on the market. An HFC-based ternary blend consisting of R-125/R-143a/R-134a (44%/52%/4%)

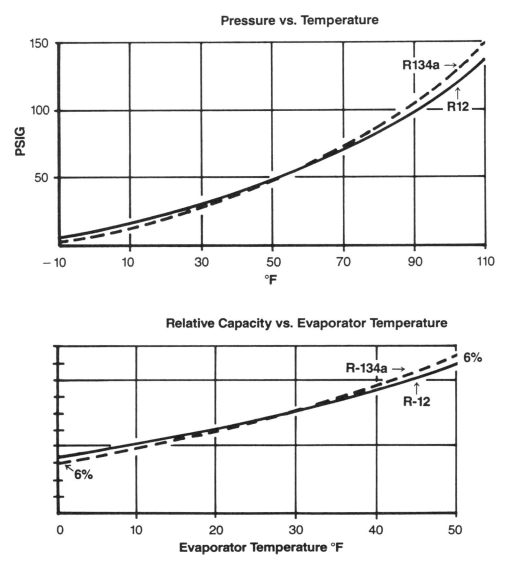

Pressure vs. Temperature

R134a →

R12

Relative Capacity vs. Evaporator Temperature

R-134a →

6%

R-12

6%

Evaporator Temperature °F

Figure 8-8. Pressure/temperature and capacity curves for R-134a (Courtesy, Tecumseh Products Company)

will be the long-term replacement for R-502 until a pure compound can be found. DuPont is marketing this blend as Suva HP62 (R-404A).

Some chlorine-free R-502 long-term replacement candidates are as follows:

- R-125/R-143a (45/55), an azeotrope

- R-125/R-143a/R-134a (44/52/4)

- R-32/R-125/R-143a (10/45/45)

- R-125 (100)

Because R-125 has such a low critical temperature, questions arise about it being a strong stand-alone replacement candidate for R-502. Appropriate high side system cooling must be incorporated if using R-125 as a stand-alone replacement to stay below refrigerant critical temperature. Some R-502 replacement refrigerant blends currently on the market are AZ-50 (R-507) (Genetron), Forane FX-10 (R-408A) (Elf Atochem), KLEA 61 (R-407B) (ICI

America), and HP80 (R-402B), 81 (R-402A), and 62 (R-404A) (DuPont).

R-12

As mentioned earlier, R-134a is the long-term high and medium temperature replacement for R-12. R-134a is not a drop in replacement for R-12. Near-azeotropic interim replacement refrigerant blends consisting of R-22/R-152a/R-124 with varying percentages for different temperature applications are also replacing R-12. These consist of constituent percentages of 53%/13%/34% for commercial refrigeration and 61%/11%/28% for automotive air conditioning. DuPont is manufacturing these refrigerants as the Suva MP (medium pressure) series. FX-56 (R-409A) by Elf Atochem is an interim R-12 replacement consisting of R-22/R-124/R-142b.

R-11

R-11, a low pressure refrigerant used in centrifugal chillers, has R-123 as an interim replacement. Because of the chlorine content in R-123, its fate has yet to be decided. As of this date, the year 2030 will be its last date of manufacture. There is some speculation that OEM high pressure chillers may one day replace low pressure chillers now that the future of R-123 is on the line.

All replacement refrigerant blends and pure compounds will require some sort of retrofit procedure, often accompanied by oil changes. Consult with the OEM for specific retrofit procedures. Tables 8-7 and 8-8 show some of the proposed acceptable refrigerant alternatives for chillers and refrigeration equipment.

REFRIGERANT OILS

Oils used in modern refrigeration compressors are not considered standard lubricants. Years of testing and research have made refrigeration oil a specialty product. Understanding the behavior of refrigeration oil requires information on its composition, properties, and application.

In any refrigeration system, oil and refrigerant are always present. The refrigerant is required for cooling, and the oil is required for lubricating. Refrigerant and oil are miscible (mixable) in one another, and the degree of miscibility depends on the type of refrigerant, temperature, and pressure. A certain amount of oil will always leave the compressor crankcase and circulate with the refrigerant. Refrigerant and oil can separate into two phases and become immiscible with one another at certain temperatures. Refrigeration oils and refrigerants are

Application	Alternatives for retrofitting equipment				Alternatives for new equipment			
	R-123	R-22, R-152a, R-124 blends	R-134a	Ammonia Systems[+]	R-123	R-22	R-134a	Lithium bromide water absorption blends
R-11 (centrifugal)	*			*	*	*	*	*
R-12 (centrifugal)			*	*	*	*	*	*
R-12 (reciprocating)			*			*	*	
R-500 (centrifugal)		*	*	*		*	*	

[+]ASHRAE Standard 15-1992 sets special requirements for systems using ammonia-containing refrigerants.

Table 8-7. Proposed acceptable refrigerant alternatives for chillers (Source: EPA)

Application	Alternatives for retrofitting equipment				Alternatives for new equipment			
	R-22	R-22, R-152a, R-124 blends	R-22, propane, R-125 blends	R-134a	Ammonia systems+	R-22	R-22, propane, R-125 blends	R-134a
R-12 (cold storage warehouses)	*	*		*	*	*		*
R-12 (commercial ice machines)		*			*	*		*
R-12 (industrial process refrigeration)	*	*		*	*	*		*
R-12 (refrigerated transport)		*		*		*		*
R-12 (retail food)	*	*		*	*	*		*
R-12 (vending machines)	*	*		*		*		*
R-500 (refrigerated transport)		*	*	*		*	*	*
R-502 (cold storage warehouses)	*		*		*	*	*	
R-502 (industrial process refrigeration)	*		*	*	*	*	*	*
R-502 (refrigerated transport)	*		*	*		*	*	*
R-502 (retail food)	*		*		*	*	*	
R-502 (commercial ice machines)					*	*		

+ASHRAE Standard 15-1992 sets special equipment room requirements for systems using ammonia-containing refrigerants.

Table 8-8. Proposed acceptable refrigerant alternatives for refrigeration equipment (Source: EPA)

usually soluble in one another over wide temperature ranges. If not soluble, the oil would not move freely around the system and oil-rich pockets would form, causing restrictions, poor heat transfer, and inadequate oil return to the compressor.

The primary function of oil is to minimize mechanical wear and reduce the effects of friction. In a refrigeration system, oil accomplishes many more tasks, such as:

- acting as the seal between the discharge and suction sides of the compressor.

- preventing excessive blow-by around the piston and through the valves in a reciprocating compressor.

- preventing blow-by in some centrifugal compressors by putting a seal around their vanes.

- acting as a noise dampener, which reduces internal mechanical noise within a compressor.

- performing heat transfer tasks by sweeping away heat from internal rotating and stationary parts.

Bearing performance is affected by both boundary and hydrodynamic lubrication. *Boundary lubrication* occurs in the reciprocal motion of the compressor, especially when a compressor starts. Metal rubbing against metal during start-up is an example of boundary lubrication. Load carrying ability or film strength are terms often associated with boundary lubrication. Boundary lubrication may also occur during high load situations, which is undesirable. *Hydrodynamic lubrication* involves separating bearing elements with an oil film while they are moving. Theoretically, mechanical parts in an operating compressor should never touch one another; there should always be an oil film dividing or supporting compressor parts to prevent wear. Hydrodynamic lubrication depends on the viscosity or thickness of the oil and involves the oil passing load tests.

Oil Properties

Refrigeration oil mixes with the working refrigerant, which requires the refrigeration oil to be specially designed and formulated. The following definitions and discussions are some of the most important properties that must be considered when designers select an oil for a specific application:

- **Viscosity** - The resistance that the oil offers to flow. High viscosity means a thick oil, and low viscosity means a thin oil. It is also defined as the body of the oil or its ability to perform a lubricating function. It is sometimes referred to as the thickness of the oil, although this can be misleading. Oil viscosity is measured in Saybolt Seconds Universal (SSU or SUS). SSU is the time (in seconds) that it takes for a known sample of oil at a controlled temperature (usually 100°F) to flow by gravity from a container through a capillary tube of known length and diameter. If an oil at 100°F takes 150 seconds to flow through the capillary tube, the oil has a viscosity of 150 SSU. If it takes 300 seconds, the oil has a viscosity of 300 SSU. The SSU value is usually printed on the oil container. The more time it takes for the oil to flow through the tube, the higher the SSU value, the higher the viscosity, and the thicker the oil. Viscosity is affected by temperature and by the amount of refrigerant dissolved in the oil, Figure 8-9. The top line in Figure 8-9 represents pure oil and shows how its viscosity decreases as temperature increases. The other lower lines represent percentages of R-12 mixed with the oil. For a constant temperature as the percentage of refrigerant increases, the viscosity decreases and a thinning of the oil results. Increases in the temperature and percentage of refrigerant in the oil decrease the viscosity of most oils.

- **Chemical stability** - The oil's ability to lubricate for extended periods of time

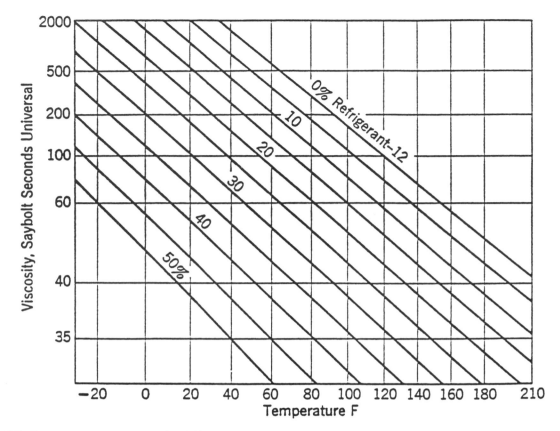

Figure 8-9. Temperature vs viscosity for different refrigerant/oil concentrations (Courtesy, ASHRAE)

without breaking down or reacting with other materials of construction. The original oil may often remain in a compressor for over 10 years. Reactions between oil, refrigerant, and oxygen may cause problems such as copper plating, varnishing, gumming, sludging, and coke formation.

- **Dielectric strength** - The measure of resistance the oil has to electric current. Dissolved metals or moisture in the oil lessen the resistance to electric current flow, giving the oil a lower dielectric strength. Oils of low dielectric strength may cause grounding of motor windings in hermetic compressors.

- **Pour point** - The lowest temperature at which oil will flow at certain test conditions. The pour point should be well

below the lowest temperature obtained in the evaporator. The amount of paraffin wax content in the oil determines the pour point. Higher pour points indicate greater wax contents. If the pour point is reached in a low temperature application, the oil will congeal (curdle) in the evaporator, causing low heat transfer and a loss in efficiency.

- **Cloud point** - The temperature at which wax begins to precipitate out of the oil and the oil becomes cloudy. The precipitated wax will plug metering devices and reduce evaporator heat transfer. Lower cloud points are preferred. This is a very important property for low temperature applications.

- **Floc point** - The temperature at which wax precipitates from a mixture of 10%

oil and 90% refrigerant. The floc point can be ignored when a non-miscible refrigerant is employed. The floc point is very important for low temperature applications.

- **Flash point** - The temperature of the oil when the oil is heated and a flame is passed over its surface causing a flash. The oil may momentarily flash but may not continue to burn. Higher flash points are preferred.

- **Fire point** - The temperature at which oil continues to burn once the flash point is reached. Higher fire points are preferred.

- **Oxidation value** - The value obtained when oil is heated in an atmosphere of oxygen and the resulting sludge is weighed. This predicts the amount of sludge an oil may produce in an operating environment if conditions are right.

- **Low temperature miscibility** - The temperature at which a mixture of oil and refrigerant will start to separate. The refrigerant and oil must mix at low temperatures, and oil and refrigerant can separate (two-phase) in the evaporator. Two-phasing can cause oil hang-up in the evaporator because of the high viscosity of cold oil. Poor heat transfer and poor oil return to the compressor are the results. High temperature miscibility is also important because temperatures often reach 300°F in refrigeration systems.

- **Foaming** - Oil foaming depends on the amount of refrigerant dissolved in the oil and is a characteristic of the oil itself. Foaming usually occurs in the crankcase at start-up but may occur during the running cycle. When a compressor is shut down or cycled off, refrigerant migrates slowly to the oil in the crankcase because of a difference in vapor pressure. An automatic pumpdown system or crankcase heater will prevent

this from happening. At start-up, the sudden pressure drop on the crankcase causes refrigerant to come out of solution rapidly, causing the oil to foam. The oil-refrigerant foam may generate excessive pressures in the crankcase and cause some of the oil to flow past the compressor clearances and be pumped into the high side of the refrigeration system. The foaming oil can cause compressor moving parts to wear rapidly from loss of lubrication. Poor motor cooling may also result, because the crankcase has less oil. Agitation of the oil by the crankshaft and its counterbalances can also cause oil foam, but this is usually much less severe.

- **Material compatibility** - This test involves soaking materials of construction in an oil and refrigerant mixture to check if the oil or refrigerant is compatible with the materials in a system. Elastomers, polyesters, copper, aluminum, etc., are found throughout the refrigeration system. The oil and refrigerant must not react, swell, shrink, deteriorate, weaken, pit, embrittle, or extrude any of these materials.

The properties just listed can become very confusing when choosing an oil to use for a certain compressor. This is why it is recommended that the compressor manufacturer specify the oil for any application. If uncertain as to what oil to use, call the compressor manufacturer or use their specification data along with their performance curve literature. An example of a compressor data sheet was shown in Figure 3-34. In this figure, the viscosity is 150 SSU for this compressor. Even though oil additives may be incorporated in the oil, nothing is mentioned about additives in the specifications. This specification data also gives the initial oil charge as 144 oz and the oil recharge of 128 oz. Every conscientious service technician should have notebooks of every major compressor manufacturer's published performance curves and specification data for their manufactured compressors. This information should be carried in the service vehicle at all times.

Oil Groups

Oils can be categorized into three basic groups: animal, vegetable, and mineral. Animal and vegetable oils cannot be refined or distilled without a change in composition. Both are considered poor lubricants in the refrigeration industry because of this changing composition. Poor stability is another disadvantage of animal and vegetable oils; these oils will form acids and gums very easily. Another problem with animal and vegetable oils in refrigeration applications is their somewhat fixed viscosity. Different viscosities for the diverse temperature applications in the refrigeration and air conditioning industries are mandatory and cannot be reached with animal or vegetable oils.

Mineral oils can be categorized into three main groups: paraffinic, naphthenic, and aromatics, Figure 8-10. The *paraffinic* group is refined from crude oil found in the eastern United States. The *naphthenic* group is refined from California and Texas crude oil and constitutes most of the mineral oils used in modern refrigeration. Naphthenic oils have a low wax content, low pour point values, and a lower viscosity index for the same temperature when compared with paraffinic oils. Paraffinic oils are recommended for electric motor lubrication. Paraffinic oils have excellent chemical stability but have somewhat poor solubility with polar refrigerants. *Aromatics* are a bit more reactive but have good boundary lubrication properties.

Synthetic oils

Because of the somewhat limited solubility of mineral oils with certain refrigerants such as R-22 and R-502, synthetic oils for refrigeration applications have been used. Ozone depletion and global warming scares have prompted much more research in synthetic oils. Three of the most popular synthetic oils are: alkylbenzenes, glycols, and ester-based oils.

Alkylbenzene is synthesized from the raw materials propylene and benzene. Its synthetic origin distinguishes it from naphthenic and paraffinic mineral oils. One advantage of an alkylbenzene is that it is not dependent on the quality of any crude oil; therefore, it always has a consistent

composition. Alkylbenzene is an aromatic hydrocarbon and consists of only one type of compound. Hydrocarbons consist of molecules that contain hydrogen and carbon atoms. Most modern refrigerants have either ethane or methane, which are both hydrocarbon molecules, as their base molecule. Aromatic hydrocarbons also make up mineral oils but in widely varying compositions.

Near-azeotropic refrigerant mixtures (narms) that consist of refrigerant mixtures or blends are being used as R-12, R-22, and R-502 replacements. Testing to date indicates that most HCFC-based blends perform best with alkylbenzene lubricants when compared to other paraffinic and naphthenic mineral oils. This is because existing mineral oils are not completely soluble in refrigerant blends. The HFC-based blends perform best with an ester lubricant.

Blends are soluble in a mixture of mineral oil up to a concentration of 20%. This indicates that mineral oil systems that are retrofitted with HCFC-based refrigerant blends will not require extensive flushing of the oil. Alkylbenzenes are used quite often in refrigeration applications (i.e., Zerol). A few manufacturers are using a synthetic ester-based lubricant for HFC-based refrigerant blends. Extensive oil flushing may be required when retrofitting with ester oils.

Some of the most popular glycol-based lubricants are *polyalkylene glycols* (PAGs), which were the first generation lubricants used with R-134a. R-134a is a polar molecule, which contributes to its low solubility in non-polar lubricants such as mineral oils and synthetic oils used presently in refrigeration applications. R-134a may replace R-12 in medium and high temperature applications. The automobile industry has accepted R-134a because of its low hose permeability and satisfactory efficiencies.

Because of the commercial availability of PAG oils, widespread testing is being done to evaluate the lubricant. However, many polyalkylene glycol lubricants tested with R-134a are not fully soluble and will two-phase. PAG oils also have a track record of being very hygroscopic (attracting and retaining moisture); however, modi-

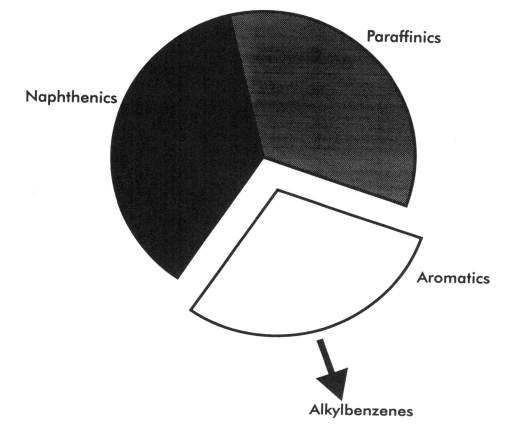

Figure 8-10. Chart showing mineral oil composition

fied PAGs are still being researched. Another disadvantage of PAG oils is they do not lubricate aluminum on steel well. This is especially important when used with some compressors in the automobile industry. PAGs have also been known to have a reverse solubility in refrigeration systems. This means that the oil may separate in the condenser instead of the evaporator. PAGs also have a very high molecular weight and can be harmful if inhaled in certain concentrations. PAGs are incompatible with chlorine, which means the field retrofitting of R-12 systems with R-134a will be very difficult and labor intensive if not impossible. Most R-134a systems will have to come from the factory as virgin systems with PAG oil if they are to be used.

Ester oils are also gaining popularity. Ester-based oils contain no wax, which results in lower pour and floc point properties. Both small and large scale testing is being done with ester-based

oils because of the hygroscopic nature of PAGs. Extensive field testing has proven that ester-based oils are the best lubricants for many applications incorporating HFC refrigerants like R-134a. R-134a is not compatible with current refrigerant mineral oils, and retrofitting an R-12 system with R-134a is complicated and labor intensive.

Mineral oils are not miscible with R-134a, so systems will have to come from the factory as virgin systems with ester oils for refrigeration use. If not, the technician will have to go through a step-by-step flushing process to rid the system of most of its mineral oil. There are a few ester-based oils that are compatible with R-134a and a low percentage of mineral oil. However, a retrofit flushing process must be followed to rid the system of most of its mineral oils. Not more than 5% of the original mineral oil can be left in the system. Flushing procedures have been established to incorporate non-ozone depleting refrigerants and ester oils in active CFC systems from 2 hp to 2000 hp that incorporate mineral oils. A new expansion valve and liquid line drier are often the only equipment change involved in the retrofit.

Polyol esters (POEs) are ester-based lubricants, which have became very popular recently. These lubricants are somewhat compatible with the minerals oils found with the common chlorine-based CFC and HCFC refrigerants. POEs do have similar miscibility and lubricity as mineral oils; however, they absorb atmospheric moisture more rapidly than mineral oils, Figure 8-11. POEs not only absorb moisture more readily, they hold it more tightly, so refrigeration system evacuation procedures may require longer times. Maximum moisture level concentrations are usually around 50 parts per million (ppm). At 100 ppm, copper plating, acid, sludge, and rusting problems will occur more easily.

Polyol esters also attract small particles, which must be filtered out, more easily than mineral oils. This phenomenon occurs because of the higher level of polarity of POEs. POEs also have a tendency to come from their suppliers with additives to improve various properties mentioned earlier. Many conventional filter driers contain alumina, which is a bonding agent and acid remover. It has been found that the alumina in conventional filter driers will strip some of these additives from the POEs. Because of these factors, many manufacturers have designed special driers to use with POE lubricants and alternative refrigerants. Newer driers for use with HFC and POE systems incorporate minimum alumina content and maximum filtration and moisture absorbing abilities. Always consult with the compressor manufacturer when changing refrigerant, oil, or drier styles.

To ensure proper handling of polyol ester lubricants, consider the following points, which were reprinted with permission from the Copeland Corporation:

- Because the moisture level of a POE can increase significantly when exposed to air, oil containers must remain sealed at all times, unless the oil is actually being dispensed. Where possible, use smaller containers to minimize leftover oil, which can become contaminated by moisture. Leaving compressors and related equipment open during work

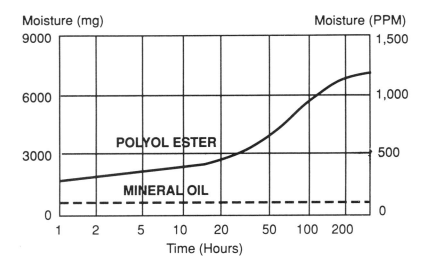

Figure 8-11. Hygroscopicity comparison of mineral oil vs polyol ester lubricant as a function of time (Courtesy, Copeland Corporation)

breaks or overnight can also lead to unacceptable levels of moisture in POE lubricants.

- Storing POE lubricants in their original metal containers will prevent contamination by moisture. Most plastic containers used to package oils are permeable to moisture. Even with the cap tightly in place, moisture can penetrate through the plastic and into the POE lubricant.

- Using the correct filters and driers keeps debris and other impurities from collecting in the POE lubricant.

- In general, proper installation and service techniques will keep problems from occurring and may even increase the life of the compressor.

To date, it has been found that POEs provide the same amount of reliability and performance as mineral oils. The high degree of flexibility they offer brings an even greater advantage to the refrigeration end-user.

OIL FLUSHING PROCEDURES

Proper handling of POEs safeguards the life of your compressor. The procedures outlined in this section are recommended by Castrol North America.

Mineral oils are completely miscible with CFC and HCFC refrigerants but are not miscible with HFC refrigerants. If mineral oil is used with an HFC refrigerant, it will separate in the condenser. Oil plugs will form, which will impede refrigerant flow, and cause the refrigerant to sputter as it passes through the metering device. Restrictions of flow will eventually decrease mass flow rate, robbing the system of capacity. The compressor will also be robbed of oil from the non-returning refrigerant. Once the oil is in the evaporator, the immiscible mineral oil will settle in the tubes, causing further restriction of refrigerant flow and a decrease in cooling capacity, Figure 8-12. This can eventu-

ally cause lack of oil return to the compressor and ruined mechanical parts.

Because mineral oils are not miscible with HFC refrigerants, a step-by-step oil flushing process must be performed to rid the system of residual mineral oils. Specific POE lubricants are miscible with both HFC refrigerants and mineral oils and therefore provide the means to flush existing CFC systems. Castrol has patented a step-by-step flushing procedure to use in the field. Figure 8-13 shows one such flow chart for oil retrofitting procedures. The procedure shown in Figure 8-13 can be used on compressors ranging from 2 hp to 2000 hp. This procedure entails isolating the CFC refrigerant from the compressor and replacing the original mineral oil with the appropriate viscosity ester-based lubricant. The totally compatible system is then run normally. The action of the CFC refrigerant passing around the system flushes the remaining mineral oil back to the compressor sump. Through repeated oil changes, the contamination level of the mineral oil within the system is reduced to an acceptable level at which it will no longer create a problem on conversion to R-134a and other HFC refrigerants.

When the drained oil contains less than 5% of the original mineral oil, a fully compatible system for HFCs is ensured. The number of lubricant changes required is determined by the size and complexity of the system involved. The mineral oil percentage contamination is reduced to approximately one-fifth with each oil change, Figure 8-14. A sizable refrigeration system requires only about three oil changes before mineral oil contamination levels are acceptable. A new expansion valve and compatible filter drier are the only equipment changes recommended.

To ensure the drained oil has less than 5% residual mineral oil, oil companies have created retrofit oil test kits. The mineral oil test kit shown in Figure 8-15 is a "go/no-go" test for converting CFC/mineral oil systems to HFC/ester-based oil systems. It enables the service technician to determine residual mineral oil amounts in the drained oil. This test kit eliminates guesswork and facilitates retrofit work. Always contact the original equipment manu-

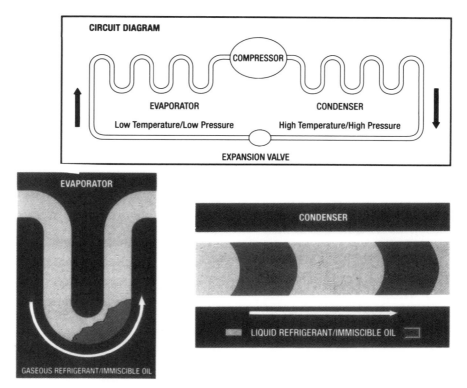

Figure 8-12. Oil causing restriction in refrigeration tubing (Courtesy, Castrol Specialty Products Division)

CASTROL CONVERSION PROCEDURE

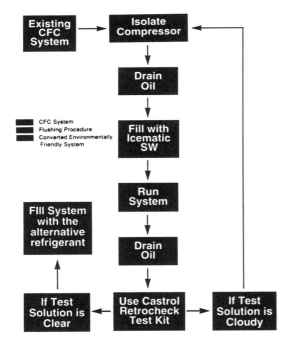

Figure 8-13. Flow chart for oil retrofitting procedure (Courtesy, Castrol Specialty Products Division)

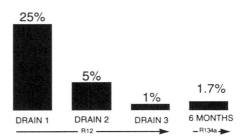

Figure 8-14. Mineral oil contamination declines as a function of oil changes (Courtesy, Castrol Specialty Products Division)

facturer and distributor of the retrofit test kit before retrofitting. Figure 8-16 shows how to use one of these retrofit oil test kits to test an oil sample.

Figure 8-17 lists many popular refrigerants and their recommended oils for stationary refrigeration equipment with direct expansion applications. The triangle suggests the maximum cross oil contamination that should be tolerated when retrofitting. Figure 8-18 shows general refriger-

RESIDUAL MINERAL OIL TEST KIT

For use only with Castrol Icematic SW Synthetic Ester Refrigeration Lubricants during CFC conversion, using the "Castrol Retrofill Procedure".

- Is a "go/no-go" test for converting to HFC refrigerant.

- Enables service engineers to determine mineral oil contamination residual.

- Facilitates retrofits...by eliminating guess work.

Figure 8-15. Mineral oil test kit (Courtesy, Castrol Specialty Products Division)

ant and oil retrofitting guidelines, and Figure 8-19 shows pressure control setting guidelines for direct expansion applications.

OIL ADDITIVES

Oil additives fall into three general groups: polymers, sulfur compounds, and chlorine compounds. Additives can lower floc and pour points, improve thermal stability, include antifoaming agents, inhibit oxidation, improve viscosities, decrease metal activation and copper plating, prevent rusting, decrease metal wear, and handle extreme pressure situations. Oil additives often help in one area but may be objectionable in others. All oil additives must be compatible with materials of construction. Oil additives can also be combined. The service technician may not even know if additives are in the oil of the serviced compressor, which is why it is so important to consult with the compressor manufacturer or use the data specified by the manufacturer when adding oil.

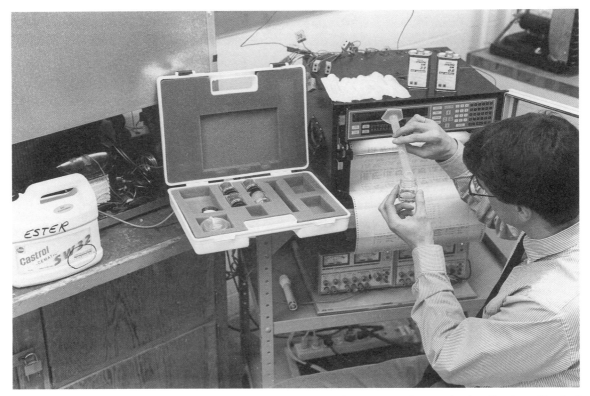

Figure 8-16. Using an oil test kit (Photo by Bill Bitzinger, Office of University Communication Services, Ferris State University)

Service technicians must realize the magnitude of the refrigerant and oil transition the refrigeration/air conditioning industry is experiencing. Refrigerants and oils have become a complex science. "Rules of thumb" that were used to match a certain viscosity with the temperature application no longer exist today. The diversification of oils and oil additives used with past and present refrigerants make these rules of thumb obsolete. Education through reading current literature is one method a technician can use to keep abreast of the new technologies and changes in the industry. A technician must always refer to manufacturer literature for each compressor to obtain information on what oil to incorporate.

Notes

[1] In this case, binary means that two refrigerants are present.

Suggested OIL Guide[1]
Stationary Refrigeration Applications
Direct Expansion Applications

Suva®
Refrigerants

Refrigerant		MO	AB	POE
R-12	②	1	2	3
R-13	②	1	2	3
R-22	③	1	2	3
R-23	④			1
• R-134a	④			1
R-500	②	1	2	3
R-502	②	1	2	3
R-503	②	1	2	3
• AC9000	④			1
• HP62	④			1
• HP80	③		1	2
• HP81	③		1	2
• MP39	③		1	2
• MP66	③		1	2

- • Suva® refrigerants
- ① *Interim data subject to change*
- ② *CFC refrigerant*
- ③ *HCFC refrigerant*
- ④ *HFC refrigerant*

CROSS OIL CONTAMINATION PERCENT GUIDELINE

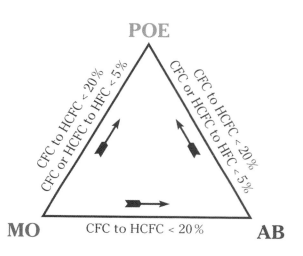

NOTE:

Where possible use OEM recommended oil type, charge size and viscosity. If oil type information is not available the perferred order of oil selection would be 1, 2, 3.

MO .... Mineral Oil
AB .... Alkylbenzene Oil
POE .... Polyol Ester Oil

DuPont Canada

Figure 8-17. Suggested oil guide (Courtesy, DuPont Canada)

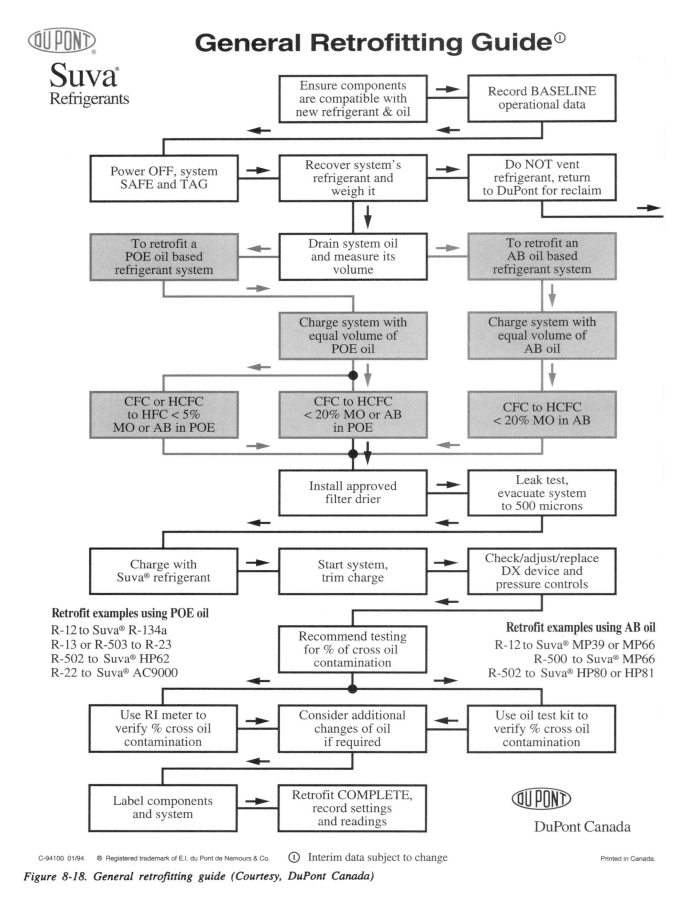

General Retrofitting Guide①

Suva® Refrigerants

Ensure components are compatible with new refrigerant & oil	Record BASELINE operational data
Power OFF, system SAFE and TAG	Recover system's refrigerant and weigh it	Do NOT vent refrigerant, return to DuPont for reclaim
To retrofit a POE oil based refrigerant system	Drain system oil and measure its volume	To retrofit an AB oil based refrigerant system
	Charge system with equal volume of POE oil	Charge system with equal volume of AB oil
CFC or HCFC to HFC < 5% MO or AB in POE	CFC to HCFC < 20% MO or AB in POE	CFC to HCFC < 20% MO in AB
	Install approved filter drier	Leak test, evacuate system to 500 microns
Charge with Suva® refrigerant	Start system, trim charge	Check/adjust/replace DX device and pressure controls
	Recommend testing for % of cross oil contamination	
Use RI meter to verify % cross oil contamination	Consider additional changes of oil if required	Use oil test kit to verify % cross oil contamination
Label components and system	Retrofit COMPLETE, record settings and readings	

Retrofit examples using POE oil
R-12 to Suva® R-134a
R-13 or R-503 to R-23
R-502 to Suva® HP62
R-22 to Suva® AC9000

Retrofit examples using AB oil
R-12 to Suva® MP39 or MP66
R-500 to Suva® MP66
R-502 to Suva® HP80 or HP81

DuPont Canada

C-94100 01/94   ® Registered trademark of E.I. du Pont de Nemours & Co.   ① Interim data subject to change   Printed in Canada.

Figure 8-18. General retrofitting guide (Courtesy, DuPont Canada)

Suva
Refrigerants

Pressure Control [1]
Setting Guide
Direct Expansion Applications

Applications		Desirable Temperature Degrees F	HP62		HP80		HP81		MP39		MP66	
			Out	In	Out	In	Out	In	Out	In	Out	In
Ice Cream	Open	-27 to -22	2	17	2	18	0	15	—	—	14"	1"
	Closed	-15 to -10	6	23	6	25	4	21	—	—	12"	3
Frozen Food	Open	-10 to -5	17	29	18	31	14	27	4"	4	2"	6
	Closed	-5 to 0	21	33	22	35	19	30	0	6	2	8
Deli Case	Serving	43 to 45	44	69	47	75	42	68	12	26	14	29
Fresh Meat	Cooler	34 to 38	34	56	36	60	32	54	6	18	8	20
	Cases	38 to 42	36	65	39	70	34	63	8	23	10	26
Dairy	Serving	50 to 60	54	69	59	75	52	68	17	26	19	29
	Storage	31 to 36	36	58	39	63	34	56	8	20	10	22
Produce	Serving	40 to 45	50	80	53	87	48	78	15	32	18	35
	Storage	35 to 40	38	69	41	75	37	68	9	26	11	29
Meat Prep	Product	28 to 32	53	69	59	75	52	68	17	26	19	29
	Room	42 to 47	34	80	36	87	32	78	6	32	8	35
Walk-In Coolers	Meat	28 to 32	53	71	57	77	51	69	17	26	19	29
	Poultry	20 to 30	52	69	56	75	50	68	16	26	18	29
	Fish	32 to 34	54	80	59	78	52	78	17	26	19	29
	Dairy	33 to 35	61	83	66	90	59	82	24	39	27	43
	Produce	36 to 42	65	85	70	92	63	84	27	40	30	44
	Bakery	35 to 40	61	83	66	90	59	82	24	39	27	43
Walk-In Freezers	Grocery	-15 to -12	11	23	12	25	9	21	9"	1	7"	3
	Meat	-5 to 0	21	33	22	35	19	30	0	6	2	8
	Bakery	0 to 5	25	37	27	40	23	35	2	8	4	10
Beverage	Milk	35 to 40	54	69	59	75	52	68	17	26	19	29
	Beer	36 to 43	56	72	60	78	54	70	18	26	20	29
Fur Storage	Shock	10 to 15	34	52	36	56	32	50	6	16	8	18
	Storage	30 to 35	64	94	69	101	62	92	23	39	26	43
Cut Flowers	Display	32 to 35	11	94	75	101	68	92	26	39	29	43
Ice Cuber	Dry Type	-5 to 0	25	48	27	52	23	46	2	14	4	17
Soda Fountain		29 to 33	56	71	60	77	54	69	18	26	20	29
Salad Bars		40 to 42	50	79	53	86	48	77	15	31	18	35
Domestic	Refrigerator	36 to 46	65	85	70	92	63	84	27	42	30	46
	Freezer	-10 to 0	17	33	18	35	14	30	4"	6	2"	8

Note: Pressure values expressed in PSIG or inches Hg.

[1] *Interim data subject to change*

DuPont

DuPont Canada

C-94105   01/94   Registered trademark of E.I. du Pont de Nemours and Company.   Printed in Canada.

Figure 8-19. Pressure control setting guide (Courtesy, DuPont Canada). Note: The following are ASHRAE designations of the refrigerants listed in the chart: HP62 (R-404A), HP80 (R-402A), HP81 (R-402B), MP39 (R-401A), MP66 (R-401B).

CHAPTER NINE

Leak Detection, Evacuation, and Clean-up Procedures

All sealed systems leak. The leak may be one pound per second or one ounce every million years. Every pressurized system leaks, because flaws exist at every joint, fitting, seam, or weld. These flaws may be too small to detect with even the best leak detection equipment. Time, vibration, temperature, and environmental stress, will cause these flaws to become larger detectable leaks, Figure 9-1.

It is technically incorrect to state that a unit has no leaks, because all equipment leaks to some degree. A sealed system that has operated for 20 years without ever needing a charge is called a *tight system*. The equipment still has leaks but not enough leakage to read on a gauge or to affect cooling performance. No pressurized machine is perfect.

A leak is not some arbitrary reading on a meter. Gas escapes at different times and at different rates, and some leaks may not be detected at the time of the leak test. Leaks may plug and then re-open under certain conditions. A leak is a physical path or hole, usually of irregular dimensions, that may be the tail end of a well fracture, a speck of dirt on a gasket, or a micro-groove between fittings.

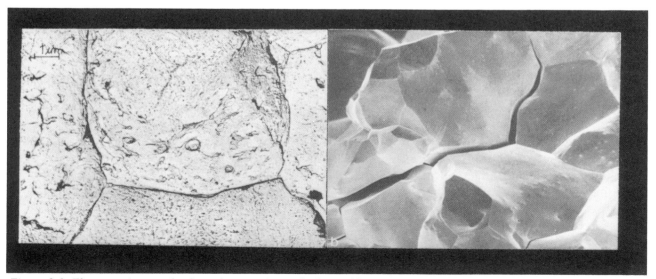

Figure 9-1. Electron micrograph of metal separation and silver solder joint (Courtesy, Refrigeration Technologies)

Refrigerant vapor can flow under layers of paint, flux, rust, slag, and pipe insulation. The refrigerant gas may show up quite a long distance from the leak sight, which is why it is important to clean the leak sight by removing loose paint, slag, flux, rust, or pipe insulation. Oil and grease must also be removed from the sight, because they will contaminate the delicate detection tips of electronic detectors.

There are six classes of leaks:

1. *Standing leaks* are leaks that can be detected while the unit is at rest (off) and fully equalized. This includes freezer evaporator coils warmed up by defrost. Standing leaks are the most common type of leak.

2. *Pressure dependent leaks* are leaks that can only be detected as the pressure builds. Nitrogen is used to pressurize the low side of the system to 150 psig and the high side to 450 psig. Carbon dioxide (CO_2) and oxygen should never be used, but helium is acceptable. A pressure dependent leak test should be performed if no leaks are discovered in the standing leak test.

3. *Temperature dependent leaks* are associated with the heat of expansion. They usually occur from high temperature ambient air, condenser blockages, or during a defrost.

4. *Vibration dependent leaks* only occur during unit operation. The mechanical strain of motion, rotation, refrigerant flow, or valve actuation are all associated with vibration dependent leaks.

5. *Combination dependent leaks* are flaws that require two or more conditions in order to induce leakage. For example, temperature, vibration, and pressure cause the discharge manifold on a semi-hermetic compressor to expand and seep gas.

6. *Cumulative microleaks* are all the individual leaks that are too small to detect

with standard tools. The total loss over many years of operation slightly reduces the initial gas charge. A system having many fittings, welds, seams, or gasket flanges will have a greater amount of cumulative microleaks.

Basic Leak Detection[1]

Successful leak detection is solely dependent upon the careful observation made by the testing technician. All refrigeration systems circulate compressor oil with the refrigerant. Oil will blow off with the refrigerant gas and oil mark the general area of leakage. Oil spots appear wet and have a fine coating of dust, Figure 9-2. The technician must determine that the area wetness is oil and not condensate. This can be accomplished by rubbing the area and feeling for oil slickness.

Figure 9-2. Oil spots and dust coating on vibration eliminator (Courtesy, Refrigeration Technologies)

Oil spotting is the technician's first quick check for leaks, but it is not always reliable for the following reasons:

- Oil is always present at Schrader valves and access ports due to the discharging of refrigerant hoses on the manifold and gauge set, Figure 9-3. These parts are often falsely blamed as the main point of leakage.

- Oil blotches can originate from motors, pumps, and other sources.

- Oil residue may be the result of a previous leak.

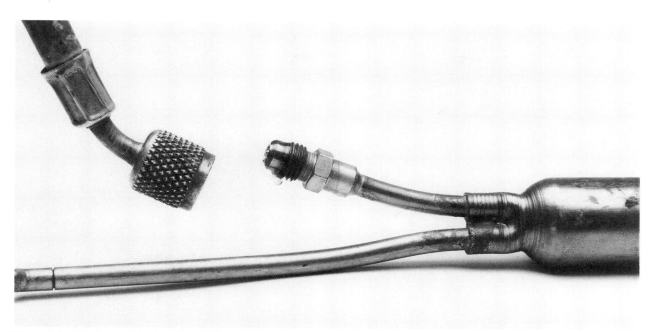

Figure 9-3. Oil shown dripping from Schrader valve (Courtesy, Refrigeration Technologies)

- Oil is not always present at every leak sight. It may take months, even years of unit operation to cause enough oil blow off to accumulate on the outer side.

- Oil may not be present with microleaks.

- Oil may not reach certain leak positions.

- Oil may not be present on new start-ups.

Testing for Evaporator Leaks

Many leaks that go undetected are in the evaporator coil. This is because most evaporators are contained in cabinets or framed into areas that do not allow easy access. In order to avoid stripping off covers, ductwork, and blower cages, an easy electronic screening method is as follows:

1. Turn off all system power including evaporator fan motors.

2. Pressurize system to equalization including defrosting of freezer coils.

3. Warm up and calibrate an electronic leak detector to its highest sensitivity.

4. Locate the evaporator drain outlet or downstream trap.

5. Position the detector probe at the drain opening, Figure 9-4. (Be careful that the detector probe does not come into contact with any water.)

6. Test a minimum of ten minutes or until a leak is sensed. Recalibrate the device and test again. Two consecutive *positive* tests confirm an evaporator leak. Two consecutive *negative* tests rule out an evaporator section leak.

Refrigerant gas is heavier than air, and gravity will cause the gas to flow to the lowest point. If the evaporator section tests positive, the technician should expose the coil and coat all surfaces with a specially formulated bubble/foamer solution, Figure 9-5. Leaks are easily pinpointed with bubble/foamer solutions. Specially formulated and patented bubble solutions form a fine foam or "cocoon" when in contact with a leak, Figure 9-6. A mild soap and water solution may also be used for bubble checking, although a soap and water solution does not have the microfoamers, coagulant base, and wet adhesives of specially formulated bubble/foamer so-

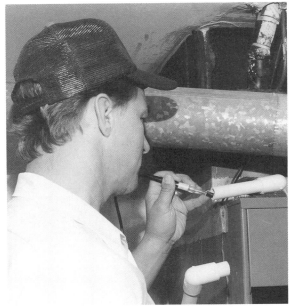

Figure 9-4. Leak checking drain opening of evaporator
(Courtesy, Refrigeration Technologies)

Figure 9-5. Bubble/foamer leak reactant solution
(Courtesy, Refrigeration Technologies)

lutions. Household detergents often contain chlorides and will pit and corrode brass and iron.

Once the solution is applied over the suspected leak area, bubbles or foam will indicate the location of the leak. Bubbles must stand for at least 10 to 15 minutes if small leaks are suspected. Bubbles may also be used with nitrogen or refrigerants pressurizing the system. Small leaks of less than a few ounces per year can often be found with specially formulated bubble/foamer solutions.

If the evaporator tests negative, continue on to the condensing unit test.

Testing for Condensing Section Leaks

If leaks are not located in the evaporator, check the condensing unit using the following procedure:

1. Calibrate an electronic leak detector to its highest sensitivity, and place the probe at the base of the unit (usually under the compressor). The unit should be fully pressurized to equalization.

Figure 9-6. Stress crack indicated by expanding foam spot
(Courtesy, Refrigeration Technologies)

2. Cover the condensing unit with a cloth tarp or bed sheet to serve as a barrier against any outside air movement and to trap any refrigerant gas, Figure 9-7. Do not use a plastic material.

3. Monitor for leakage for ten minutes or until a leak is sensed. Recalibrate and test again. Two consecutive positive

tests confirm condensing section leakage. Two consecutive negative tests rule out a detectable leak.

4. Use the electronic leak detector to check for leaks on the bellows of the pressure controls. Remove the control box cover and place the probe within the housing. Cover the control tightly with a cloth barrier and monitor for ten minutes as above.

5. If the results are positive, uncover the equipment and begin spray coating with a bubble/foamer solution. If the results are negative, continue to the suction/ liquid line leak test.

Suction and Liquid Line Leak Test

The longer the tubing run between the evaporator and condensing unit, the greater the pos-

sibility of defects. These defects may include a typical sightglass-drier connection leak or a poor solder joint hidden under pipe insulation. Check the suction and liquid line by performing the following procedure:

1. Calibrate an electronic leak detector to its highest sensitivity.

2. Tuck the probe underneath the pipe insulation, Figure 9-8. Monitor for ten minute intervals while the system is at rest and fully pressurized to equalization. It may be necessary to insert the probe at several downstream points.

3. If a leak is sensed, strip off insulation and apply a bubble/foamer solution to all surfaces. If no leak is positively screened, test the liquid line.

Figure 9-7. Leak checking a covered condenser (Courtesy, Refrigeration Technologies)

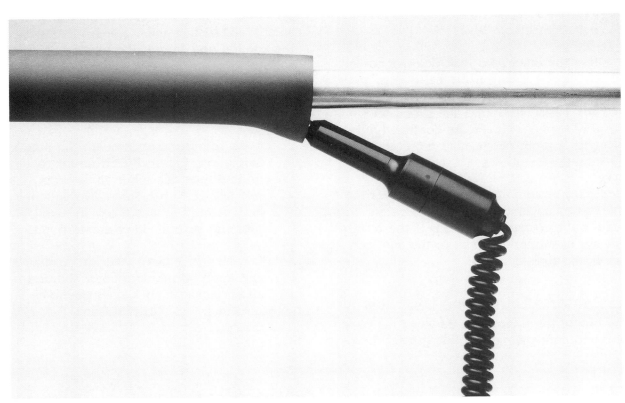

Figure 9-8. Insulated pipe being scanned for leaks (Courtesy, Refrigeration Technologies)

ADVANCED LEAK DETECTION

Under Section 608 of the Clean Air Act, Environmental Protection Agency (EPA) in its prohibition on venting allowed R-22 and nitrogen to be used as leak test gases. Part of this section reads as follows:

Mixtures of nitrogen and R-22 can be used as holding charges or as leak test gases, because in these cases, the ozone-depleting compound is not used as a refrigerant. However, a technician cannot avoid recovering refrigerant by adding nitrogen to a charged system; before nitrogen is added, the system must be evacuated to appropriate levels. Otherwise, the CFC or HCFC vented along with the nitrogen will be considered a refrigerant. This will constitute a violation of the prohibition on venting. The use of a CFC as a trace gas is not permitted.

The R-22 and nitrogen mixture is not an easy procedure to use, because nitrogen and R-22 should be mixed ahead of time and then introduced into the refrigeration system. This prevents the gas that is introduced second from simply pushing the first gas farther into the refrigeration system. It takes time for the two gases to mix, and the system may sit idle for some time until the gases completely mix. This mixing of the two gases can take days for complete mixture.

A method commonly used to ensure the two gases have mixed properly is to bleed a minute amount of gas from the end of the system opposite to the point at which they were initially introduced. Use an electronic leak detector to detect trace gas. If it is not detected, the system must idle longer.

The amount of R-22 introduced into the system need not be any more than that required for the detector to register its presence. This may vary from detector to detector. R-22 is gradually being phased out of production and will be completely phased out of production by the year 2030. This means it will become more and more expensive. The cost, pre-mixing, and time of mixing may encourage technicians to use an alternative type of pressurized leak detection. Nitrogen alone with a bubble/foamer solution may be the alternative to this type of leak checking. *Warning: Never use pure oxygen or air to raise the pressure in a refrigeration system. Pure air contains about 20% oxygen. The pure oxygen and/or the oxygen in the air can combine with refrigerant oil and cause an explosive mixture. Even some refrigerants when mixed with air or oxygen can become explosive under pressure. Pure oxygen and the oxygen in the air will oxidize system oil rapidly. In a closed system, pressure from the oxidizing oil can build up rapidly and may generate pressures to a point of exploding.*

Testing for Pressure Dependent Leaks

Test for pressure dependent leaks by performing the following procedure:

1. Pressurize the low side to 150 psig and the high side to 450 psig using dry nitrogen or helium. (The equipment rating plate usually states the maximum pressure permissible.) Make sure that valving and other components can take these pressures whether or not they are original equipment. If the high side and low side cannot be split by isolation valves, pressurize the entire system to about 350 psig if permissible.

2. Always conduct proper bubble testing by thoroughly saturating all surfaces with a bubble/foamer solution. Allow up to 15 minutes of reaction time for the microfoamer to expand into visible white "cocoon" structures, Figure 9-9. Use an inspection mirror to view undersides and a light source for dark areas. Saturate surfaces as follows:

   a. Starting at the compressor, coat all suspected surfaces. Continue to coat

Figure 9-9. White cocoon structure pinpoints leaking area (Courtesy, Refrigeration Technologies)

all suction line connections back to the evaporator section.

b. Spray coat all fittings starting at the discharge line at the compressor to the condenser coil. Spray coat all of the soldered condenser coil U-joints.

c. From the condenser, continue to spray coat all liquid line connections including the receiver, valves, seams, pressure taps, and any mounting hardware. Continue the liquid line search back to the evaporator section.

d. Spray coat any control line taps to the sealed system the entire length of their run all the way back to the bellow device.

e. Expose the evaporator section and coat all connections, valves, and U-joints.

Notice that the first sequence of searching starts with the compressor and suction line due to their large surface areas. Next comes the discharge line, condenser, and liquid line connection. The evaporator section is the last and least desirable component to pressure test in the field.

Testing for Temperature Dependent Leaks

All mechanical connections expand when heated. The connections on a refrigeration system are usually of soft metals such as copper, brass, or aluminum. These metals actually warp when heated, then contract and seal when heat is removed. Test for temperature dependent leaks by using the following procedure:

1. Place the unit in operation, and raise the operating temperature by partially blocking the condenser air intake. Warm water may also be used for system pressurization.[2] An electronic leak detector may be used while the system is running; however, a running system usually causes fast air currents from

fans and motors, which may interfere with electronic detection. Covering the unit with a blanket or sheet helps to collect escaping gases. The leaking refrigerant is easier to detect with an electronic detector if it can collect somewhere, instead of being dissipated by air currents.

2. Spray coat all metal connections with a bubble/foamer solution one at a time and observe for leakage. Re-wet extremely hot surfaces with water to keep the fluid from evaporating too quickly.

3. When testing evaporator components, induce heat by placing the unit into defrost.

Testing for Vibration Dependent Leaks

Leaks that only occur while the unit is in operation are very rare. These leaks open and close from physical shaking; however, studies show that certain components and piping on refrigeration units will develop vibration leaks. To check for vibration dependent leaks, use an electronic detector or a bubble/foamer solution while the unit is running. Again, drafts must be minimized when the unit is running. If an electronic detector is used first, use a blanket or sheet to help collect escaping gases and minimize air currents. If a bubble/foamer promoter is used, place the unit in operation and spray coat the following areas:

- All compressor bolts and gasket edges

- Suction line connection at compressor

- Suction line connection at evaporator

- Discharge line connection at compressor

- Discharge line connection at condenser

- Vibration eliminators

- Any joint or fitting on unsupported pipe runs

- Expansion and solenoid valves

- Capillary tube connections

- Sightglass

Testing for Combination Dependent Leaks

Combination dependent leak checking involves overlapping the procedures already mentioned. At least two or three procedures should be merged into one procedure. This type of testing requires a high order of skills and observation techniques. Each suspected component must be isolated and tested in the following manner:

1. Subject valve or fitting to high pressure

2. Spray coat valve or fitting

3. Tap the component repeatedly with a rubber mallet to induce vibration. If no leakage, gently add heat to the component. If no leakage, continue on to another component.

Testing for Cumulative Microleaks

Cumulative microleaks are measured using a helium mass spectrometer. Such superfine leak testing is beyond the normal operations of the service technician. Microleaks are considered an acceptable amount of leakage in the industry at this point in time. For more information on acceptable leakage rates, see Chapter Ten under EPA's *Stratospheric Ozone Protection - The Final Rule Summary*.

MODERN LEAK DETECTORS

The newer refrigerants have caused leak checking technology to advance rapidly. Some of the alternative refrigerants have very few or no chlorine atoms in their molecules. Many older leak checking devices depended on the chlorine molecule for detection. These devices must now be made to detect either a smaller amount of the chlorine atom or the more elusive fluorine atom.

Since the fluorine atom is harder to detect than the chlorine atom, many conventional leak detection methods are just not effective on the new, alternative HFC refrigerant R-134a. Many conventional leak detectors like the halide torch will not work on any HFC refrigerant. This is because halide torches search only for chlorine, which changes the color of the flame. HFC refrigerants do not contain chlorine. Newer electronic leak detectors will detect HFCs, but they are much more sensitive than halide torches and can be prone to false alarm if sensitivity is off.

Electronic Leak Detection

Electronic leak detection has been around the hvac/r industry for years. More sensitive instrumentation has recently been developed to detect both fluorine and chlorine atoms in the refrigerant molecule. Electronic leak detectors are some of the most sensitive leak detection instruments on the market. The four types of electronic leak detectors are: *corona discharge*, *thermistor*, *dielectric*, and *ion*.

Corona discharge

In a *corona discharge* leak detector, a small motor driven micropump (fan assembly) mounted in the sensing probe handle draws air into the sensing tip and vents it through the back of the handle. Detection of halogen gases occurs as a result of *electron affinity*. In the presence of an ionized (electrically conductive) atmosphere, some gases (halogens in particular) have the ability to steal or capture electrons from the ionization current flow. In the electronic detector, this results in a decrease in the current within the sensing tip. It is this change that causes an increase in the beeping or ticking of the detector.

Because of the growing concern over ozone depletion, chlorine-free refrigerants (HFCs) such as R-134a have been developed. HFCs contain fluorine as their halogen and not chlorine. Manufacturers have rushed to produce a detector able to respond to a gas with no chlorine content. All previous halogen refrigerants contained chlorine, which is the most easily detected component using corona discharge technology. Detectors must now respond to fluorine instead of chlorine. Detecting fluorine is much more difficult than detecting chlorine, so major changes in tip sensitivity and circuit gain were required. However, the main problem with high-gain amplifiers is their susceptibility to electronic noise and their amplification of unwanted signals. To solve this problem, one leak detector manufacturer designed a separate circuit with a manual switch for variable gains to sense CFC, HCFC, and HFC refrigerants in one detector. Figure 9-10 shows an electronic detector being used on a domestic refrigeration system after a complete retrofit procedure from R-12 to MP39 (R-401A).

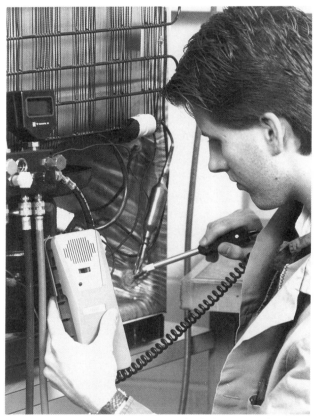

Figure 9-10. Using an electronic leak detector (Photo by Bill Bitzinger, Office of University Communication Services, Ferris State University)

Thermistor

A *thermistor-type* leak detector is a variable resistor that varies its resistance in response to temperature changes. This type of leak detector is based on a change of temperature as gases pass over its tip.

Dielectric

A *dielectric* type of leak detector responds to the different heat conductivity and/or the dielectric differences between gases. As gases are detected between plates of capacitors, their different dielectric strengths, which act as dielectric insulators within the capacitors, change the capacitance of capacitors. A solid state electronic circuit board with oscillators puts out different frequencies, indicating leak position.

Ion

The *ion* type of detector ionizes or charges the gas as it is drawn over the tip of the detector. A transistorized electronic circuit board differentiates the gases.

Electronic Leak Detector Operation

Most electronic detectors have small fans or pumps that actually "sniff" or draw a sample of gas over their sensing tips. This sensing tip is then passed over the suspected leaking source to sample the air. If a halogen refrigerant is sensed, the detector either causes an alarm to sound or lights to flash.

If the environment is contaminated with many refrigerant leaks, an ambient compensation device will compensate for the contaminated environment. Ambient compensation automatically recalibrates leak detector sensitivity to the ambient temperature surrounding the tip. Ambient compensation makes the detector less sensitive, allowing the source of the leak to be pinpointed in contaminated ambients without nuisance false alarms. Some electronic leak detectors have an ambient reset switch built into the handle. These detectors may also have light emitting diodes (LEDs) that indicate the size of the leak and can also be seen when working in poorly lighted areas.

Because refrigerants are heavier than air, the sensing probe should be positioned below the suspected leaking area. Drafts should be minimized along with air movement from fans. Excess air movement will dilute the local leak area with air from surrounding ambient temperatures.

Most electronic detectors have filters beneath their sensing tips. These filters must be changed on a regular basis in order to keep the proper flow through the tip. When leak checking, move the tip about one inch per second across the suspected leak area. In dirty, linty, or oily environments, avoid touching surfaces with the tip. This will help keep the filter and sensing tip clean and dry.

Ultraviolet (UV) Fluorescent Leak Detection

The UV fluorescent leak detection method uses special fluorescent tracer additives and a high intensity black-light (ultraviolet) lamp to pinpoint the exact source of the refrigerant leak, Figure 9-11. It does this quickly and easily with a bright yellow-green glow. This clean method is not to be confused with the messy red visible dyes that stain everything. UV leak detection can be used on CFC, HCFC, and HFC refrigerant systems and refrigerant blends using ester, alkylbenzene, or PAG lubricants. Special formulas are also available for oil-less and low temperature hvac/r systems.

To use a UV leak detector, a special refrigerant- and lubricant-specific tracer is infused as a safe mist into an operating air conditioning and refrigeration system. This mist mixes and circulates with the system refrigerant and oil. A small amount of the tracer is deposited at any leakage point. A quick scan with the high intensity UV lamp pinpoints the exact source of all leaks, including multiple ones and very small leaks less than 0.125 oz per year, Figure 9-12. This system of leak detection is used and approved by major compressor manufacturers and is accepted for recovery and recycling.

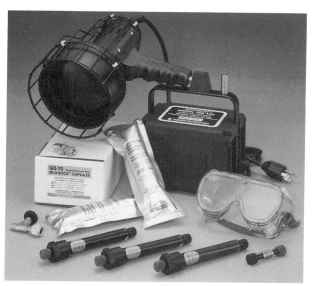

Figure 9-11. Ultraviolet leak testing kit (Courtesy, Spectronics Corporation)

The Spectronics Corporation suggests the following procedure when using one of their UV leak detectors:

1. Pour the appropriate pre-measured additive into the refillable Mist Infuser™. As an alternative, pre-measured disposable GLO-STICK™ capsules may be used, which will not require the Mist Infuser™, Figure 9-13.

2. Connect the Mist Infuser™, Figure 9-14, or GLO-STICK™ capsule, Figure 9-15, between the low-pressure gauge, Figure 9-16, and the refrigeration system service port.

3. Open the valves slowly to gradually add the additive to the unit. Allow enough time for the additive to circulate. The actual time depends on the size of the system.

4. Trace the unit with the UV lamp. A bright yellow-green glow will show the precise leak locations, Figure 9-17. Since the additives safely remain in the system, repaired leaks can be confidence-checked to prevent reworks.

Figure 9-12. High intensity ultraviolet lamp illuminating a leak (Courtesy, Spectronics Corporation)

Preventive maintenance inspections will reveal leaks and reduce refrigerant losses.

Choosing a UV Lamp

Choosing the right UV lamp is important but not difficult if you follow these guidelines.

- Avoid low intensity lamps, as they do not have sufficient power to locate most leaks.

- Select a UL-listed (CSA-approved in Canada) high intensity UV lamp that provides a steady-state intensity of at least 4500 microwatts/cm² at 12 inches. Lamps rated in peak intensity misrepresent their power.

Halide Torch Leak Detection

When refrigerant molecules contact a copper element in the presence of a flame, a colored flame will result. As refrigerants are drawn into the flame, the refrigerant decomposes and copper salts form. If a chloride salt is formed, a

Figure 9-13. GLO-STICK™ capsules (Courtesy, Spectronics Corporation)

greenish flame is produced. This is the principle behind the halide torch. Halide torches can detect leaks up to as small as 0.7 oz/year if conditions are ideal.

In a halide torch, a propane or acetylene tank in conjunction with a burner usually supplies the

Figure 9-14. Attach Mist Infuser™ to test manifold with a charging hose or permanently attached, as shown, with an optional adapter (Courtesy, Spectronics Corporation)

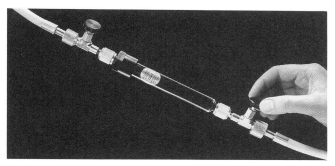

Figure 9-15. Installation of GLO-STICK™ capsule (Courtesy, Spectronics Corporation)

Figure 9-17. Bright green glow showing position of GLO-STICK™ capsule (Courtesy, Spectronics Corporation)

flame. A hollow, rubber tube is connected to the burner, Figure 9-18, and the other end of the rubber tube "sniffs" for leaks. A natural draft is created when the burner is lit, drawing any refrigerant leak through the burner and into contact with the flame and heated copper element. The copper element must reach a slightly glowing, dull red color before it is ready to detect refrigerant gases. There is always a delay before the air sample reaches the heated element. The technician must keep an eye on both the hose and the flame simultaneously. In the presence of a chlorinated refrigerant, the flame will glow with a light green color. If the flame goes out, it could be an indication of high levels of refrigerant gases.

False readings may be detected if the halide torch comes into contact with insulation that has been blown with halogenated refrigerants such as R-11 and other CFCs. R-11 will no

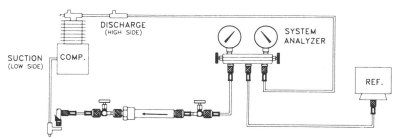

Figure 9-16. System diagram showing position of GLO-STICK™ capsule (Courtesy, Spectronics Corporation)

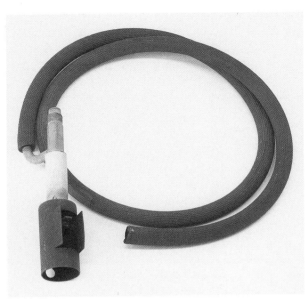

Figure 9-18. Halide leak detector (Photo by Bill Bitzinger, Office of University Communication Services, Ferris State University)

longer be manufactured by the end of 1995, and other non-ozone depleting propellants are replacing it.

As stated earlier, HFC refrigerants do not contain chlorine. Because of this, HFC refrigerants cannot currently be detected with the halide leak detector. It is known that fluorine is the most electronegative of all elements and that there is no oxidizing agent strong enough to liberate it from its ions. This is one of the main reasons why fluorine does not affect the ozone layer.

Care should be taken when using a halide leak detector in explosive environments or when flammable materials or gases are nearby. The open flame may cause fires or explosions. Some refrigerant gases may also form poisonous compounds when exposed to a flame. Material Safety Data Sheets (MSDS) should be checked to see if the refrigerant being exposed to the open flame of the halide torch will form poisonous vapors. R-123 is an example of a refrigerant that will produce a poisonous gas or dangerous decomposition products when exposed to an open flame or heating element. Figure 9-19 shows two pages from the MSDS sheet for R-123.

Refrigerant Dyes

In the past, certain red dyes were added to the refrigerant in a system. When the system developed a leak, a bright red dye would stain the source of the leak. More modern and sensitive methods of leak detection are replacing dyes. Buying refrigerant with dye is becoming more expensive. Most of the time, the entire refrigerant charge must be replaced with refrigerant containing the dye additive. Some major compressor manufacturers will void warranties if refrigerant with dyes are used in their compressors. Always check the warranty before adding any dyed refrigerant to the system.

Most refrigerant dyes contain a mineral oil base. Many HFC-based alternative refrigerants that have entered the market require the use of ester-based or alkylbenzene oils. Because of this, refrigerant dyes should not be used in ester-based or alkylbenzene oil systems.

Standing Pressure

Dry nitrogen can be used to pressurize a system between 125 and 150 psi or higher if components permit once system leaks are repaired. Dry nitrogen can also be used when all of the refrigerant leaks out of the system and the leak must be located. A microfoamer solution may then be used to leak test the system for remaining leaks. A mixture of R-22 and nitrogen can be used as a holding charge or as leak test gas, because in these cases, the ozone-depleting compound of R-22 is not used as a refrigerant. When R-22 is used as a trace gas with dry nitrogen, it allows the use of an electronic leak detector or microfoamer solution to find leaks. Figure 9-20 shows the placement of regulated dry nitrogen into a refrigeration system followed by the use of a microfoamer solution. *Warning: Never use compressed air with refrigerants for leak checking. The approximately 20% oxygen in standard air could cause an explosive mixture when mixed with many refrigerants if exposed to a flame.*

Some advantages of dry nitrogen include its high pressure generation abilities, its inexpensiveness, and the fact that it is readily available almost anywhere. *Warning: Dry nitrogen must*

MATERIAL SAFETY DATA SHEET
MSDS

```
                          "SUVA" CENTRI-LP
2190FR                 Revised 12-MAR-1994      Printed 25-APR-1994
```


CHEMICAL PRODUCT/COMPANY IDENTIFICATION

Material Identification

   "SUVA" is a registered trademark of DuPont.

   Corporate MSDS Number   : DU002798
   Formula   : CHCl2CF3

Company Identification

   MANUFACTURER/DISTRIBUTOR
                 DuPont
                 1007 Market Street
                 Wilmington, DE 19898

   PHONE NUMBERS
      Product Information : 1-800-441-9442
      Transport Emergency : CHEMTREC: 1-800-424-9300
      Medical Emergency   : 1-800-441-3637

COMPOSITION/INFORMATION ON INGREDIENTS

Components

```
Material                             CAS Number    %
*                                    306-83-2
ETHANE, 2,2-DICHLORO-1,1,1-TRIFLUORO- (HCFC 123)    100
```

* Regulated as a Toxic Chemical under Section 313 of Title III of the
Superfund Amendments and Reauthorization Act of 1986 and 40 CFR part
372.

Figure 9-19. R-123 MSDS information, continued on next page (Courtesy, DuPont Company)

DuPont
Material Safety Data Sheet

(ACCIDENTAL RELEASE MEASURES - Continued)

Accidental Release Measures

Dike spill. Prevent liquid from entering sewers, waterways or low areas.

Ventilate area. Collect on absorbent material and transfer to steel drums for recovery/disposal. Comply with Federal, State, and local regulations on reporting releases.

Du Pont Emergency Exposure Limits (EEL) are established to facilitate site or plant emergency evacuation and specify airborne concentrations of brief durations which should not result in permanent adverse health effects or interfere with escape. EEL's are expressed as airborne concentration multiplied by time (CxT) for up to a maximum of 60 minutes and as a ceiling airborne concentration. These limits are used in conjunction with engineering controls/monitoring and as an aid in planning for episodic releases and spills. For more information on the applicability of EEL's, contact Du Pont.

The Du Pont Emergency Exposure Limit (EEL) for HCFC-123 Refrigerant is 1000 ppm for up to 60 minutes with a 1 minute not-to-exceed ceiling of 2500 ppm.

HANDLING AND STORAGE

Handling (Personnel)

Avoid breathing high concentrations of vapor. Provide adequate ventilation for storage, handling, and use, especially for enclosed or low spaces. Avoid contact of liquid with eyes and prolonged skin exposure. Do not allow product to contact open flame or electrical heating elements because dangerous decomposition products may form. *

Storage

Clean, dry area. Do not heat above 125 deg F.

EXPOSURE CONTROLS/PERSONAL PROTECTION

Engineering Controls

Normal ventilation for standard manufacturing procedures is generally adequate. Local exhaust should be used when large amounts are released. Mechanical ventilation should be used in low or enclosed places.

Figure 9-19. Continued from previous page

Figure 9-20. Using pressurized nitrogen and a microfoamer solution to leak check (Photo by Bill Bitzinger, Office of University Communication Services, Ferris State University)

Ultrasonic Leak Detection

Ultrasonic leak detection relies on the sounds or ultrasonic frequencies that gases make when leaking. These sounds or frequencies can be caused either by gas leaking out of or into an area. Gases that leak in (vacuum leaks) are not as easy to detect as gases that leak out. Leaks normally cannot be detected by the human ear.

Each detector has a sensitivity adjustment for small to large leaks. Some readouts may be in decibels (analog or digital), while others have a "peeping" sound that increases in intensity and frequency closer to the leak.

Ultrasonic detectors are reliable in drafty situations especially in outdoor usage, and some even come with headphones to dampen background noise and amplify leak noises. They can also be safely used when combustible gases are present. Ultrasonic detectors can be used on any type of refrigerant, because they rely on the sounds of a leaking gas.

be used in conjunction with a hand shut-off valve, a downstream pressure regulator, and a safety blow-off valve set for the appropriate pressures.

Standing Vacuum

Once an evacuation procedure is performed and the vacuum pump is isolated from the refrigeration system, simply letting a deep vacuum of about 500 microns stand for a period of 10 to 20 minutes can detect leaks. If the micron level keeps creeping up but stops at 0 psig (atmospheric pressure), a leak exists. If the micron level keeps creeping up past atmospheric pressure and never levels off, either residual refrigerant or moisture is still vaporizing in the system and more vacuum time is needed. This method can be used if the system does not stand the chance of having water entering the evaporator or condenser from a leak when water-cooled components are used on the condenser or the system evaporator is chilling water.

LEAK PINPOINTING AND AREA MONITORING[3]

Refrigeration service personnel have used leak detection equipment for years when servicing refrigeration equipment. In the past, leak detection was typically only concerned with leak pinpointing, which was performed with an ultrasonic device, soap bubbles, dyes, thermal conductivity, or other electronic devices. Current refrigerant leak detectors not only pinpoint leaks, they also monitor entire areas on a continual basis. This continuous monitoring helps conserve expensive refrigerants, protect valuable refrigeration equipment, reduce fugitive emissions, and protect employees.

Leak detectors can be placed into two broad categories: leak pinpointers and area monitors. Due to the different nature of their applications, these two broad classes of detectors each have their own specific set of requirements and specifications. *Leak pinpointers* are used to check individual joints or components of a refrigeration system for leaks. *Area monitors* are used to check the level of refrigerant vapor present in an equipment room or other location where human exposure can occur. Stationary area monitors usually let personnel know when a leak occurs within a given space. The leak is then tracked down with a portable leak pinpointer.

Several criteria should be considered before purchasing a monitor or pinpointer, including (but certainly not limited to) sensitivity, detection limits, and selectivity. These terms are not all independent from each other. Other factors to consider when choosing a detector include cost, ruggedness, and ability to be calibrated.

Sensitivity

The *sensitivity* of any device is defined as the amount of input (material being measured) necessary to generate a certain change in output signal. For leak detection, the material is the vapor concentration being measured and the output is the reading from a panel meter, voltage output, or other display device. Detectors with good (high) sensitivity require very little material to generate a large change in output signal, while detectors with poor (low) sensitivity require a larger amount of material to change the output signal. For example, a detector with high sensitivity may be able to accurately discriminate concentration levels of 1 or 2 parts per million (ppm) of vapor, while a low sensitivity detector may only be able to discriminate in increments of 20 ppm or higher.

The sensitivity of a device is determined by a number of factors, including the method of detection and the material being detected. For example, an ionization detector that demonstrates high sensitivity for R-12 may have poorer sensitivity for R-123 and R-134a. Sensitivity differences of 100x to 1000x may exist when comparing R-12 to R-134a with some ionization-based detectors. In this case, the variation in sensitivity would be due to less chlorine, which is very easily ionized and detected. On the other hand, an infrared-based area monitor will show roughly the same sensitivity to the CFC, HCFC, and HFC compounds mentioned above.

Detection Limit

Certain analytical techniques have well-defined sensitivity values, but these do not exist for leak detectors. The most common measure of how sensitive a detector can be is the *detection limit*, which is usually defined as the minimum amount of material a unit can sense that gives a signal at least two times the background noise level. A sensitive device does not necessarily mean a low detection limit (it could have a high background electronic noise level), although the two measures of performance usually tend to coincide.

Detection limits for monitors are measured in either oz/yr for pinpointing applications or ppm (parts per million) for area monitoring. Portable leak pinpointers typically have detection limits of 0.25 oz/yr, while area monitors have detection limits as low as 1 ppm, although a more typical value is 3 to 4 ppm for most compounds.

Since sensitivity can vary greatly with different compounds, the detector must be matched to the intended application. For example, an ionization detector that claims a detection limit of 0.25 oz/yr with R-12 does not work very well for R-134a detection. Conversely, an ionization detector made specifically for R-134a may be too sensitive for R-12 leak pinpointing. Some manufacturers are now considering an option that would allow the operator to choose various sensitivity settings on a single instrument based on the application.

Selectivity

For leak detection applications, *selectivity* can be defined as the ability to detect only the refrigerant of interest without interference from other compounds that may be present in the area. Selectivity is not as important for leak pinpointers, because once the leak is pinpointed, its identity is known. As the specificity of the monitor increases, complexity and cost also increase.

While selectivity requirements for area monitoring will vary with each specific installation, some general statements can still be made:

- Area monitors must work on a continuous basis and are exposed to more potential interfering compounds than a leak pinpointer, which is usually used for only minutes at a time.

- Due to the larger number of potential compounds (and wider range of concentrations over time), selectivity is more important for area monitors.

- Selectivity is a required feature of an area monitor if there are other compounds present with vastly different threshold limit values (TLVs). For example, many equipment rooms with R-123 chillers (AEL = 10 ppm) also have chillers with R-11 (TLV = 1000 ppm). Without being able to distinguish between the two compounds, a nonselective detector will alarm when 10 ppm of either refrigerant is detected. This can lead to concern about excessive R-123 exposure when there may be no exposure to that compound and only inconsequential exposure to the R-11. This can also lead to many false alarms and eventual complacency toward alarms. Despite this fact, some operators prefer nonspecific detection so they can be alarmed when any refrigerant is detected. The identity of the refrigerants will be discovered once the leak is pinpointed.

Using selectivity as a criterion, leak detectors can be placed into one of three categories: nonselective, halogen selective, or compound specific.

Nonselective detectors

Nonselective detectors are those that will detect any type of emission or vapor present, regardless of its chemical composition. Typical detectors in this category are based on electrical ionization, thermal conductivity, ultrasonics, or metal-oxide semiconductors. These detectors are typically quite simple to use, rugged, inexpensive (normally less than $500), and almost always portable, which makes them ideal for leak pinpointing applications. However, they cannot be calibrated, they will drift over the long term, and they lack selectivity and sensitivity (detection limits usually between 50 and 100 ppm for R-134a), which limits their use for area monitoring.

Halogen-selective detectors

Halogen-selective detectors use a specialized sensor that allows the monitor to detect compounds containing fluoride, chloride, bromide, and iodide without interference from other compounds. The major advantage of such a detector is a reduction in the number of false alarms, which are caused by the presence of some compound in the area other than refrigerant.

These detectors are typically easy to use, feature higher sensitivity than nonselective detectors (typically <5 ppm when used as an area monitor and <0.05 oz/yr when used as a leak pinpointer), and are very durable. These instruments can also be calibrated easily.

As an area monitor, halogen-selective detectors are best suited for use in moderately clean equipment rooms in which only one refrigerant must be monitored. Their lack of response to nonhalogenated compounds allows their use in storage areas or areas where other (nonhalogenated) compounds may be present. As a leak pinpointer, the benefit of better sensitivity must be weighed against the higher cost of the product.

Compound-specific detectors

The most complex, expensive detectors are the *compound-specific detectors*. These units are typically capable of detecting the presence of a single compound without interference from other compounds. Compound specific detectors are typically infrared-based (IR), although some of the newer types are infrared-photocoustic (IR-PAS).

The IR and IR-PAS detectors normally have detection limits around 1 ppm, depending on the compound being detected. There are also several IR detectors on the market that have detection limits of approximately 10 ppm. These detectors typically have a much lower price per unit and are less complex than those with lower detection limits. For refrigerants other than R-123, these units probably will yield acceptable performance.

Due to recent improvements in technology, the price of compound-specific detectors has dropped by about 50% to 60%. Several years ago, IR-based detectors could be purchased for approximately $10,000 per unit. Units with comparable performance can now be purchased for $3500 to $4000.

Fluorescent Dyes

As stated earlier, fluorescent dyes have been used in refrigeration systems for several years. These dyes, invisible under ordinary lighting but visible under ultraviolet (UV) light, are used to pinpoint leaks in systems. The dyes are typically placed into the refrigeration lubricant when the system is serviced. Leaks are detected by using a UV light to search for dye that has escaped from the system. When subjected to UV light, the color of the dye is normally bright green or yellow.

As a leak pinpointer, fluorescent dyes work very well, because large areas can be rapidly checked by a single individual. The recent introduction of battery-powered UV lights has made this task even simpler. Dyes can detect leak rates of less that 0.25 oz/yr. The only drawback to the use of dyes is that some areas may be hidden due to cramped spaces. *Caution: The compatibility of the specific dye with the lubricant and refrigerant should be tested prior to use.*

Choosing an Area Monitor

Deciding which area monitor to purchase and install is a complicated issue. Many factors, both instrumental and application oriented, must be considered before purchasing a monitor, including the following:

- Location of the refrigeration room and the monitor

- Other chemical compounds in the room

- Degree of specificity required

- Number of detectors required

- Amount of capital to be invested

If the equipment room is located and ventilated so that few outside vapors enter the room, a halogen-specific instrument could probably be chosen as an area monitor. However, if the equipment room has vapors coming in through the ventilation system that could be detected by a halogen-specific system, the use of a compound-specific instrument would probably be best. Installation examples of the first type (fairly isolated) may include newer office buildings, supermarkets, etc. Examples of the second type (less isolated) may include older installations in which the ventilation system pulls hydrocarbon or other vapors into the building, chemical plants, processing areas, manufacturing sites, etc.

The presence of other chemical compounds in the equipment room and the desired degree of specificity also play an important part in the decision. Many equipment rooms have multiple refrigeration systems present, often with two or more refrigerants being used. If the entire equipment room uses only one refrigerant (such as a supermarket using only R-22), use of a halogen-specific detector may be acceptable. If the equipment room has multiple refrigerants present, the halogen-specific detector will work well if the TLVs of the refrigerants are fairly similar. If multiple refrigerants are present and their TLVs are not very close, use of a compound-specific detector is recommended. Obviously, a compound-specific detector that can be switched either manually or automatically to search for other compounds would be best. Units of this type are just beginning to be introduced.

Two other related factors to consider when purchasing an area monitor are the number of monitors required and the amount of money that can be spent. Only the smallest equipment rooms can get by with a single detector - most rooms should have at least two. With the price of detectors typically running more than $2000, purchasing multiple detectors may be cost prohibitive. Several instrument manufacturers have addressed this issue by making sample manifolds that allow air from several locations to be sampled sequentially by a single detector. The price of this manifold is normally around $500, which is a much less expensive option than purchasing several detectors.

Other factors to consider when purchasing an area monitor include the following:

- How and where alarms are indicated?

- Who has access to the instrument?

- Who maintains the detector?

- Is the detector fail-safe?

Table 9-1 presents a comparison of the various types of leak detectors. Table 9-2 illustrates points to consider when purchasing an area monitor. While not inclusive, this table should be enough to get started when considering which area monitor to purchase.

SYSTEM EVACUATION USING THE DEEP VACUUM METHOD

When installing a vacuum pump on a refrigeration or air conditioning system, technicians ask themselves questions such as the following:

- Why am I really pulling a vacuum on this system?

- What time span should I pull the vacuum for?

- Should I use the single- or double-stage vacuum pump?

These three questions should be thoroughly understood in order to obtain good results from a system evacuation.

Why Pull a Vacuum?

The main reason for pulling a vacuum on a system is to rid the system of unwanted gases, which consist mainly of air and water vapor. Air is a noncondensible and will become trapped in the high side of a refrigeration system due to the condenser liquid seal (subcooled liquid) at its bottom, preventing the passage of air. System head pressure, condensing temperature, discharge temperature, and compression ratio will elevate, causing unwanted inefficiencies because of the reduced condensing surface area. At these elevated temperatures, oxygen in the air will react with the refrigeration oil to form organic solids. This reaction of oil and oxygen usually occurs at the discharge valve, because it is the hottest part of the refrigeration system.

Removing air or other noncondensible gases from a system with a vacuum pump is known as *degassing* a system. Removing water vapor from a system is known as *dehydration*. In the hvac/r industry, the process of removing both air and water vapor is referred to as *evacuation*:

Degassing + Dehydration = Evacuation

Water vapor causes more serious problems. All air contains water vapor. If air is present in the system, water vapor will also be present. Water vapor can cause freeze-ups at expansion devices, as well as sludging problems due to the formation of acids, which cause more serious system problems.

Water is considered to be the universal solvent and is usually thought of as a relatively harmless liquid; however, it can be a service technician's worst nightmare if not handled properly. Water vapor in the air is commonly measured in terms of specific humidity or relative humidity. Water vapor can enter a refrigeration system by way of a leak, poor service practices, or chemical reaction of oil, refrigerant, and excessive heat. In the last case, the high temperature arc from a hermetic motor burnout can cause the refrigerant and oil mixture to break down into water and corrosive

	Nonselective	Halogen-Selective	Compound-Specific	Fluorescent Dyes
Advantages	• Price ($250 to $1500) • Simplicity • Ruggedness	• Simple/rugged • Can be calibrated • Good sensitivity • Low maintenance	• Virtually interference-free • Can be calibrated • Good sensitivity	• Low price • Little specialized equipment required • Good detection limits • Rapid detection • Interference-free
Disadvantages	• Poor detection limits • Cross-sensitive to other species • Most cannot be calibrated	• Price ($280 to $2500) • Not compound specific • Detector lifetime • Stability	• Price (~$10,000; lower through chiller manufacturers); IR-PAS may be lower priced • Complexity/maintenance • Stability	• Potential compatibility problems • Cannot be automated • Cannot be calibrated • Some areas not observable
Vendors	• More than six for leak pinpointers • Three to four for area monitors	• Three for leak pinpointers • Two for area monitors	• Several for IR • Several working on IR-PAS	Two currently exist
Applications *Leak Pinpointing*	R-134a, R-123, blends	All HCFCs, R-134a, blends	Not recommended due to high price	Works with most systems
Area Monitoring	Not recommended due to cross-sensitivity and poor detection limits	All HCFCs, R-134a, blends	All HCFCs, R-134a, blends	Not applicable
Other	None	When only one refrigerant is used or in moderately clean equipment rooms	"Dirty" environments or multi-refrigerant environments	None

Table 9-1. Comparison of various types of leak detectors

acids. This moisture is transported through the system by the refrigerant until it experiences the sudden pressure drop at the expansion device. The water crystallizes and gradually builds frozen layers until the expansion device is restricted enough to restrict or stop the flow of refrigerant, causing a reduction or complete loss of cooling. This restriction opens the low pressure control, freezestat, internal overload or other protection device and interrupts power to the cooling unit. Once the unit cycles off by some protection device, the ice layers on the expansion device orifice will melt. If the protection device is not manually reset, the cooling unit will cycle back on and start a short cycling routine until the problem is corrected. Short cycling is not good for motors or controls. Whether or not expansion device freeze-

Category/Concern	Features to Consider
Selectivity	• Is only one refrigerant present? • Are other materials (nonrefrigerants) stored in the equipment room? • Will the ventilation system bring potential interfering vapors into the equipment room?
Operation	• Is battery backup required? Is there access to UPS? • Is fail-safe operation required? • Who has access to instrument controls and alarms? • What is the desired maintenance schedule? • What are the anticipated temperature and humidity ranges?
Instrument Parameters	• Is multi-port capability required? • What is the desired response time? • What is the baseline stability? • Is the instrument auto-zeroed? • What is the minimum detection level? • How selective is the monitor?
Output/Alarm Functions	• Where is the alarm indicated (local vs area alarms)? • Is the alarm latching or nonlatching? • Is the alarm single or multilevel? • Can the instrument show rate-of-change of refrigerant concentration? • How are the concentrations displayed? • Is the computer interface present or necessary? • Who has the capability to shut off alarms?
Miscellaneous	• What is the cost? • What type of enclosure is required? • Is an intrinsically safe detector required?

Table 9-2. Points to consider when purchasing an area monitor

up occurs depends on both the amount of water vapor present and the size of the ice crystals. Even if freeze-up does not occur, corrosion, acids, and sludges from the mixture of water vapor, excessive heat, oil, and refrigerant can seriously damage a cooling system.

When moisture, heat, and refrigerant are present in a refrigeration system, acids will start to form after a short period of time. Refrigerants such as R-12, R-502, and R-22 contain chlorine and will hydrolyze with water, forming hydrochloric and hydrofluoric acids and more water. Once acids form, metal corrosion occurs. Heat is the catalyst in this complex chemical reaction, and as more heat is generated, acid formations are accelerated, Figure 9-21. When this acid travels through the system and mixes with system oil, globules of sludge form along with organic acids and oil decomposition products. Sludge

is a tightly bound mixture of water, acid, and oil:

Moisture + Acid + Oil = Sludge

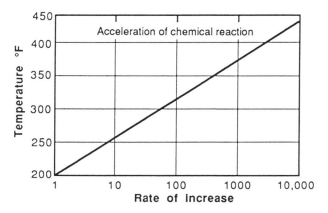

Figure 9-21. Rate of chemical reaction versus temperature compared to that at 200°F (Courtesy, Refrigeration Service Engineers Society)

Refrigeration oil has a high affinity for water vapor and refrigerant because of the low vapor pressure of oil. This means that refrigerant and water vapor will be attracted to the oil, and all three will mix and be soluble in one another. Sludge breaks down the oil and reduces its lubricating abilities, causing serious mechanical damage to the system. If any mechanical parts are corroded from acids, small particles from the corroded parts will be carried in the sludge. These sludges and solids tend to build up in the hottest points in the system, which are the discharge valve seats. The valves will no longer seat properly and *wiredrawing* occurs, which is when vapor is forced through a very small orifice at high speeds, creating friction and elevating temperatures up to 1000°F. Sludge can exist as minute, slimy, or sticky solids, powdery solids, or thick, slimy, and oily liquids.

To avoid corrosion and sludging problems in cooling systems, moisture must be kept out through good service practices and an effective preventive maintenance program. Sludging and corrosion cause expansion devices, filter driers, and strainers to plug and malfunction. The only sure way to rid a cooling system of moisture is to employ good evacuation procedures through the use of a high vacuum pump. Once sludge is formed, standard clean-up procedures using oversize driers specified for sludging problems must be followed (discussed later in the chapter). Vacuum pumps are not designed to remove solids such as sludge. Deep vacuum procedures will not take the place of liquid line or suction line driers, because the vacuum pump cannot remove solids.

Moisture always needs heat to vaporize, and the closest heat source is in the moisture itself. Because of this, moisture may freeze to ice before it completes its vaporization process. Sublimation will then occur, in which the ice changes state directly to a gas (water vapor) without passing through the liquid phase. Heating the system can prevent moisture from freezing and will save evacuation time.

Deep Vacuums

A common belief of many service technicians is that a vacuum pump pulls out liquid moisture particles from the system. This is incorrect. Even if moisture exists as both liquid and vapor in the cooling system, the vacuum pump is not capable of drawing out liquid water. When a vacuum pump is used, the system pressure is reduced to the boiling point of water at normal temperatures. For example, water boils at 212°F at an atmospheric pressure of 14.696 psia (0 psig). In order to vaporize any liquid moisture from the cooling system at atmospheric pressure, the system must be at a temperature of 212°F, which is not a normal temperature. In order to vaporize (boil) water at lower temperatures, the internal pressure of the cooling system must be reduced. If the internal pressure of the cooling system can be significantly lowered, liquid water will vaporize and be drawn through the vacuum pump and expelled to the atmosphere. The lower the internal pressure on the system, the lower the boiling point temperature of the moisture in the system, Table 9-3.

Table 9-3 shows that if the internal pressure of a system is reduced to 1.066 psia (27.75 in. Hg), water in the system will boil at 104°F. The system will still have to be exposed to a temperature of 104°F in order for the vaporization process to occur. A compound gauge set installed on this system will indicate a vacuum of 27.75 in. Hg if the gauge is calibrated. If the internal system pressure is brought down to 28.67 in. Hg, any moisture in the system would vaporize at a temperature of 86°F, which is a more reasonable temperature for a system without the addition of artificial heat. However, the difference between 27.75 and 28.67 in. Hg is too small for a compound gauge to accurately differentiate and measure. At best, a compound gauge has an accuracy of +/-1 in. Hg, which is why compound gauges are impractical for accurate vacuum measurements. This small difference must be measured, because moisture will not be removed if the vacuum is 27.75 in. Hg and the system is below 104°F. This is where the use of a good quality micron gauge comes into play.

Micron Gauges

All deep vacuums should be measured with an electronic micron gauge, which can be digital, analog, or light emitting diode. Figure 9-22 shows an analog micron gauge that will check deep vacuums down to 10 microns. One micron is equivalent to one millionth (0.000001) of a meter or one thousandth (0.001) of a millimeter. There are 25.4 millimeters to an inch, so there are 25,400 microns to an inch. Referring to Table 9-3, a perfect vacuum, which exists only on theory, is either 29.92 inches of mercury, 0 psia, or 0 microns. From 25,400 microns to 0 microns would be the equivalent

Temperature (°F)	Microns	In. Hg Vacuum	Pressure (psia)
212	759,968	00.00	14.696
205	535,000	4.92	12.279
194	525,526	9.23	10.162
176	355,092	15.94	6.866
158	233,680	20.72	4.519
140	149,352	24.04	2.888
122	92,456	26.28	1.788
104	55,118	27.75	1.066
86	31,750	28.67	.614
80	25,400	28.92	.491
76	22,860	29.02	.442
72	20,320	29.12	.393
69	17,780	29.22	.344
64	15,240	29.32	.295
59	12,700	29.42	.246
53	10,160	29.52	.196
45	7620	29.62	.147
32	4572	29.74	.088
21	2540	29.82	.049
6	1270	29.87	.0245
-24	254	29.91	.0049
-35	127	29.915	.00245
-60	25.4	29.919	.00049
-70	12.7	29.9195	.00024
-90	2.54	29.9199	.000049
----	0.00	29.9200	.000000

One inch = 25,400 microns
Theoretical perfect vacuum = 29.92 in. Hg = 0 microns = 0 psia

Table 9-3. Boiling point of water at various vacuum levels

Figure 9-22. Analog micron gauge for deep vacuum measurements (Courtesy, Thermal Engineering Company)

to pulling a vacuum on the last inch of mercury (28.92 in. Hg to 29.92 in. Hg) when dealing with a compound gauge. Deep vacuum is the only sure method to ensure the system is dry and free of noncondensibles.

In the hvac industry, vacuum is often measured with closed end manometers, compound gauges, and electronic thermistor or thermocouple gauges. The closed end manometer is only accurate in millimeters of mercury, not microns. The closed end manometer becomes impractical at about the 100 micron level. The compound gauge is only designed to read inches of mercury, not millimeters or microns. Thermocouples have the disadvantage of rusting or corroding at their junctions if made of ferrous metals. Using noble metal type thermocouples will prevent this corroding or rusting. When thermocouple vacuum gauges are operating properly, they can read microns of mercury. However, the most practical and accurate method of measuring deep vacuum today is with a thermistor micron gauge.

The only difference between today's deep vacuum techniques and yesterday's triple evacuation or blotter method is that vacuum results are known when using a thermistor micron vacuum gauge. If the triple evacuation method is used, the technician must use a micron gauge on the final vacuum to be sure of the results. With the rising cost of refrigerants and their

ozone depleting potential, technicians must think twice before using the triple evacuation method. Whatever evacuation technique is chosen, a 500 micron vacuum should be pulled and held on all systems for at least 10 minutes once the vacuum pump is isolated from the system. It is no longer advisable to guess when dehydration and degassing have been adequately completed.

The thermistor micron gauge employs a thermistor to sense vacuum. A thermistor is a temperature sensitive resistor that will exhibit a change in its electrical resistance with a change in its temperature. Thermistors used in vacuum measurements are negative coefficient thermistors, meaning that as their temperature increases their resistance decreases and vise versa. Electronic thermistor vacuum gauges are specially designed for high vacuum pumping applications. Thermistor gauges are rugged and portable, so the service technician can use them on the job.

Thermistor vacuum gauges can be accurate to 1.0 micron. Electronic thermistors are either mounted somewhere in the vacuum line or a sensing tube is mounted in the vacuum line. An electronic circuit senses total vapor pressure as a function of thermal conductivity of a gas. Thermistors are sensitive to water vapor and other condensibles or noncondensibles and can give a very accurate measurement of the vacuum in the system.

Thermistor gauges are heat sensing devices in that the sensing element (thermistor) generates heat. The heat rate changes as surrounding vapors are removed, which results in the decrease of surrounding vapor pressure. As gases are removed from the system, less heat is dissipated from the thermistor, raising its temperature and decreasing its resistance because of its negative temperature coefficient. Since the output of the thermistor changes as a function of its heat dissipation rate, a change in output will be indicated on a meter calibrated in microns of mercury. As soon as all the moisture in the system is vaporized, the vapor pressure and heat dissipation rate of the system will decrease, decreasing the micron measurement.

Note: It is this low micron measurement from the meter that tells the service technician when the evacuation is complete. The more vaporization of water or degassing of unwanted gases from the system, the more time it will take to reach a low micron level when using the same size vacuum pump. Again, it is the measurement from the meter that determines when an evacuation is complete.

Vacuum Pump

Deep vacuums require the use of a rotary vacuum pump, Figure 9-23. Rotary vacuum pumps do not require any head clearance like piston type pumps. This results in rotary vacuum pumps having much higher volumetric efficiencies when compared to piston type pumps. It is for these reasons that the rotary type vacuum pump is superior for deep vacuums. These pumps are usually two-stage vacuum pumps.

Vacuum pump capacity is measured in cubic feet per minute (cfm). Vacuum pump capacity has little to do with evacuation time because of the internal restrictions in cooling systems. Metering devices, tubing length, return bends, and service valve orifices all offer restrictions during evacuation. The only way to increase flow through a fixed orifice is to increase the pressure difference across that orifice. When evacuating the system, the source of pressure in and out of the system on a 3/16 inch gauge orifice is not much to work with, which is why vacuum pumps in the 1 cfm to 4 cfm range should handle 95% of the work. If the vacuum pump is too large and overpowers the system, there may be a low micron level reading before all of the moisture is removed. However, the micron level will soon creep up after the vacuum pump is isolated from the system. Many technicians use the general rule that one cfm of vacuum pump capacity will be needed for every 7 tons of system capacity.

Reducing the pressure drop of the system will increase the evacuation time. This can be accomplished by pulling a vacuum through both the high and low sides of the system, provided the connecting hoses and manifold do not add too many additional restrictions. It is better to have larger vacuum hoses and manifold because of less pressure drop. Another method to reduce pressure drop is to use two vacuum pumps at two different locations in the system. On larger systems, two smaller pumps at different locations are more advantageous than one large vacuum pump at one location, because each is pulling against only half of the system restrictions.

Single-stage vacuum pumps

Single-stage vacuum pumps are generally less expensive and smaller than a two-stage vacuum pump of equal cfm capacity, Figure 9-24. Because single-stage vacuum pumps discharge directly into the atmosphere, the added backpressure of approximately 14.7 psi (atmospheric pressure) prevents them from pulling a vacuum as low as that of a two-stage vacuum pump. Manufacturers claim that single-stage vacuum pumps will pull vacuums to about 50 microns under ideal conditions. However, there are many service applications in which single-stage vacuum pumps will not perform the evacuation task because of the added atmospheric backpressure or ignored maintenance schedules.

Two-stage vacuum pumps are two single-stage vacuum pumps in series, Figure 9-25. They are capable of pulling lower vacuums because of their second stage, which experiences a much lower intake pressure. This lower pressure results from the exhaust of the first-stage vacuum pump being exhausted into the intake of the second-stage vacuum pump instead of to atmospheric pressure. This gives the first-stage vacuum pump less backpressure and more efficiency. The second-stage vacuum pump begins pulling at a lower pressure, so it pulls a higher vacuum. Two-stage vacuum pumps are capable of pulling vacuums of about 1 micron under ideal conditions. In field situations, two-stage pumps will almost always pull down to 20 microns. Two-stage vacuum pumps have the best track record, because they consistently pull lower vacuums and are much more efficient when removing moisture.

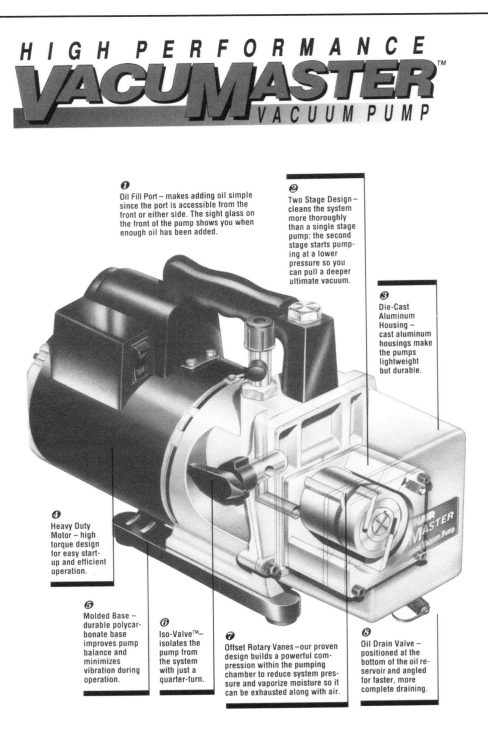

HIGH PERFORMANCE
VacuMASTER™
VACUUM PUMP

❶ Oil Fill Port – makes adding oil simple since the port is accessible from the front or either side. The sight glass on the front of the pump shows you when enough oil has been added.

❷ Two Stage Design – cleans the system more thoroughly than a single stage pump: the second stage starts pumping at a lower pressure so you can pull a deeper ultimate vacuum.

❸ Die-Cast Aluminum Housing – cast aluminum housings make the pumps lightweight but durable.

❹ Heavy Duty Motor – high torque design for easy start-up and efficient operation.

❺ Molded Base – durable polycarbonate base improves pump balance and minimizes vibration during operation.

❻ Iso-Valve™– isolates the pump from the system with just a quarter-turn.

❼ Offset Rotary Vanes –our proven design builds a powerful compression within the pumping chamber to reduce system pressure and vaporize moisture so it can be exhausted along with air.

❽ Oil Drain Valve – positioned at the bottom of the oil reservoir and angled for faster, more complete draining.

Figure 9-23. Rotary vacuum pump (Courtesy, Robinair Division, SPX Corporation)

The limitation of any vacuum pump is referred to as its *blank-off pressure*. When the inlet of the pump is blanked off or closed tightly, the blank-off pressure should be quickly reached. Piston-type pumps rarely reach 28 in. Hg, while rotary vacuum pumps will blank-off at a very low micron level. The vapor pressure of their lubricating and sealing oil is sometimes their limit.

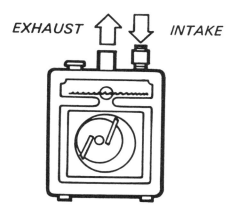

EXHAUST ⬆ ⬇ INTAKE

Figure 9-24. Single-stage vacuum pump (Courtesy, Robinair Division, SPX Corporation)

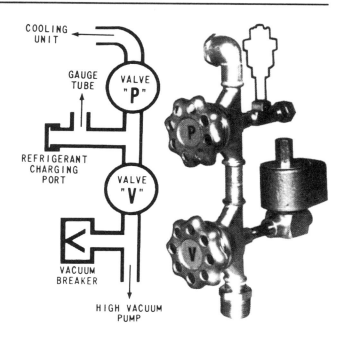

COOLING UNIT ◀

GAUGE TUBE

VALVE "P"

REFRIGERANT CHARGING PORT

VALVE "V"

VACUUM BREAKER

HIGH VACUUM PUMP

Figure 9-26. Two valve test manifold (Courtesy, Refrigeration Service Engineers Society)

EXHAUST ⬆ ⬇ INTAKE

SECOND   FIRST

Figure 9-25. Two-stage rotary vacuum pump (Courtesy, Robinair Division, SPX Corporation)

Vacuum test manifold with two valves

A two valve test manifold consists of a P valve and a V valve, along with a vacuum breaker port and a micron gauge tube access, Figure 9-26. When a vacuum is pulled, both P and V valves are open, Figure 9-27a. Once the desired vacuum level is reached on the micron gauge, valve V can be closed, which isolates the system from the vacuum pump, Figure 9-27b. The micron gauge is still open to the refrigeration system, so it is possible to watch for signs of a leak. The vacuum pump may then be vented to the atmosphere so as not to hydraulically load the cylinder on the next start-up. If the micron gauge rises slowly and then stabilizes at atmospheric pressure (0 psig), a leak is indicated. If the micron gauge pressure increases above atmospheric pressure, either residual re-

frigerant or moisture is still vaporizing in the system and more vacuum time is needed. Remember, if performing a standing vacuum test and a leak is found, gases such as air and water vapor may enter the system through the leak. The standing vacuum test only tells the technician that the system does leak; it does not pinpoint the leak. Other leak checking methods must be employed to locate the exact location of the leak. Once the leak is found, a deep vacuum must be performed on the system.

If the desired vacuum level cannot be reached and the system has been leak checked very thoroughly, close valve P. This arrangement will blank-off the vacuum pump. The micron gauge will still be able to service the vacuum pump but not the system, Figure 9-27c. If the micron gauge immediately goes to the lower blank-off pressure reading, the pump is in good condition. If the micron gauge does not go immediately to blank-off pressure, a pump problem, contaminated oil, or pump leak exists; it is not the fault of the refrigeration system.

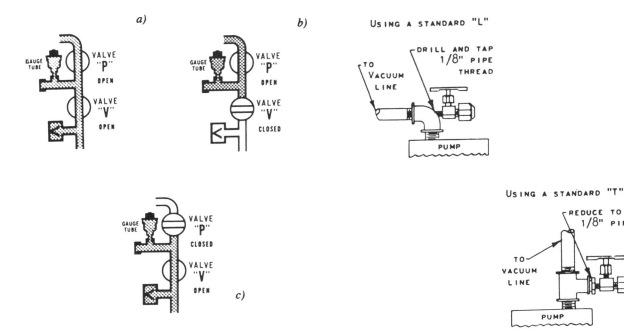

Figure 9-27. a) P and V valves open; b) P valve open, V valve closed; c) P valve closed, V valve open (Courtesy, Refrigeration Service Engineers Society)

Figure 9-28. Vacuum breaker valve (Courtesy, Refrigeration Service Engineers Society)

Vacuum breaker valves

When a vacuum pump is shut off the lowest pressure is in its cylinder, because it is the driving force that creates the pressure difference to make a vacuum possible. The lower pressure area in the cylinder attracts vacuum pump oil, which means when the pump is off, the cylinder will be full of oil. Upon start-up, hydraulic loading of the cylinder will occur, which may seriously damage mechanical and electrical components of the vacuum pump. To prevent this from happening, vent the cylinder to the atmosphere while shutting down the pump. The cylinder area will be at a relatively higher pressure (atmospheric) and will not accept any vacuum pump oil because of this higher pressure. Atmospheric venting before shutdown prevents hydraulic cylinder loading during shutdown and pump damage during the next start-up. Make sure the vacuum pump is isolated from the system through the manifold gauge set before venting a pump to the atmosphere, otherwise, the vacuum that was just pulled on the system will be ruined. Figure 9-28 shows the construction of a vacuum breaker valve that has been added to a vacuum system.

One manufacturer has an atmospheric vent (purge) built into the vacuum pump, which is initiated by the shut-off switch, Figure 9-29. This prevents the technician from forgetting to vent the vacuum pump to atmosphere at shut off. Again, make sure the vacuum pump is isolated from the system before shutting off the vacuum pump.

Pressure drop

Whenever there is a pressure difference from high to low inside a refrigeration system, there will be fluid flow. Just as a temperature difference is the driving potential for heat transfer, a pressure difference is the driving potential for fluid (gas or liquid) flow. When an operating vacuum pump is connected to a system through a manifold and gauge set, the lowest pressure in the system will be in the vacuum pump cylinder, the next highest pressure will be in the manifold gauge set, and the highest pressure will be in the system itself. Figure 9-30 shows pressure drop as gases travel from the refrigeration system to the vacuum pump. The pressure drop from the system to the vacuum

Figure 9-29. Vacuum breaker built into shut-off switch (Courtesy, TIF Instruments, Inc.)

pump is caused by restrictions as the gas flows to the vacuum pump. If the micron gauge is located on the vacuum pump, vacuum pump pressure will be monitored, not system pressure. Remember that system pressure will be a little higher.

Once the desired vacuum level is reached and the vacuum pump is isolated from the system, there may be a slight rise in pressure on the micron gauge. The gauge should then stop rising and stabilize or level off at a desired low micron level, Figure 9-31a. This desired low micron level should be at least 500 microns. This slight rise in pressure is referred to as system equalization. The higher system pressure is equalizing with the lower vacuum pump pressure, giving an intermediate pressure on the micron gauge. If the micron gauge rises and levels off at a high micron reading, the system is still too wet and needs additional dehydration time.

If the micron gauge is located on the refrigeration system, there may be a decrease in pressure when the vacuum pump is isolated from the system. This is caused by the lower vacuum pump pressure equalizing with the higher system pressure, giving an intermediate pressure. If the micron gauge keeps rising and does not stabilize once the pump is isolated from the

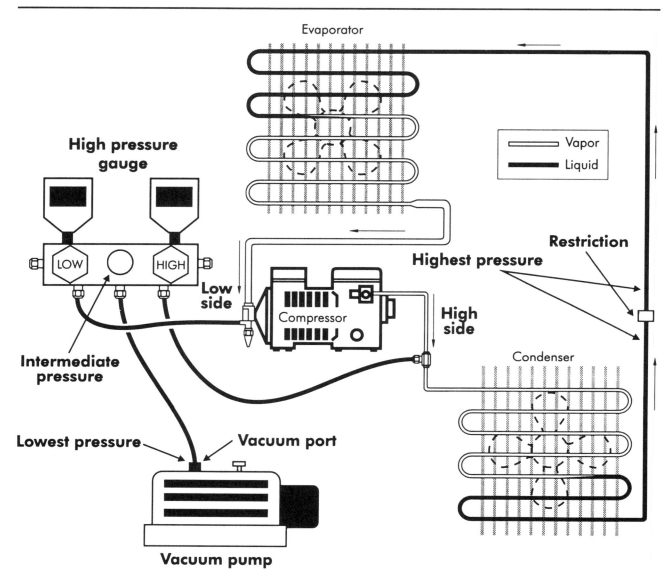

Figure 9-30. Pressure drop as gas travels to vacuum pump

system, there is probably a system leak letting in atmospheric pressure at the leak site, Figure 9-31b. The leak may not be in the system; it may be in the hook-up of the vacuum equipment. Fittings coming out of the vacuum pump may be coated with a clear sealant made especially for deep vacuum equipment. This sealant must have a vapor pressure of 1 micron when in liquid form. The liquid is painted over the threads and joints and allowed to dry. Once dry, it will remain in the plastic state.

Figure 9-32 shows a state-of-the-art vacuum pump that has a micron gauge with light emit-

ting diodes built into it. The number of lights that are on and off indicate the vacuum level; when all lights are off, the desired vacuum level is reached. This vacuum pump also has an atmospheric purge or vent built into the power switch to prevent hydraulic loading of the cylinders; a built-in blank-off valve to check for system leaks; an oil contamination light; and a test button to test the electronic circuitry in the system. This type of system eliminates the need for a two valve test manifold, because it is built into the vacuum pump.

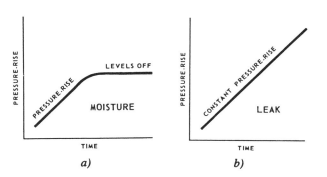

Figure 9-31. Pressure vs time graphs for standing vacuum leak checking (Courtesy, Refrigeration Service Engineers Society)

Vacuum pump oil

Vacuum pumps require a different oil than that used in standard refrigeration systems. Vacuum pump oil has a much lower boiling point than regular refrigeration oil; for example at 100°F, vacuum pump oil vapor pressure must be no greater than 5 microns. Standard refrigeration oil vapor pressure is much higher and can vary considerably; as such, it cannot be controlled closely enough for use in deep vacuum applications. Low vapor pressure is a very important quality of vacuum pump oil. A vacuum pump cannot be expected to pull a vacuum less than the vapor pressure of the oil that is lubricating and sealing the pump.

Because used oil contains system contaminants, the vacuum pump oil must be changed once the evacuation is completed. It is important to drain the oil while the vacuum pump crankcase is still warm from use, Figure 9-33. The oil will be much thinner when warm, making the draining process easier and more efficient. Do not expose any new oil to the atmosphere for over a minute or two, because it will readily absorb atmospheric moisture, causing loss of lubrication and sealing abilities. It is a good idea to first run the vacuum pump for one to two minutes with the new oil, then drain the oil again and refill with fresh oil, Figure 9-34. This process ensures all contaminants are removed from the pump and oil. When storing the vacuum pump, always close off its inlet and outlet to ensure no atmospheric moisture will contaminate the oil or rust the internals of the pump.

Gas ballasts

Pump efficiency depends on the cleanliness and dryness of its sealing and lubricating oils. Moisture in the oil takes away the sealing and lubricating properties of oil. Larger amounts of air and moisture are evacuated from the refrigeration system during the initial stages of any vacuum procedure. It is at these times that moisture in the air may condense back into liquid and become compressed as it moves through the vacuum pump. This condensation will fall into the pump oil and degrade its sealing and lubricating abilities, which is why some vacuum pump manufacturers have incorporated a gas ballast into the design of their vacuum pumps.

The *gas ballast* is an opening in the second stage pumping chamber that the technician can open or close when desired, Figure 9-35. It is usually opened during the initial stages of a vacuum procedure to let relatively dry atmospheric air into the second stage pumping chamber. (The gas ballast is usually in the second stage pumping chamber because of its deeper vacuum abilities.) The atmospheric air is considered dry relative to the compressed air and water vapor in the second stage chamber that has been drawn from the system. This atmospheric air will mix with the chamber air and water vapor and decrease its relative humidity. This will lessen the chance of moisture condensation when it is compressed in the second stage of the vacuum pump.

The gas ballast helps prevent condensation, which keeps the sealing and lubricating oil dry and the system running at maximum efficiency. Once the initial stages of the vacuum procedure are completed and a greater percentage of moisture and air have been removed from the system, the gas ballast can be closed. This will allow the vacuum pump to reach its desired micron levels and finish the evacuation procedure. Remember, it is only during the first stages of the vacuum procedure that there is a lot of air and moisture being removed from the system, which requires the opening of the gas ballast. The gas ballast must be closed during the final stages of evacuation, or the desired vacuum level will never be met.

Figure 9-32. Vacuum pump (Courtesy, TIF Instruments, Inc.)

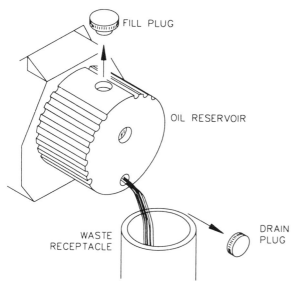

Figure 9-33. Draining hot vacuum pump oil (Courtesy, TIF Instruments, Inc.)

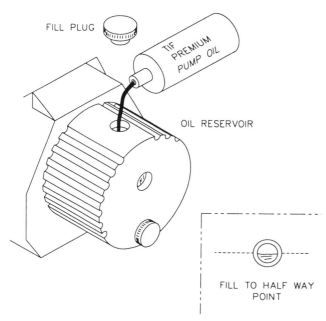

Figure 9-34. Filling vacuum pump with fresh oil (Courtesy, TIF Instruments, Inc.)

SYSTEM BURNOUT

When moisture, heat, and refrigerant are present in a refrigeration system, hydrochloric and hydrofluoric acids are sure to form in time. Air and moisture enter a refrigeration system through leaks or poor service practices when charging or installing gauges. Air and moisture are detrimental to a refrigeration system. Air

The Importance of the Gas Ballast

The efficiency of rotary vane vacuum pumps relies to a great extent on the quality and purity of the vacuum pump oil. The gas ballast feature helps to prevent moisture from contaminating the oil and reducing its viscosity.

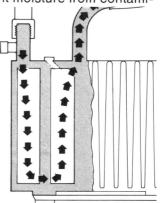

As air and moisture vapor are exhausted from the system through the pump, the vapor has a tendency to condense back to a liquid as it moves through the pump. The gas ballast lets in enough relatively dry atmospheric air to prevent the condensation, keeping the oil cleaner. Using the gas ballast will keep the vacuum pump running at maximum efficiency.

Figure 9-35. The importance of the gas ballast (Courtesy, Robinair Division, SPX Corporation)

remains in the condenser, causing high head pressures, high compression ratios, and higher discharge temperatures. Air trapped in a compressor during installation is impossible to remove by purging; a deep vacuum pump must be used. New and replacement compressors are shipped with a dry nitrogen charge that must be removed and then evacuated before using.

Refrigerants containing chlorine will hydrolyze with water, forming hydrochloric and hydrofluoric acids and more water. Once acids are formed, most metals in the system will corrode. As stated previously, when acids travel throughout the system and mix with the system oil, globules of sludge form along with organic acids and oil decomposition products. Sludge tends to build up in the hottest point in the system, which is the discharge valve seat.

Sludge and corrosion will cause expansion devices, filter driers, and strainers to plug and malfunction. The only sure way to rid moisture from the system is to use good evacuation procedures in conjunction with a deep vacuum pump. Once sludge is formed, standard clean-up procedures using oversize driers specified for sludge problems must be followed. Vacuum pumps are not designed to remove sludge, which is why deep vacuum procedures do not take the place of liquid line or suction line driers.

System Clean-Up

When a compressor motor burns out for some reason, the high temperature arc causes the refrigerant/oil mixture to break down into sludge, acids, and more water. If the contamination from a burned-out and/or sludged compressor reaches the replacement compressor crankcase, the replacement compressor will probably burn out.

The old method of flushing a system with R-11 is now obsolete, because R-11 is a CFC, which has a high ozone depleting potential. A new method of system cleaning involves the use of approved filter driers. Filter driers incorporate an adequate desiccant in both the liquid and suction lines. This method is very economical if the system refrigerant is recovered using certified recovery devices.

To clean up a system in which burnout has taken place, perform the following steps (Reprinted with permission from Copeland Corporation): *Warning: Whenever working with a burned out system, extreme caution must be used when opening the system. Adequate ventilation, rubber gloves, and safety glasses must be worn to protect the technician from acids. The oil from a burnout can cause serious skin irritation and possible burns. In some cases, the fumes are toxic.*

1. If possible, front seat (close) compressor service valves to isolate the compressor from the system. *Note: If the compressor does not have service valves, see Step 6.* To avoid losing refrigerant to the atmosphere, recover the refrigerant

using standard refrigerant recovery equipment and procedures. Remove the inoperative compressor, and install the replacement compressor.

2. Take a sample of oil from the replacement compressor. Seal this sample in a glass bottle for comparison purposes after the cleaning procedure is complete.

3. With the compressor isolated from the rest of the system, evacuate the compressor. After evacuation, open the compressor service valves and close the liquid line valve and any other available shut-off valves. This will minimize the amount of refrigerant to be handled during pumpdown. Pump down the system. Some contaminants will be returned to the compressor during the pumpdown procedure, but the compressor will not be harmed by the short period of operation required for pumpdown. These contaminants will be removed after the installation of the filter driers.

4. Inspect all system controls such as expansion valves, solenoid valves, check valves, reversing valves, and contactors. Clean or replace if necessary. Remove or replace any filter driers previously installed in the system, and clean or replace any filters or strainers. Install a good quality moisture indicator if the system does not have one.

5. Install the recommended filter drier size in the suction line. Install an oversize filter drier in the liquid line. Both liquid and suction lines will require a special temporary filter drier manufactured especially for system burnout cleaning procedures.

    Note: If compressor has service valves, go to Step 7.

6. For systems without service valves, recover the refrigerant and then evacuate the system. Perform the inspection

and filter drier changes listed in Steps 4 and 5. Charge recovered refrigerant back into the system through a filter drier. Add additional refrigerant as necessary.

7. Start the compressor and run the system. As contaminants in the system are filtered out, the pressure drop across the filter driers will increase. Observe the pressure differential across the filter driers for a minimum of 4 hours. Use the same gauge to measure the pressure on each side of the driers to minimize error in gauges (driers come with pressure taps on both of their ends). If the pressure drop exceeds the maximum limit shown on the curves in Figure 9-36, replace the filter drier and restart the system. Note that the curves in Figure 9-36 show both permanent and temporary filter driers. During the clean-up procedure, technicians will encounter oversize filter driers for both the liquid and suction lines.

8. Allow the unit to operate for 48 hours. Check the odor (smell cautiously), and compare the color of the oil with the sample taken in Step 2. If an acid test kit is available, test for acid content. If the oil is discolored, acidic, has an acrid odor, or if the moisture indicator indicates a high moisture content in the system, change the filter driers. The compressor oil may also be changed. Allow the system to operate for an additional 48 hours, and recheck as before. Repeat until the oil remains clean, odor free, and the color approaches that of the original sample.

9. Replace the liquid line filter drier with one of normal size and for normal applications. Remove the suction line filter drier, and replace it with a normal size, permanent type suction line filter only.

10. After the cleaning procedure is completed, recheck in approximately two weeks to ensure that the system condition and operation are completely satisfactory.

Standard Suction Line Filters

It is essential that all foreign material be removed from a system at the time of the original installation. Filings, shavings, dirt, solder, flux, metal chips, bits of steel wool, sand from sandpaper, and even wire from cleaning brushes may all contaminate the refrigeration system, ending up in the compressor. Compressors returned to the factory for replacement indicate that many such replacements could have been prevented if all contaminants had been removed from the refrigeration system at the time of field installation. Returned compressors may be reported as motor failures; in reality, the return is due to damaged bearings, reed valves, or connecting rods caused by contaminants, which then caused motor damage. Many contaminants are small enough to pass through fine mesh screens. In addition, metal fragments may be rotating, due to gas velocity, and cut or break the usual compressor suction screen.

Due to the possibility of contaminants being in a system after the initial installation, a heavy duty suction line filter is recommended for every new field installation. Suction line filters are filters only, not driers, and come with reasonable pressure drops so as not to affect capacity by any measurable amount. These suction filters are placed between the evaporator and the compressor in the suction line and give maximum protection for the most vulnerable part of the system. A pressure fitting should be provided ahead of the filter to facilitate checking pressure drop. An additional benefit of suction line filters is the added protection they provide if a burnout should occur; any small particles will be picked up by the filter and prevented from traveling to the compressor.

Combination Acid and Moisture Test Kits

Good preventive maintenance includes periodically checking refrigeration and air conditioning systems for acid and moisture. Acid test

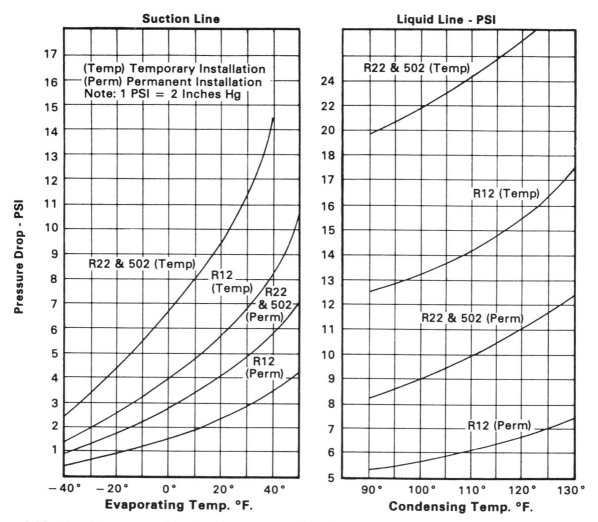

Figure 9-36. Filter drier pressure drop (maximum recommended) (Courtesy, Copeland Corporation)

kits are available to test compressor oil for acid content. However, obtaining a large enough oil sample usually means pumping down the compressor, obtaining the sample, and then evacuating the compressor, which can be very time consuming.

One test kit on the market allows the technician to test for acid and moisture simultaneously, without obtaining an oil sample. This same test kit can be used to test any refrigerant gas source that is in the vapor phase, including any recovered refrigerant that may be in a refrigerant storage cylinder. This test kit is simple, ready to use, and provides quick, accurate test

results. The system consists of a detector tube, Figure 9-37, which is inserted into a metering device, Figure 9-38. The assembly is then attached to a low side gauge hose and manifold set. The low side valve is opened, allowing flow through the entire manifold and hoses. Pressure will register on both low and high side gauges. When the low side gauge reaches a predetermined termination pressure, Figure 9-39, the detector tube is ready to be read. The detector tubes are calibrated to ARI purity standards, Figure 9-40. Figure 9-41 shows how to use the kit in more detail.

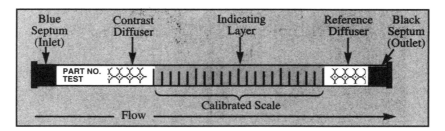

Septum ends will automatically be pierced when fully assembled; then self seal upon disassembly. Exposure to counter indicating atmospheric moisture is nil by this process.

Diffusers assure even gas flow over the indicating reagent layer. Indicating layer will react with a graphic color change. The reacted layer is read against a scale calibrated to ARI standards. The threshold limit is always the center mark. The reference diffuser is a sample color change of a positive test.

A detector tube can only be used once even if the test is negative.

Figure 9-37. Detection tubes (Courtesy, Refrigeration Technologies)

The Metering Device

The CHECKMATE BODY is a precision engineered in-line gas sampling device built to endure a lifetime of usage.

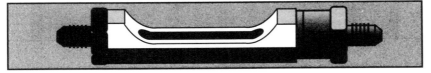

- A short extension hose provides adequate clearance for connection to system or cylinder.

- Blue inlet fitting containing a regulating Needle-Nozzle.

- Recessed track providing in-line viewing of Detector Tube.

Figure 9-38. Metering device (Courtesy, Refrigeration Technologies)

Termination Pressure

The CHECKMATE method is based on sound principals of analytical chemistry where a **specific volume** of gas is allowed to pass through the detection tube. Each detection tube will react in a **proportional amount** to the contaminant. The termination pressure is respectfully different for all Refrigerant Gases - **Knight takes King** - CHECKMATE!

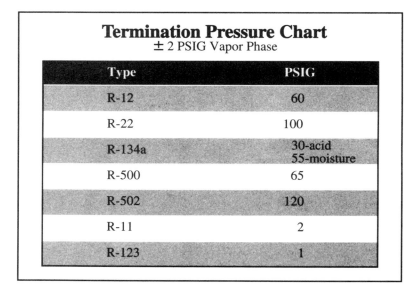

Termination Pressure Chart
± 2 PSIG Vapor Phase

Type	PSIG
R-12	60
R-22	100
R-134a	30-acid 55-moisture
R-500	65
R-502	120
R-11	2
R-123	1

Figure 9-39. Termination pressure (Courtesy, Refrigeration Technologies)

Contaminant	Allowable Limit by Weight
Moisture	10 ppm
Oil	100 ppm
Acid	1.0 ppm
Chlorides	none

Figure 9-40. ARI Standard 700-88, Purity for Reclaimed and Virgin Refrigerant

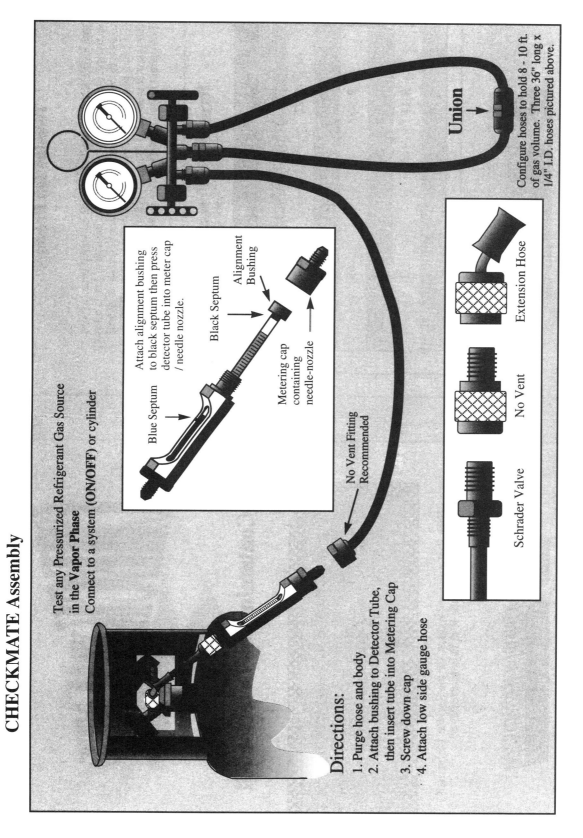

CHECKMATE Assembly

Test any Pressurized Refrigerant Gas Source in the **Vapor Phase**
Connect to a system (ON/OFF) or cylinder

Attach alignment bushing to black septum then press detector tube into meter cap / needle nozzle.

Blue Septum

Black Septum

Alignment Bushing

Metering cap containing needle-nozzle

No Vent Fitting Recommended

Schrader Valve

No Vent

Extension Hose

Union

Configure hoses to hold 8 - 10 ft. of gas volume. Three 36" long x 1/4" I.D. hoses pictured above.

Directions:

1. Purge hose and body
2. Attach bushing to Detector Tube, then insert tube into Metering Cap
3. Screw down cap
4. Attach low side gauge hose

Figure 9-41. Assembling and disassembling the test kit, continued on next page (Courtesy, Refrigeration Technologies)

Taking the Test

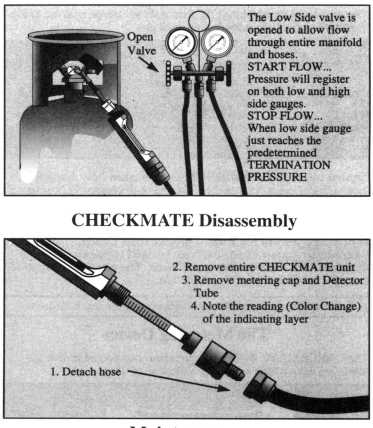

Open Valve

The Low Side valve is opened to allow flow through entire manifold and hoses.
START FLOW...
Pressure will register on both low and high side gauges.
STOP FLOW...
When low side gauge just reaches the predetermined TERMINATION PRESSURE

CHECKMATE Disassembly

2. Remove entire CHECKMATE unit
3. Remove metering cap and Detector Tube
4. Note the reading (Color Change) of the indicating layer

1. Detach hose

Maintenance

Keep your CHECKMATE Unit clean and free of contaminants with a purge of dry gas, especially after a positive reading. Never use soap, water or solvent on your unit. If a solvent is desired, rinse with rubbing alcohol and allow to dry.

Figure 9-41. Continued from previous page

NOTES

[1] Refrigeration Technologies and TIF Instruments provided most of the information on leak detection found in this chapter.

[2] Water chillers are usually pressurized using controlled warm water. When dealing with chillers, valve off the condenser and evaporator water circuits. Then introduce controlled warm water to the evaporator tube bundle. This causes the rate of vaporization of the refrigeration to increase, causing higher pressures in the evaporator. A gradual amount of warm water must be introduced to avoid temperature shock to the evaporator. The rupture disc on the evaporator may open if the pressures are raised too high. There are special fittings available from the chiller manufacturer to equalize pressure inside and outside of the rupture disc to prevent rupture. Consult with the chiller manufacturer before attempting to service or leak check any chiller.

[3] This section on leak pinpointing is by DuPont. Reprinted with permission.

Ozone Depletion and Global Warming

Ozone (O_3) is a gas that is found in both the stratosphere and troposphere, Figure 10-1. Ozone is a molecule that consists of three oxygen atoms, Figure 10-2, instead of the standard two atom oxygen or diatomic oxygen, Figure 10-3. Stratospheric ozone is considered "good" ozone, because it shields the earth from harmful ultraviolet (UV-B) radiation. This three atom structure enables stratospheric ozone to absorb harmful ultraviolet light from the sun. Tropospheric ozone is considered "bad" or unwanted ozone, because it is a pollutant.

Stratospheric ozone, or "good" ozone, resides in the stratosphere, which is between 7 and 30

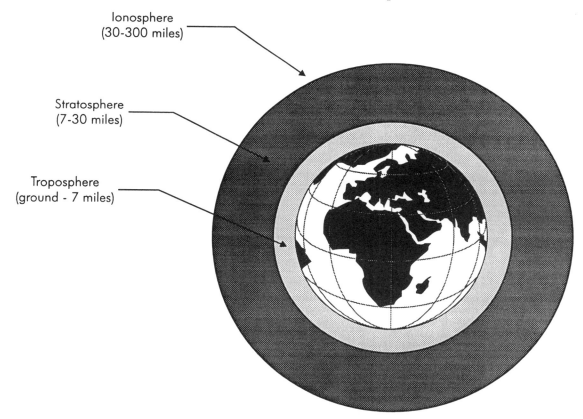

Figure 10-1. Atmospheric regions

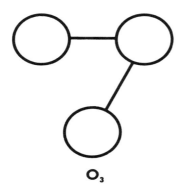

Figure 10-2. Ozone molecule

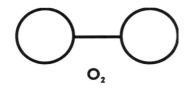

Figure 10-3. Diatomic oxygen

miles above the earth's surface. This ozone accounts for over 90% of all ozone; however, it is rapidly being depleted by man-made chemicals containing chlorine, including refrigerants such as CFCs and HCFCs. Figure 10-4 shows the annual production of fluorocarbons.

Tropospheric ozone, or "bad" ozone, resides in the troposphere, which extends from ground level to about 7 miles. This ozone accounts for only about 10% of all ozone. The troposphere contains 90% of the atmosphere and is well mixed by weather patterns. Tropospheric ozone is formed by reactions between hydrocarbons and oxides of nitrogen in sunlight. It has a bluish color when seen from the surface of the earth and may be irritating to the mucous membranes when inhaled. A popular term for tropospheric

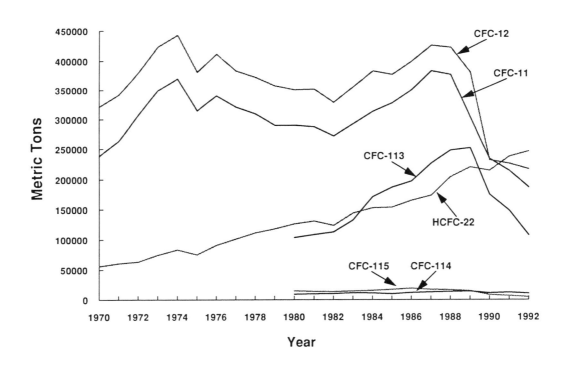

NOTES: "Production" (as defined in the Montreal Protocol) does not include feedstock uses.
Annual data are not available for CFC-113, CFC-114 and CFC-115 prior to 1980.

Figure 10-4. Annual production of fluorocarbons reported to AFEAS (Courtesy, Alternative Fluorocarbons Environmental Acceptability Study)

ozone pollution is smog. Tropospheric ozone contributes to the greenhouse effect, or global warming, which will be covered later in this chapter.

plants) and krill (tiny shrimp), which are both at the bottom of the food chain in the ocean. Almost all ocean dwellers rely on a form of photoplankton and krill for survival.

RADIATION

The sun emits three types of ultraviolet (UV) radiation: UV-A, UV-B, and UV-C radiation. UV-A radiation is not absorbed by the stratospheric ozone molecule at all; it is not biologically reactive and is of no worry to life on earth. UV-B radiation is preferentially absorbed by the stratospheric ozone molecule; it is biologically reactive and will affect life on earth. UV-B radiation has wavelengths in the range of 280 to 320 manometers. It represents the portion of solar radiation reaching the earth's surface that is most efficiently filtered and controlled by stratospheric ozone. UV-C radiation never makes it to the earth.

Under a clear sky at noon, up to 0.5% of the energy reaching the surface of the earth from the sun consists of UV-B radiation. The intensity is primarily controlled by the angle of the sun, so that during winter or in the morning and evening, it is only a small fraction of that at noon in the summer. Clouds absorb and scatter a significant amount of UV-B radiation, as do trace gases (e.g., sulfur dioxide and nitrogen dioxide) and atmospheric particulates. A complicating factor is that ozone in the lower atmosphere, such as that associated with air pollution, can absorb UV-B that has passed through the stratospheric ozone layer.

Excess ultraviolet radiation causes the lens of the eye to cloud up with cataracts, which may cause blindness if left untreated. Ultraviolet radiation can also lead to skin cancers, including the often deadly melanoma. Increased exposure to ultraviolet radiation affects the body's ability to fight off disease. High doses of UV radiation also reduce crop yields worldwide. UV-B is the most dangerous of the ultraviolet radiations and can penetrate many meters below the surface of the ocean. This radiation has been known to kill photoplankton (one-celled

STRATOSPHERIC OZONE DEPLETION

Oxygen molecules (O_2) have been generated for millions of years through a process called *photosynthesis*, which takes place in the oceans and on the earth. Oxygen molecules travel to the stratosphere where ultraviolet radiation from the sun breaks them apart into individual, free oxygen (O) atoms. The free oxygen (O) atoms are very unstable and need to bond to other atoms or molecules. Some of these free oxygen (O) molecules bond to other oxygen (O_2) molecules, which is how ozone (O_3) is formed in the stratosphere, Figure 10-5. It is this ozone in the stratosphere that shields us from the sun's harmful UV-B radiation. Once ozone is formed, much of it is easily destroyed by the sun's ultraviolet radiation. The sun's radiation will break up the ozone molecule into diatomic oxygen (O_2) and free elemental oxygen (O), Figure 10-6.

Ozone is constantly being created and destroyed in the stratosphere; the balance has been going on for millions of years. However, man-made chemicals containing chlorine, such as CFCs and HCFCs, have knocked this delicate process out of balance in the last few decades. Volcanic eruptions and other man-made chemicals also contain chlorine, but their contributions are almost negligible when compared to the amount of chlorine emitted into the atmosphere by CFCs and HCFCs. The result is that ozone is being depleted faster than it is being generated, and too much harmful ultraviolet radiation (UV-B) is reaching the earth.

CFC and a few HCFC molecules emitted into the troposphere reach the stratosphere by tropospheric winds. CFC molecules are very stable and have a long life that may extend over 100 years. In fact, if production and use of CFCs were to cease today, their effect on the ozone layer would not disappear until the year 2050

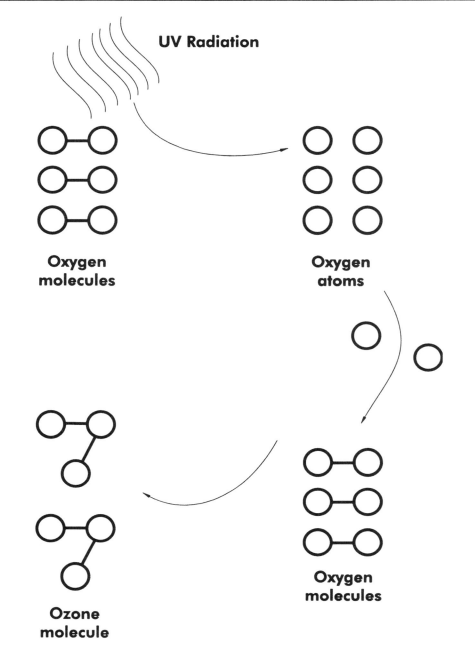

**Oxygen
molecules**

**Oxygen
atoms**

**Oxygen
molecules**

**Ozone
molecule**

Figure 10-5. The formation of stratospheric ozone

and beyond. HCFC molecules have much shorter lives and few ever reach the stratosphere.

Once in the stratosphere, ultraviolet radiation from the sun breaks off a chlorine atom (Cl) from the CFC or HCFC molecule. This chlorine atom is unstable and floats around looking for another molecule with which to bond. Upon finding an (O_3) molecule, the chlorine atom breaks it up into diatomic oxygen (O_2) and

chlorine monoxide (ClO), Figure 10-7. This is the process by which ozone is depleted. An oxygen molecule is normally a diatomic molecule and wants to exist in pairs to be stable. A free oxygen (O) atom is very unstable and wants to combine with the unstable chlorine atom. The two molecules formed by this break-up are chlorine monoxide (ClO) and diatomic oxygen (O_2).

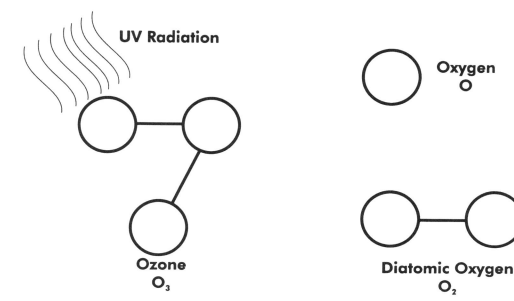

UV Radiation

Oxygen
O

Ozone
O_3

Diatomic Oxygen
O_2

Figure 10-6. Breaking up an ozone molecule

When the sun's ultraviolet radiation breaks up diatomic oxygen (O_2) into unstable, individual, free oxygen (O) atoms, these unstable oxygen atoms float around the stratosphere looking to bond with another atom or molecule. Upon finding a chlorine monoxide (ClO) molecule, the free oxygen atom breaks off the chlorine atom from the molecule and bonds to the now free oxygen atom to form a stable diatomic molecule of oxygen (O_2). This reaction happens because diatomic oxygen (O_2) is more stable than chlorine monoxide (ClO). The free chlorine atom is left to float around the stratosphere looking for another ozone (O_3) molecule to attack and break up. This starts the ozone depletion process all over again. The chlorine atom is actually a catalyst in these ozone depleting reactions. It enters a reaction and comes out unchanged and unharmed, ready to start other reactions. It is believed that one chlorine atom can destroy up to 1,000,000 ozone molecules, which is why chlorine in the stratosphere is so destructive to the ozone layer.

Ozone depletion is most pronounced over Antarctica. This is because of a polar vortex or swirling air mass, which has collected CFCs from industrialized nations over time. It is this swirling polar vortex that isolates the Antarctic atmosphere from the rest of the world. Antarctica has very cold air, which causes tiny clouds of ice crystals to form in its stratosphere even though the humidity is low. When ultraviolet radiation breaks down the CFC molecules, chlorine monoxide (ClO) and other inorganic forms of chlorine cling to the ice crystals. Spring sunlight melts the ice crystals, releasing chlorine monoxide and other chlorine compounds to further react with ozone and deplete it in large quantities. These ozone-depleted masses of air are then spread throughout the stratosphere by stratospheric winds, which dilute ozone-rich masses of air in other parts of the stratosphere. This causes ozone depletion in other parts of the stratosphere, creating a global problem. Data from NASA shows that the ozone layer over the northernmost parts of the United States, Europe, Russia, and Canada is depleted in the early spring as much as 35% in some years.

An index called the ozone depletion potential (ODP) has been adopted for regulatory purposes under the United Nations Environment Programme (UNEP) Montreal Protocol. The ODP of a compound is a measure of the ability of a chemical to destroy ozone molecules. The ODP shows relative effects of comparable emissions of various compounds. Table 10-1 shows some refrigerants with their ozone depletion potentials.

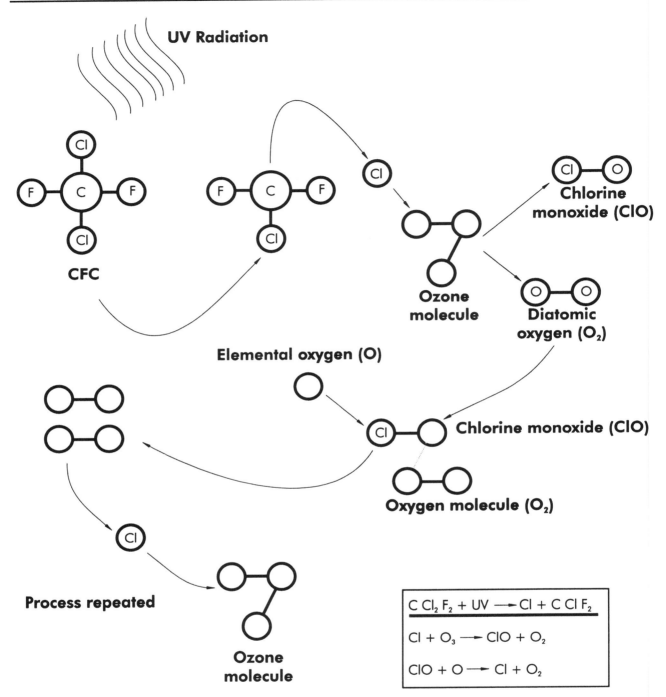

UV Radiation

Cl

F C F

CFC

F C F

Cl

Cl

Ozone
molecule

Cl O
Chlorine
monoxide (ClO)

O O
Diatomic
oxygen (O₂)

Elemental oxygen (O)

O

Cl O Chlorine monoxide (ClO)

Oxygen molecule (O₂)

Process repeated

Cl

Ozone
molecule

$C\ Cl_2\ F_2 + UV \longrightarrow Cl + C\ Cl\ F_2$
$Cl + O_3 \longrightarrow ClO + O_2$
$ClO + O \longrightarrow Cl + O_2$

Figure 10-7. Chlorine atom attacking and breaking up ozone molecule

When stratospheric ozone intercepts ultraviolet light, heat is generated, Figure 10-8. This generated heat causes stratospheric winds, which are the main forces behind weather patterns on earth. By changing the amount of ozone, or even its distribution in the stratosphere, the temperature of the stratosphere can be affected, which can seriously affect weather on earth.

GLOBAL WARMING

Global warming, or *greenhouse effect*, is a completely different phenomenon than ozone depletion. Global warming refers to the physical phenomenon that may lead to the heating of the earth. While global warming and ozone depletion are two completely different issues,

Refrigerant	ODP
R-11	1.0
R-12	0.93
R-114	0.71
R-115	0.38
R-22	0.055
R-123	0.016
R-32	0.0
R-125	0.0
R-134a	0.0
R-143a	0.0
R-152a	0.0
R-32	0.0
CO_2	0.0
HP80 (R-402A)	0.030
HP81 (R-402B)	0.020
HP62 (R-404A)	0.0
AC9000 (R-407C)	0.0

Table 10-1. Ozone depletion potentials

they can be caused by the same man-made chemicals such as refrigerants.

Most of the sun's energy reaches the earth as visible light. After passing through the atmosphere, part of this energy is absorbed by the earth's surface and is converted into heat energy. The earth, warmed by the sun, radiates heat energy back into the atmosphere toward space. Naturally occurring gases and lower atmospheric (tropospheric) pollutants such as CFCs, HCFCs, HFCs, carbon dioxide, carbon monoxide, water vapor, and many other chemicals absorb, reflect, and/or refract the earth's infrared radiation and prevent it from escaping the lower atmosphere. This process slows the heat loss, making the earth's surface warmer than it would be if this heat energy had passed unobstructed through the atmosphere into space. The warmer earth surface then radiates more heat until a balance is established between incoming and outgoing energy. This warming process, which is caused by the atmosphere's absorption of the heat energy radiated from the earth's surface, is called global warming or the greenhouse effect. Increased concentrations of gases from man-made sources (e.g., carbon dioxide, methane, and CFCs) that absorb the heat radiation could lead to a slow warming of the earth, Figure 10-9.

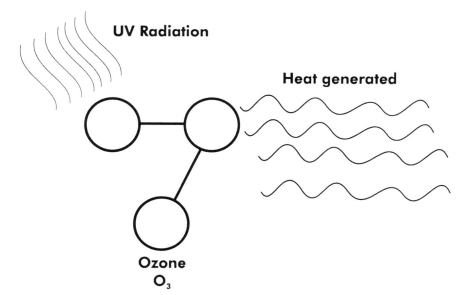

Figure 10-8. Heat generated from the break-up of an ozone molecule

METHANE IN THE AIR

Methane and global warming

*C*olorless and odorless, methane (CH_4) is a flammable gas that rises into Earth's atmosphere from numerous man-made and natural sources. Together with carbon dioxide, chlorofluoro-carbons (CFCs) and other substances, it contributes to global warming.

Atmospheric methane has doubled since the early 1800s, but the rate of increase is slowing. Some scientists predict it will reach its global peak by 2006.

Methane's impact still causes concern. During the 1980s, carbon dioxide was responsible for about half of the increase in global warming. Methane accounted for another 20 to 25 percent.

Methane from lakes and ponds may be a bigger natural culprit than previously suspected.

Methane increase

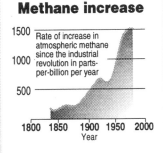

1500 — Rate of increase in atmospheric methane since the industrial revolution in parts-per-billion per year
1000
500

1800 1850 1900 1950 2000
Year

Major natural sources of atmospheric methane

Selected methane sources by the millions of tons – megatons – they add to the atmosphere each year:

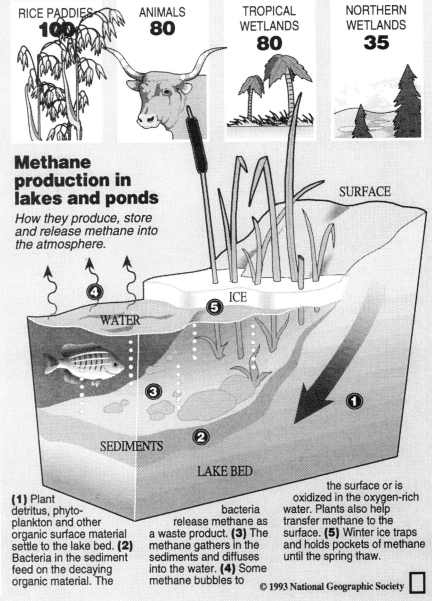

RICE PADDIES **100**   ANIMALS **80**   TROPICAL WETLANDS **80**   NORTHERN WETLANDS **35**

Methane production in lakes and ponds

How they produce, store and release methane into the atmosphere.

SURFACE
ICE
WATER
SEDIMENTS
LAKE BED

(1) Plant detritus, phytoplankton and other organic surface material settle to the lake bed. **(2)** Bacteria in the sediment feed on the decaying organic material. The bacteria release methane as a waste product. **(3)** The methane gathers in the sediments and diffuses into the water. **(4)** Some methane bubbles to the surface or is oxidized in the oxygen-rich water. Plants also help transfer methane to the surface. **(5)** Winter ice traps and holds pockets of methane until the spring thaw.

© 1993 National Geographic Society

SOURCES: Ecos magazine, Lesley K. Smith and William M. Lewis, Jr., University of Colorado   **BILL PITZER/** National Geographic News Service

Figure 10-9. Methane in the air (Courtesy, Bill Pitzer, National Geographic News Service)

Global warming concerns infrared radiation, which is less powerful and has a much lower frequency when compared to ultraviolet radiation, Figure 10-10. Re-radiated infrared radiation may cause a gradual increase in the average temperature of the earth. Over 70% of the earth's fresh water supply is either in ice cap or glacier form. Scientists are concerned that these ice caps or glaciers will melt if the average temperature rises too much, causing increased water levels Global warming may also cause decreased crop yields and added smog levels on earth.

Considerable uncertainty exists about the climate change response to greenhouse gas emissions. This is due to the incomplete understanding of the following:

• Interactive feedback mechanism between clouds, oceans, and polar ice.

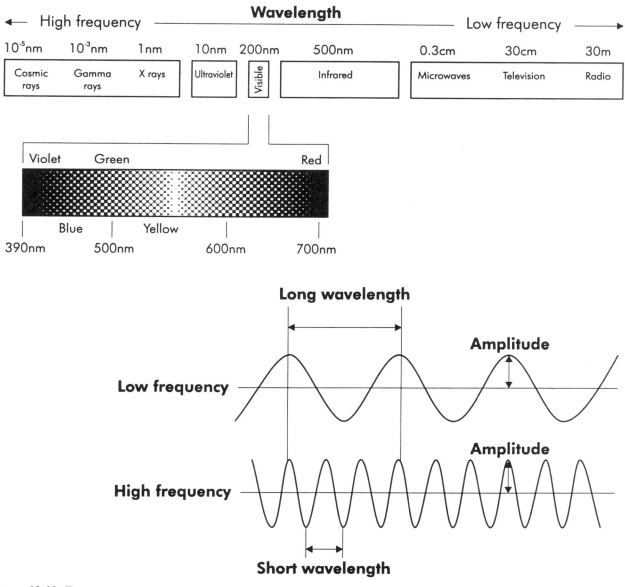

Figure 10-10. Energy waves

- Size and nature of the sources and sinks of the greenhouse gases.

Direct Effects

Chemicals that are emitted directly into the atmosphere and are not caused by any other related secondary process are direct emissions. These direct emissions are measured by an index referred to as the global warming potential (GWP). The GWP measures how much of a direct effect these emissions have on global warming. Refrigerant leaking from a refrigeration or air conditioning system is a good example of a direct emission, which has a direct effect on global warming. Refrigerants are measured compared to R-11, which has a global warming potential (GWP) of 1. Table 10-2 shows the GWPs for some popular refrigerants.

Refrigerant	GWP
R-11	1.0
R-502	4.10
R-12	3.10
R-22	0.34
R-123	0.02
R-134a	0.27
MP39 (R-401A)	0.22
MP66 (R-401B)	0.24
HP62 (R-404A)	0.94
HP80 (R-402A)	0.63
HP81 (R-402B)	0.52
AC9000 (R-407C)	0.28

Table 10-2. Global warming potentials (direct effects)

Carbon dioxide, CFCs, HCFCs, and HFCs are purged from the atmosphere at very different rates. Carbon dioxide in the atmosphere decays very slowly. CFCs also decay very slowly, with rates varying from a 55-year lifetime for R-11 to a 550-year lifetime for R-115. Decay rates of HCFCs and HFCs also vary, but their lifetimes are much less than those of CFCs, Table 10-3. The decay of an HFC compound with a mid-range lifetime is shown in Figure 10-11.

Indirect Effects

Carbon dioxide (CO_2) is the number one contributor to global warming. Humanity has boosted levels of carbon dioxide in the atmosphere to over 25% the usual amount. Most of this CO_2 increase is caused by the combustion of fossil fuels. The combustion of fossil fuels, resulting in increased CO_2, is what causes the indirect effects of global warming. Fossil fuel combustion is required for electricity, which is used to power refrigeration and air conditioning equipment. More efficient hvac/r equipment requires less electrical energy, which means a decreased need for the combustion of fossil fuels.

Indirect effects of global warming relate to the energy efficiency of the equipment. For example, refrigeration or air conditioning equipment that contains a relatively small charge of refrigerant that never leaks may still have a great impact on global warming. This is because the equipment may be undercharged or overcharged. The equipment is very inefficient under these conditions, and the carbon dioxide (CO_2) generated from longer run times contributes more to global warming than leaking refrigerant. This is an example of an indirect effect of global warming.

In the hvac/r industry, scientists are mainly concerned with these indirect effects. Most of the newer refrigerants are more energy efficient than their predecessors; however, some are not as efficient. Just because newer refrigerants may not contribute to ozone depletion does not mean they do not contribute to the direct and indirect effects of global warming. An example of this is the newer refrigerant R-134a. It has a zero ozone depletion index but does contribute to global warming directly and indirectly. Table 10-4 shows the indirect effects of refrigerants on global warming potentials.

Compound	Estimated Atmospheric Lifetime	GWPs for Various Integration Time Horizons		
		20 years	100 years	500 years
Carbon Dioxide	†	1	1	1
CFC-11	55	4500	3400	1400
CFC-12	116	7100	7100	4100
CFC-113	110	4600	4500	2600
CFC-114	220	6100	7000	5900
CFC-115	550	5500	7000	8500
HCFC-22	15.8	4200	1600	540
HCFC-123	1.7	330	90	30
HCFC-141b	10.8	1800	580	200
HCFC-225ca[1]	2.7	610	170	60
HCFC-225cb[1]	7.9	2400	690	240
HFC-32[2]	6.2	2440	720	260
HFC-125	40.5	5200	3400	1200
HFC-134a	15.6	3100	1200	400
HFC-143a	64.2	4700	3800	1600
HFC-152a	1.8	530	150	49
Methane[3]	10.5	35	11	4

† The decay of carbon dioxide concentrations cannot be reproduced using a single exponential decay lifetime. Thus, there is no meaningful single value for the lifetime that can be compared directly with other values in this table.

[1] The HCFC-225ca/cb values were calculated by Atmospheric and Environmental Research, Inc. and are based on rate constant measurements reported by Z. Zhang et al., Geophysical Research Letters, Vol. 18, January 1991, and the infrared energy absorption properties measured at AlliedSignal Central Research Laboratory.

[2] The HFC-32 values were calculated by Atmospheric Environmental Research, Inc. and are based on the recommended value for the rate constant reported in Chemical Kinetics and Photochemical Data for Use in Stratospheric Modeling, Evaluation Number 10, JPL Publication 92-20, August 1992.

[3] The GWP values include the direct radiative effect and the effect due to carbon dioxide formation, but do not include any effects resulting from tropospheric ozone or stratospheric water formed as methane decomposes in the atmosphere.

Table 10-3. Integration time horizon (Courtesy, Alternative Fluorocarbons Environmental Acceptability Study)

Refrigerant	GWP
CO_2	1.00
R-22	570
R-123	28
R-11	1300
R-12	3700
R-114	6400
R-115	13,800
R-134a	400

Table 10-4. Indirect effects of refrigerants on global warming potentials

Total Equivalent Warming Impact

The total equivalent warming impact (TEWI) takes into consideration both the direct and indirect effects of refrigerants on global warming. Refrigerants with the lowest global warming and ozone depletion potentials have the lowest TEWI. Using HFC and HCFC refrigerants in the place of CFCs reduces the TEWI. All of the newer refrigerant alternatives introduced have a much lower TEWI.

Figure 10-12 shows that carbon dioxide from energy generation by a household refrigerator

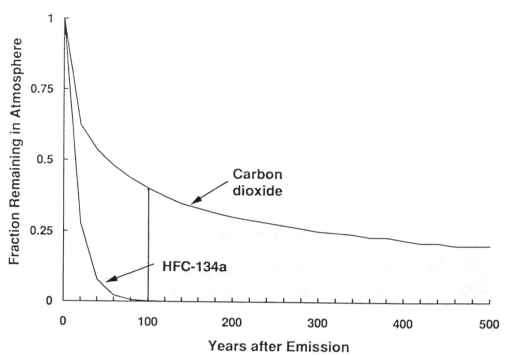

Figure 10-11. Carbon dioxide and R-134a fractions in the atmosphere as a function of time (Courtesy, Alternative Fluorocarbons Environmental Acceptability Study)

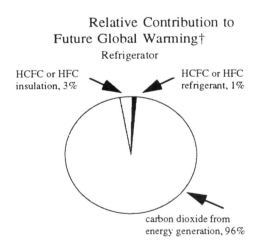

Relative Contribution to
Future Global Warming†
Refrigerator

HCFC or HFC
insulation, 3%

HCFC or HFC
refrigerant, 1%

carbon dioxide from
energy generation, 96%

†Based on a 100-year ITH for the GWP value.

Figure 10-12. Relative contribution to future global warming (Courtesy, Alternative Fluorocarbons Environmental Acceptability Study)

using HCFCs or HFCs would account for 96% of the contribution to global warming. If the HCFC or HFC refrigerant is recaptured at the end of the refrigerator's useful life, the direct contribution to global warming is eliminated. For refrigeration applications, the energy factor is so dominant that the difference between HCFC and HFC alternatives is insignificant, as long as the refrigerator efficiency is not compromised over its useful life. Figure 10-13 shows another example of TEWI.

Figure 10-14 compares baseline CFC refrigerants to HCFC/HFC alternatives and HCFC/HFC alternatives with deducted losses when applied to commercial chillers. TEWI consists of direct and indirect effects, fluorocarbon emission levels, global warming potentials, and carbon dioxide emissions coming from the equipment.

CLEAN AIR ACT

The Clean Air Act (CAA), which was signed November 15, 1990, included requirements for controlling ozone depleting substances. These substances were generally consistent with those contained in the Montreal Protocol as revised in 1990. The main objective of the CAA was

An Example of the *Total Equivalent Warming Impact*
from Replacing CFCs with Alternative Fluorocarbons *

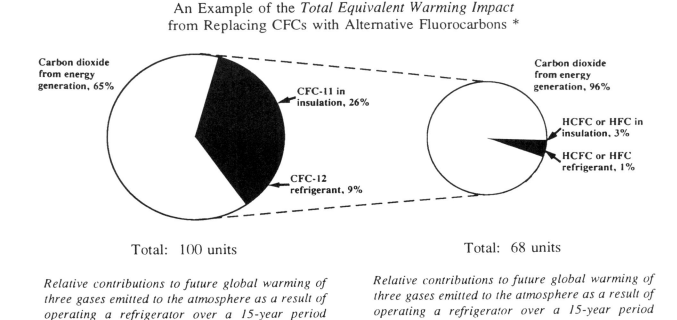

Total: 100 units   Total: 68 units

Relative contributions to future global warming of three gases emitted to the atmosphere as a result of operating a refrigerator over a 15-year period beginning in 1990.

Relative contributions to future global warming of three gases emitted to the atmosphere as a result of operating a refrigerator over a 15-year period beginning in 2000.

* Based on identical energy efficiency of the systems compared and a 100-year integration time horizon.

Figure 10-13. An example of TEWI (Courtesy, Alternative Fluorocarbons Environmental Acceptability Study)

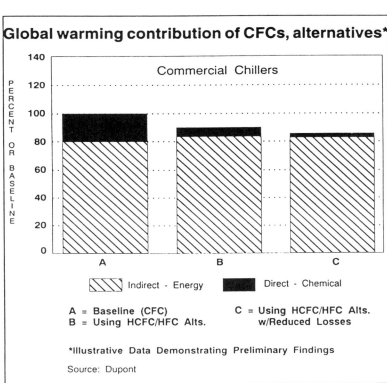

Figure 10-14. Global warming contribution of CFCs, alternatives (Courtesy, DuPont Company)

to reduce the emission, production, and use of ozone depleting compounds and to promote the recapture and recycling of these substances. The substances affected by the CAA were divided into two main classes: Class I consisted of certain ozone depleting CFCs, Halon-1211, Halon-1301, Halon-2402, carbon tetrachloride, and methyl chloroform; Class II substances consisted of most HCFCs.

Title VI of the CAA first called for a phaseout of CFCs by January 1, 2000. It also called for use restrictions and a production freeze of HCFCs in 2015 with a phaseout in 2030. In addition to the phaseout of ozone depleting substances, Title VI included a variety of other provisions intended to reduce emissions of ozone depleting substances. The CAA also prohibited the export of technology and provided technical and financial help to nations that were members of the Montreal Protocol. These monies and research would help finance research on a global scale.

New scientific evidence presented by NASA showed chlorine monoxide levels to be many times greater than expected, so the phaseout of CFCs was moved to December 31, 1995. The U.S. Environmental Protection Agency (EPA) has the right to accelerate any phaseout date if scientific evidence exists. This will cause major users of refrigerants to initiate a massive transition from CFCs to HCFCs and HFCs. Figure 10-15 shows how the transition from CFCs to HCFCs will reduce chlorine concentrations in the stratosphere.

TESTING ALTERNATIVES TO CFCs

To test alternatives to CFCs for their environmental, health, and safety characteristics, 17 chemical companies around the world joined together to form the Alternative Fluorocarbons Environmental Acceptability Study (AFEAS) and the Programme for Alternative Fluorocarbon Toxicity Testing (PAFT). These two programs were set up to provide research on the potential effects of CFC alternatives on the environment and human health. They did this through international cooperation with independent scientists, government research programs, and member companies.

By combining their resources and cooperating with academic and government research programs, AFEAS and PAFT believe that the usual period of time for environmental and toxicity testing of new chemicals will be substantially reduced. This will allow for the rapid phaseout of CFCs.

Alternative Fluorocarbons Environmental Acceptability Study (AFEAS)

The initial phase of AFEAS began in 1988 with a complete review by leading scientists of data pertinent to the environmental acceptability of alternative fluorocarbons. The review was published as a separate volume of the *UNEP/WMO Scientific Assessment of Stratospheric Ozone: 1989*. The results indicated that all of the proposed alternatives were significantly more environmentally acceptable than current CFCs. The alternatives have little or no ozone depleting potentials and make a minimal contribution to global warming. In 1990, AFEAS was expanded into a three-year research program to identify and help resolve gaps in knowledge regarding the potential environmental effects of alternative fluorocarbons. The program was extended by another three years in 1992 to focus on key issues. Funding for these three phases of AFEAS exceeds $8 million (U.S.).

The AFEAS research program has two overall goals:

- To identify and help resolve uncertainties regarding potential environmental effects of HCFCs and HFCs.

- To stimulate prompt dissemination of scientific information to the research community, government decision makers, affected industries, and the general public.

The proposed alternatives are hydrochlorofluorocarbons (HCFCs) and hydrofluorocarbons

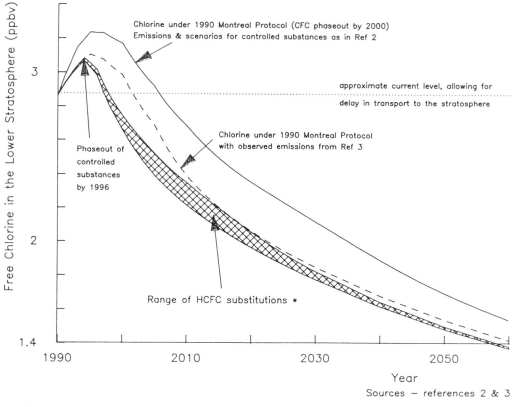

Figure 10-15. CFC to HCFC transition can reduce chlorine (Courtesy, Alternative Fluorocarbons Environmental Acceptability Study)

(HFCs). The presence of hydrogen in their structure means that, unlike CFCs, the compounds are largely removed in the lower atmosphere by natural processes. Based on assessments of expert scientists who contributed to the first AFEAS report, it is expected that the atmospheric breakdown products will not have a serious environmental impact. Subsequent research has verified the original assessment.

Because HCFCs and HFCs are destroyed more rapidly in the lower atmosphere by natural processes than CFCs, the alternative compounds will not accumulate in the atmosphere to the same extent. This means they will have only a small potential to contribute to the greenhouse effect. The direct contribution to global warming from alternative fluorocarbon emissions is usually minor compared to the indirect contribution of carbon dioxide emissions resulting

from the energy required to operate the system over its normal lifetime.

An important component of the AFEAS program is the *Global Warming and Energy Efficiency Study*, co-funded with the U.S. Department of Energy. The total equivalent warming impact (TEWI) from both direct and indirect contributions to global warming has been quantified for each alternative, including non-chemical alternatives, in the major application areas (e.g., refrigeration, air conditioning, insulation, and solvents). This analysis has shown that in a number of cases, the use of HCFCs and HFCs can lead to better energy efficiency and a smaller indirect contribution than non-fluorocarbon technology.

Programme for Alternative Fluorocarbon Toxicity Testing (PAFT)

The first of the PAFT program sectors was launched at the end of 1987 to address the toxicity of R-134a and R-123. Later sectors covered R-141b, R-124, R-125, R-225ca/cb, and R-32. Each program sector takes several years to complete. The testing of R-134a is complete and that of R-123 and R-141b is virtually complete. With well over 100 individual studies and the involvement of more than a dozen testing laboratories on three continents, the total PAFT investment is approximately $35 million (U.S.).

This level of effort shows the total commitment of PAFT member companies to ensuring the prompt availability of safe alternatives. Indications so far give encouragement to producers and users alike to continue their commercial and technical developments in the expectation that most products will prove acceptable for human health and the environment. By initiating the AFEAS and PAFT programs, the chemical industry has ensured that a global phaseout of CFCs will be possible in a timely manner and that the alternatives will be safe in use and safe to the environment.

Breakdown Products of Alternative HCFCs and HFCs

Studies funded by AFEAS, the Commission of the European Communities, and the U.S. EPA indicate that most alternative fluorocarbons can be expected to break down in the lower atmosphere to simple organic species, including carbon oxides, hydrogen halides, and carbonyl halides. Trifluoroacetyl halides are also likely to be formed by R-123 and R-124 and, to a lesser extent, R-134a. These breakdown products will not persist in the atmosphere. Hydrogen chloride and fluoride are removed from the atmosphere by dissolution in cloud water and subsequent deposition in rain, on average, in less than two weeks. Carbonyl and trifluoroacetyl halides are not expected to persist in the atmosphere for longer than a few months. They will be absorbed by cloud water or by ocean and land surfaces and be subse-quently hydrolyzed to carbon dioxide and trifluoroacetic acid, respectively, together with the hydrochloric and hydrofluoric acids.

Even if hundreds of thousands of tons of new fluorocarbons were released each year, studies show that they would not contribute appreciably to the current chloride or fluoride burden of the biosphere. Any trifluoroacetic acid formed will be present at such low concentrations in rain or sea water (parts per billion levels or less) that adverse effects are not expected. Nevertheless, the ultimate environmental fate of these compounds is being evaluated.

MONTREAL PROTOCOL

In 1974, Dr. Sherwood Roland, a chemist at the University of California at Irvine, and his colleague, Mario Molina, predicted that man-made bromine and chlorofluorocarbon (CFC) molecules would not break up very quickly in the lower region of the atmosphere referred to as the troposphere. They predicted that tropospheric winds would spread these molecules throughout the troposphere and eventually into the upper atmospheres (stratosphere). This would contribute to ozone depletion.

There was not much reaction to the Roland-Molina theory until 1978, when the United States and a few other countries banned the use of CFCs in hair sprays and other aerosol products. This banning of aerosols containing CFCs sparked public concerns and caused a lot of debate and controversy throughout the scientific community. However, it was not until the Antarctic ozone hole was confirmed in 1985 that the rest of the world started to become serious about reducing the use of CFCs. By then, about 20 million tons of CFCs had been released into the atmosphere.

In response to the discovery of the ozone hole over Antarctica and the growing evidence that chlorine and bromine could destroy stratospheric ozone on a global basis, many members of the international community came to the conclusion that an international agreement to reduce global production of ozone depleting substances was needed. Because releases of CFCs from all areas affect stratospheric ozone globally, efforts to reduce emissions from specific products by only a few nations could quickly be offset by increases in emissions from other nations. EPA evaluated this risk and concluded that an international approach was necessary to effectively safeguard the ozone layer. To control the future release of ozone depleting substances, the United States and 22 other countries signed the *Montreal Protocol on Substances that Deplete the Ozone Layer*. This Protocol now controls the production and consumption of substances that can cause ozone depletion.

Global cooperation for the protection of the stratospheric ozone layer began with negotiations of the Vienna Convention, which concluded in 1985. The Montreal Protocol was then signed in September 1987 and became effective in 1989. It contains provisions for regular review and control measures that are based on assessment of evolving scientific, environmental, technical, and economic information.

The Montreal Protocol froze the production of R-11, R-12, R-113, R-114, and R-115 back to 1986 levels starting July 1, 1989. Halon-1211, Halon-1301, and Halon-2402, which are used for fire suppression and have a very high ozone depleting potential, also had a production freeze back to 1986 levels, starting in January 1, 1992. As of May 1993, over 90 nations representing approximately 95% of the world's production capacity for CFCs and halons had signed the Montreal Protocol.

Updates and Amendments to the Protocol

Research on ozone depletion is a continuous process. Under Article 6 of the Montreal Protocol, member nations are required to assess the scientific, economic, and alternative technologies related to the protection of the ozone layer every two years. The first scientific assessment in 1989 examined the atmospheres from land-based monitoring stations and a satellite. A global ozone depletion over the north-

ern hemisphere as well as over the southern hemisphere was found. Also discovered was a 3% to 5% decrease in stratospheric ozone levels in the northern hemisphere that occurred between 1969 and 1986. This depletion occurred during the winter months and could not be attributed to known natural processes.

A second meeting of the Protocol was held in London, England in 1990. Fifty-seven members of the Protocol passed amendments and adjustments for a full phaseout of the already regulated CFCs and halons by the year 2000; a phaseout of other CFCs and carbon tetrachloride by the year 2000; and a phaseout of methyl chloroform by the year 2004. The members also passed a non-binding resolution regarding the use of hydrochlorofluorocarbons (HCFCs). HCFCs were identified as a temporary substitute for CFCs, because they did not contribute as much chlorine to the stratosphere as fully halogenated CFCs. However, there was continued concern that rapid use of HCFCs over time would pose a threat to the ozone layer. As a result, the resolution called for the phaseout of HCFCs by the year 2020 if feasible and no later than 2040 in any case.

Another meeting of the Protocol was held in Copenhagen, Denmark in 1992. In this meeting, member nations accelerated the phaseout schedule of CFCs, halons, carbon tetrachloride, and methyl chloroform. HCFCs and methyl bromide were also added to the list of chemicals to be controlled under the Montreal Protocol. The following adjustments to the phaseout schedules of previously controlled substances were adapted at the Copenhagen meeting:

- Freeze methyl bromide consumption to 1991 levels starting in 1995.

- Phase out CFCs by January 1, 1996. Essential uses of CFCs may still be used if agreed upon by the Protocol.

- Encourage member nations to recover, recycle, and reclaim all controlled substances mentioned in the Protocol.

- Allow for technology transfer from developed countries to developing countries through the use of a multilateral fund.

- Freeze production of HCFCs beginning in 1996 to a baseline ceiling. HCFC production to be capped at a percentage of the historic usage starting in 1996. This cap to decrease over time until the total production phaseout of HCFCs in the year 2030.

The cap or capacity limit discussed in the last point was established as a base for the HCFC phaseout. The cap is based on an ozone depletion potential unit concept. The base of this cap is determined by the following formula:

$$\text{Total ODP weighted cap} = (1989 \text{ CFC production}) \ (\text{ODP}) \ (3.1\%) \\ + \\ (1989 \text{ HCFC production}) \ (\text{ODP})$$

The HCFC phaseout schedule is as follows:

- January 1, 2004 - 65% of the cap

- January 1, 2010 - 35% of the cap

- January 1, 2015 - 10% of the cap

- January 1, 2020 - 0.5% of the cap

- January 1, 2030 - 0.0% of the cap

Table 10-5 shows the CFC phaseout schedule with a percent of baseline as it applies to the Montreal Protocol, U.S. Clean Air Act Amendments, and European Community Schedule.

LEGISLATION AND REGULATIONS

An overview of EPA's *Stratospheric Ozone Protection Final Rule Summary for Complying with the Refrigerant Recycling Rule* established the following regulations:

- Required service practices that maximize recycling of ozone depleting compounds (both CFCs and HCFCs)

during the servicing and disposal of air-conditioning and refrigeration equipment.

- Set certification requirements for recycling and recovery equipment, technicians, and reclaimers.

- Restricted sales of refrigerant to certified technicians.

- Required persons servicing or disposing of air-conditioning and refrigeration equipment to certify to EPA that they have acquired recycling or recovery equipment and are complying with the requirements of the rule.

- Required the repair of substantial leaks in air-conditioning and refrigeration equipment with a charge of greater than 50 pounds.

- Established safe disposal requirements to ensure removal of refrigerants from goods that enter the waste stream with the charge intact (e.g., motor vehicle air conditioners, home refrigerators, room air conditioners, etc.).

To enforce the rules and regulations, EPA performs random inspections, responds to tips, and pursues potential cases against violators. Under the CAA, EPA is authorized to assess fines of up to $25,000 per day for any violation of these regulations.

Prohibition on Venting

As of July 1, 1992, Section 608 of the CAA prohibited individuals from knowingly venting ozone depleting compounds used as refrigerants into the atmosphere while maintaining, servicing, repairing, or disposing of air-conditioning or refrigeration equipment. Only four types of releases are permitted under the prohibition:

- "De minimis" quantities of refrigerant released in the course of making good faith attempts to recapture, recycle, and/ or safely dispose of refrigerant.

CFC Phaseout Schedules – Allowed Production and Consumption
(percent of baseline)

	1987 Original Montreal Protocol	1990 London Montreal Protocol	1992 Copenhagen Montreal Protocol	1990 U.S. Clean Air Act Amendments	1992 European Community Schedule
1990	100%				
1991	100%	100%		85%	
1992	100%	100%		80%	
1993	80%	80%		75%	50%
1994	80%	80%	25%	25%*	15%
1995	80%	50%	25%	25%*	0%
1996	80%	50%	0%	0%*	
1997	80%	15%			
1998	80%	15%			
1999	50%	15%			
2000	50%	0%			

* Proposed rule for accelerated phase out schedule.

Table 10-5. CFC phaseout schedules (Courtesy, Alternative Fluorocarbons Environmental Acceptability Study)

- Refrigerants emitted in the course of normal operation of air-conditioning and refrigeration equipment (as opposed to during the maintenance, servicing, repair, or disposal of this equipment), such as from mechanical purging and leaks. However, EPA does require the repair of substantial leaks.

- Mixtures of nitrogen and R-22 that are used as holding charges or as leak test gases, because in these cases, the ozone depleting compound is not used as a refrigerant. A technician may not avoid recovering refrigerant by adding nitrogen to a charged system; before nitrogen is added, the system must be evacuated to the appropriate level. Otherwise, CFCs or HCFCs vented along with the nitrogen will be considered a refrigerant. Pure CFCs or HCFCs released from appliances will be presumed to be refrigerants, and their release will be considered a violation of the prohibition on venting.

- Small releases of refrigerant that result from purging hoses or from connecting/disconnecting hoses to charge or service appliances will not be considered violations to the prohibition on venting. However, recovery and recycling equipment manufactured after November 15, 1993, must be equipped with low-loss fittings.

Regulatory Requirements

Beginning July 13, 1993, technicians were required to evacuate air-conditioning and refrigeration equipment to established vacuum levels. If the technician's recovery or recycling equipment was manufactured any time before November 15, 1993, the air-conditioning and refrigeration equipment must be evacuated to the levels described in the first column of Table 10-6. If the technician's recovery or recycling equipment was manufactured on or after November 15, 1993, the air-conditioning and refrigeration equipment must be evacuated to the levels described in the second column of Table 10-6. The recovery or recycling equipment must also be certified by an EPA-approved equipment testing organization.

Technicians repairing small appliances, such as household refrigerators, household freezers, and water coolers, are required to recover 80% to 90% of the refrigerant in the system, depending on the status of the system compressor.

EPA has established limited exceptions to its evacuation requirements for repairs to leaky equipment and for non-major repairs that are not followed by an evacuation of the equipment to the environment. If leaks either prevent evacuation to the levels shown in Table 10-6 or would substantially contaminate the refrigerant being recovered, the person opening the appliance must:

- isolate leaking from non-leaking components wherever possible.

- evacuate non-leaking components to the levels in Table 10-6.

- evacuate leaking components to the lowest level that can be attained without substantially contaminating the refrigerant. This level cannot exceed 0 psig.

If evacuation of the equipment to the environment is not to be performed when repairs are complete and if the repair is not major, the appliance must either be evacuated to at least 0 psig before it is opened if it is a high- or very high-pressure appliance or be pressurized to 0 psig before it is opened if it is a low-pressure appliance. Methods that require subsequent purging (e.g., nitrogen) cannot be used. Major repairs are those involving removal of the compressor, condenser, evaporator, or auxiliary heat exchanger coil.

Type of Appliance	Inches of Mercury Vacuum* Using Equipment Manufactured:	
	Before Nov. 15, 1993	On or After Nov. 15, 1993
R-22 appliance** normally containing less than 200 lb of refrigerant	0	0
R-22 appliance** normally containing 200 lb or more of refrigerant	4	10
Other high-pressure appliance** normally containing less than 200 lb of refrigerant (R-12, R-500, R-502, R-114)	4	10
Other high-pressure appliance** normally containing 200 lb or more of refrigerant (R-12, R-500, R-502, R-114)	4	15
Very high-pressure appliances (R-13, R-503)	0	0
Low-pressure appliances (R-11, R-123)	25	25 mm Hg absolute

* Relative to standard atmospheric pressure of 29.9 in. Hg
** Or isolated component of such appliance

Table 10-6. Required levels of evacuation for applications, except for small appliances, motor vehicle air conditioning (MVAC), and MVAC-like appliances (Source: EPA)

EPA has also established that recovered and/or recycled refrigerant may be returned to the same system or other systems owned by the same person without restriction. If refrigerant changes ownership, however, that refrigerant must be reclaimed (cleaned to the ARI 700 standard of purity and chemically analyzed to verify that it meets this standard). This provision expires in May 1995, when it may be replaced by an off-site recycling standard.

Equipment Certification

EPA has established a certification program for recovery and recycling equipment. Under the program, EPA requires that equipment manufactured on or after November 15, 1993 be tested by an EPA-approved testing organization to ensure that it meets EPA requirements. Recycling and recovery equipment intended for use with air-conditioning and refrigeration equipment besides small appliances must be tested under the ARI 740-1993 protocol. Recovery equipment intended for use with small appliances must be tested under either the ARI 740-1993 protocol or Appendix C of the Final Rule. EPA recovery efficiency standards vary depending on the size and type of air-conditioning or refrigeration equipment being serviced. For recovery and recycling equipment intended for use with air-conditioning and refrigeration equipment, excluding small appliances, the standards are the same as those in the second column of Table 10-6. Recovery equipment intended for use with small appliances must be able to recover 90% of the refrigerant in the small appliance when the small appliance compressor is operating and 80% of the refrigerant in the small appliance when the compressor is not operating.

Equipment manufactured before November 15, 1993, including home-made equipment, will be grandfathered if it meets the standards in the first column of Table 10-6. Third-party testing is not required for equipment manufactured before November 15, 1993, but equipment manufactured on or after that date, including home-made equipment, must be tested by a third-party.

Refrigerant Leaks

Owners of equipment with charges greater than 50 lb are required to repair substantial leaks. A substantial leak is a 35% annual leak rate for the industrial process and commercial refrigeration sectors. An annual leak rate of 15% is established for comfort cooling chillers and all other equipment with a charge over 50 lb other than industrial process and commercial refrigeration equipment. Owners of air-conditioning and refrigeration equipment with more that 50 lb of charge must keep records of the quantity of refrigerant added to their equipment during servicing and maintenance procedures.

Mandatory Technician Certification

EPA has established a mandatory technician certification program, which has four types of certification:

- Type I - Servicing small appliances

- Type II - Servicing or disposing of high- or very high-pressure appliances, except small appliances and MVACs

- Type III - Servicing or disposing of low-pressure appliances

- Universal - Servicing all types of equipment

Persons removing refrigerant from small appliances and MVACs in order to dispose of these appliances do not have to be certified.

Technicians must pass an EPA-approved test given by an EPA-approved certifying organization in order to become certified under the mandatory program. Technicians must be certified by November 14, 1994. EPA plans to "grandfather" individuals who have already participated in training and testing programs provided the testing programs are approved by EPA and that they provide additional, EPA-approved materials or testing to individuals to ensure they have the required level of knowledge.

Although any organization may apply to become an approved certifier, EPA plans to give priority to national organizations able to reach large numbers of people. EPA encourages smaller training organizations to make arrangements with national testing organizations to administer certification examinations at the conclusion of their courses.

Refrigerant Sales Restrictions

Under Section 609 of the Clean Air Act, sales of R-12 in containers smaller than 20 lb are restricted to technicians certified under EPA's MVAC regulations. Persons servicing appliances other than MVACs may buy containers of R-12 larger than 20 lb until November 14, 1994. After November 14, 1994, the sale of refrigerant in any size container is restricted to technicians certified either under the program described earlier or under EPA's MVAC regulations.

Certification by Owners of Recycling and Recovery Equipment

EPA requires persons who service or dispose of air-conditioning and refrigeration equipment to certify that they have acquired (built, bought, or leased) recovery or recycling equipment and that they are complying with the applicable requirements. This certification had to be signed by the owner of the equipment or another responsible officer and sent to the appropriate EPA regional office by August 12, 1993, Figure 10-16. Although owners of recycling and recovery equipment are required to list the number of trucks based at their shops, they do not need to have a piece of recycling or recovery equipment for every truck.

Reclaimer Certification

Reclaimers are required to return refrigerant to the purity level specified in ARI Standard 700-1988 (an industry-set purity standard) and to verify this purity using the laboratory protocol set forth in the same standard. In addition, reclaimers must release no more than 1.5% of the refrigerant during the reclamation process and dispose of wastes properly. Reclaimers had

to certify by August 12, 1993 to the Section 608 Recycling Program Manager at EPA headquarters that they were complying with these requirements. The certification had to include the name and address of the reclaimer and a list of equipment used to reprocess and to analyze the refrigerant.

EPA encourages reclaimers to participate in third-party reclaimer certification programs, such as those operated by the Air-Conditioning and Refrigeration Institute (ARI). Third-party certification can enhance the attractiveness of a reclaimer's product by providing an objective assessment of its purity.

MVAC-like Appliances

Some of the air conditioners that are covered by this rule are identical to motor vehicle air conditioners (MVACs). They are not covered by the MVAC refrigerant recycling rule (40 CFR Part 92 Subpart B), because they are used in vehicles that are not defined as "motor vehicles." These air conditioners include many systems used in construction equipment, farm vehicles, boats, and airplanes. Like MVACs in cars and trucks, these air conditioners typically contain 2 or 3 lb of R-12 and use open-drive compressors to cool the passenger compartments of vehicles. (Vehicle air conditioners utilizing R-22 are not included in this group and are therefore subject to the requirements outlined earlier for R-22 equipment.) EPA defines these air conditioners as "MVAC-like appliances" and applies the MVAC rules for certification and use of recycling and recovery equipment to them. That is, technicians servicing MVAC-like appliances must properly use recycling or recovery equipment that has been certified to meet the standards in Appendix A to 40 CFR Part 82, Subpart B. In addition, EPA allows technicians who service MVAC-like appliances to be certified by a certification program approved under the MVAC rule.

Safe Disposal Requirements

Under EPA's rule, equipment that is typically dismantled on-site before disposal (e.g., retail food refrigeration, cold storage warehouse re-

THE UNITED STATES ENVIRONMENTAL PROTECTION AGENCY (EPA)
REFRIGERANT RECOVERY OR RECYCLING DEVICE
ACQUISITION CERTIFICATION FORM

EPA regulations require establishments that service or dispose of refrigeration or air conditioning equipment to certify by August 12, 1993 that they have acquired recovery or recycling devices that meet EPA standards for such devices. To certify that you have acquired equipment, please complete this form according to the instructions and **mail it to the appropriate EPA Regional Office. BOTH THE INSTRUCTIONS AND MAILING ADDRESSES CAN BE FOUND ON THE REVERSE SIDE OF THIS FORM.**

PART 1: ESTABLISHMENT INFORMATION

Name of Establishment

Street

(Area Code) Telephone Number

City   State   Zip Code

Number of Service Vehicles Based at Establishment

County

PART 2: REGULATORY CLASSIFICATION

Identify the type of work performed by the establishment. **Check all boxes that apply.**

☐ Type A - Service small appliances
☐ Type B - Service refrigeration or air conditioning equipment other than small appliances
☐ Type C - Dispose of small appliances
☐ Type D - Dispose of refrigeration or air conditioning equipment other than small appliances

PART 3: DEVICE IDENTIFICATION

Name of Device(s) Manufacturer	Model Number	Year	Serial Number (if any)	Check Box if Self-Contained
1.				☐
2.				☐
3.				☐
4.				☐
5.				☐
6.				☐
7.				☐

PART 4: CERTIFICATION SIGNATURE

I certify that the establishment in Part 1 has acquired the refrigerant recovery or recycling device(s) listed in Part 2, that the establishment is complying with Section 608 regulations, and that the information given is true and correct.

Signature of Owner/Responsible Officer   Date   Name (Please Print)   Title

Public reporting burden for this collection of information is estimated to vary from 20 minutes to 60 minutes per response with an average of 40 minutes per response, including time for reviewing instructions, searching existing data sources, gathering and maintaining the data needed, and completing the collection of information. Send comments regarding ONLY the burden estimates or any other aspects of this collection of information, including suggestions for reducing this burden to Chief, Information Policy Branch; EPA, 401 M St., S.W. (PM-223Y), Washington, DC 20460, and to the Office of Information and Regulatory Affairs, Office of Management and Budget, Washington, DC 20503, marked "Attention: Desk Officer of EPA". DO NOT SEND THIS FORM TO THE ABOVE ADDRESSES. ONLY SEND COMMENTS TO THESE ADDRESSES.

Figure 10-16. Recovery or recycling equipment acquisition certification form (Source: EPA)

frigeration, chillers, industrial process refrigeration, etc.) must have the refrigerant recovered in accordance with EPA's requirements for servicing. However, equipment that typically enters the waste stream with the charge intact (e.g., motor vehicle air conditioners, household refrigerators and freezers, room air conditioners, etc.) is subject to special safe disposal requirements. Under these requirements, the final person in the disposal chain (scrap metal recycler or landfill owner) is responsible for ensuring that refrigerant is recovered from equipment before the final disposal of the equipment. However, persons "upstream" can remove the refrigerant and provide documentation of its removal to the final person if this is more cost-effective.

The equipment used to recover refrigerant from appliances prior to their final disposal must meet the same performance standards as equipment used prior to servicing, but it does not need to be tested by a laboratory. This means that self-built equipment is allowed as long as it meets the performance requirements. For MVACs and MVAC-like appliances, the performance requirement is 102 mm of Hg vacuum. For small appliances, recovery equipment performance requirements are 90% efficiency when the appliance compressor is operational and 80% efficiency when the appliance compressor is not operational.

Technician certification is not required for individuals removing refrigerant from appliances in the waste stream. The safe disposal requirements became effective July 13, 1993. The equipment had to be registered or certified with EPA by August 12, 1993.

Major Record-keeping Requirements

The major record keeping requirements are as follows:

- *Technicians* servicing appliances that contain 50 or more pounds of refrigerant must provide the owner with an invoice that indicates the amount of refrigerant added to the appliance. Technicians must also keep a copy of their proof of certification at their place of business.

- *Owners* of appliances that contain 50 or more pounds of refrigerant must keep service records documenting the date and type of service, as well as the quantity of refrigerant added.

- *Wholesalers* who sell CFC and HCFC refrigerants must retain invoices that indicate the name of the purchaser, date of sale, and quantity of refrigerant purchased.

- *Reclaimers* must maintain records of the names and addresses of persons sending them material for reclamation and the quantity of material sent. This information must be maintained on a transactional basis. Within 30 days of the end of the calendar year, reclaimers must report to EPA the total quantity of material sent to them for reclamation, the mass of refrigerant reclaimed that year, and the mass of waste products generated that year.

Hazardous Waste Disposal

If refrigerants are recycled or reclaimed, they are not considered hazardous under federal law. In addition, used oils contaminated with CFCs are not hazardous provided they are:

- not mixed with other waste.

- subjected to CFC recycling or reclamation.

- not mixed with used oils from other sources.

Used oils that contain CFCs after the CFC reclamation procedure are subject to specification limits for used oil fuels if these oils are destined for burning. Individuals with questions regarding the proper handling of these materials should contact EPA's RCRA Hotline at 800-424-9346 or 703-920-9810.

Planning for the Future

Observing the refrigerant recycling regulations is essential in order to conserve the existing stock of refrigerants, as well as to comply with Clean Air Act requirements. However, owners of equipment that contains CFC refrigerants should look beyond the immediate need to keep existing equipment in working order. EPA urges equipment owners to prepare for the phaseout of CFCs, which will be completed by January 1, 1996. Owners are advised to begin the process of converting or replacing existing equipment with equipment that uses alternative refrigerants.

To assist owners, suppliers, technicians, and others involved in comfort chiller and commercial refrigeration management, EPA has published a series of short fact sheets and expects to produce additional material. Copies of material produced by the EPA Stratospheric Protection Division are available by calling them at 800-296-1996.

DEFINITIONS

The following are definitions that must be known in order to comply with EPA's rules and regulations:

Appliance - Any device that contains and uses a Class I (CFC) or Class II (HCFC) substance as a refrigerant and is used for household or commercial purposes, including any air conditioner, refrigerator, chiller, or freezer. EPA interprets this definition to include all air-conditioning and refrigeration equipment except that designed and used exclusively for military purposes.

Major maintenance, service, or repair - Maintenance, service, or repair that involves removal of the appliance compressor, condenser, evaporator, or auxiliary heat exchanger coil.

MVAC-like appliance - Mechanical vapor compression, open-drive compressor appliances used to cool the driver or passenger compartment of a non-road vehicle, including agricul-

ture and construction vehicles. This definition excludes appliances using R-22.

Reclaim - To reprocess refrigerant to at least the purity specified in ARI Standard 700-1988 and to verify the purity using the analytical methodology prescribed in the Standard.

Recover - To remove refrigerant in any condition from an appliance and store it in an external container without necessarily testing or processing it in any way.

Recycle - To extract refrigerant from an appliance and clean the refrigerant for reuse without meeting all of the requirements for reclamation. In general, recycled refrigerant is refrigerant that is cleaned using oil separation and single or multiple passes through devices, such as replaceable core filter driers, which reduce moisture, acidity, and particulate matter.

Self-contained recovery equipment - Recovery or recycling equipment that is capable of removing refrigerant from an appliance without the assistance of components contained in the appliance.

Small appliance - Any of the following products that are fully manufactured, charged, and hermetically sealed in a factory with 5 lb or less of refrigerant: refrigerators and freezers designed for home use, room air conditioners (including window air conditioners and packaged terminal air conditioners), packaged terminal heat pumps, dehumidifiers, under-the-counter ice makers, vending machines, and drinking water coolers.

System-dependent recovery equipment - Recovery equipment that requires the assistance of components contained in an appliance to remove the refrigerant from the appliance.

Technician - Any person who performs maintenance, service, or repair that could reasonably be expected to release Class I (CFC) or Class II (HCFC) substances into the atmosphere, including but not limited to installers, contractor employees, in-house service personnel, and in some cases, owners. Technician also means any person disposing of appliances except for small appliances.

Table 10-7 is a summary of the major recycling rule compliance dates.

Event	Date
Owners of equipment containing more than 50 lb of refrigerant with substantial leaks must have such leaks repaired.	June 14, 1993
Evacuation and recovery/recycling requirements go into effect.	July 13, 1993
Owners of recovery and recycling equipment must certify to EPA that they have acquired such equipment and are complying with the rule. Reclamation requirement also goes into effect.	August 12, 1993
All newly manufactured recycling and recovery equipment must be certified by an EPA-approved testing organization to meet the requirements in the second column of Table 10-6.	November 15, 1993
All technicians must be certified and sales restrictions go into effect.	November 14, 1994
Reclamation requirement expires.	May 14, 1995

Table 10-7. Major recycling rule compliance dates (Source: EPA)

APPENDIX "A"

Alternative Refrigerant and Oil Retrofit Guidelines

This appendix consists of the refrigerant changeover guidelines that are recommended by the Copeland Corporation. These guidelines are reprinted with permission.

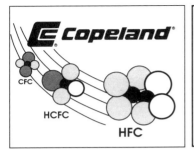

Refrigerant Changeover Guidelines
CFC-12 to HFC-134a

Leading the Industry with Environmentally Responsible Refrigerant Solutions

Copeland does not advocate the wholesale changeover of CFC refrigerants to HCFCs or HFCs. If a system is not leaking refrigerant to the atmosphere, and is operating properly, there is no technical reason to replace the CFC refrigerant. Changing the refrigerant may void the U.L. listing of the unit. However, once the decision has been made to make the change from CFC-12 (R-12) to HFC-134a (R-134a), the following guidelines are recommended.

CONSIDERATIONS

1. Retrofitting systems that employ compressors manufactured prior to 1973 is not recommended. This is due to the different materials used in motor insulation that have not been evaluated for compatibility with the new refrigerants and lubricants. Failure to heed this advice will violate the proposed U.L. Standard For Field Conversion/Retrofit Of Alternate Refrigerants In Refrigeration and Air Conditioning Equipment (U.L. 2170-2172).

2. Copeland's lubricant recommendation for use with HFC-134a is a Polyol Ester (POE), Mobil EAL™ Arctic 22 CC or ICI EMKARATE RL 32CF. The use of any other POE lubricant may void the compressor warranty.

3. R-134a should be used only in systems where the saturated suction temperature is maintained at -10°F or higher. **It should not be mixed with any other refrigerant!**

4. The expansion valve may need to be changed. The existing R-12 valve when used with R-134a will have approximately 15% more capacity. Oversized expansion valves can result in hunting and refrigerant floodback. Consult with the thermostatic expansion valve manufac-

turer for the correct valve and size.

5. Filter-driers must be changed at the time of conversion. This is proper air conditioning/refrigeration practice.

   a. The recommended drier for use with all HFC refrigerants is Alco UltraFlow.

   b. Solid core driers such as ALCO ADK are compatible with either R-12 or R-134a.

   c. Compacted bead type driers can use XH6 or XH9 molecular sieve material such as found in the ALCO EK or EKH series.

   d. If a loose fill type drier is to be used, XH9 molecular sieve is required.

6. R-134a exhibits marginally higher pressures than R-12 at normal condensing temperatures. We do not believe this will require readjustment of safety controls; however, you should verify this with the system manufacturer or component suppliers.

7. Systems that use a low pressure controller to maintain space temperature may need to have the cut-in and cut-out points changed due to the difference in Pressure/Temperature relationships.

8. Systems using R-134a may have a lower system pressure drop than with R-12. Because of the lower pressure drop, check with the manufacturer of any pressure regulators and pilot operated solenoid valves used in the system to be sure that they will operate with the lower pressure drop. It is possible that these controls may have to be downsized in order to operate properly.

9. Mineral oil lubricants, such as 3GS, **must not** be used as the compressor lubricant. Polyol Ester (POE) lubricant,

WARNING: Use only Copeland approved refrigerants and lubricants in the manner prescribed by Copeland. In some circumstances, other refrigerants and lubricants may be dangerous and could cause fires, explosions or electrical shorting. Contact Copeland Corp., Sidney, Ohio.

Mobil EAL Arctic 22 CC, or ICI EMKARATE RL 32CF are the only lubricants that can be used in a Copeland compressor when using R-134a.

Before starting the changeover, it is suggested that at least the following items be ready:

1. Safety glasses
2. Gloves
3. Refrigerant service gauges
4. Electronic thermometer
5. Vacuum pump capable of pulling 250 microns
6. Thermocouple micron gauge
7. Leak detector
8. Refrigerant recovery unit including refrigerant cylinder
9. Proper container for removed lubricant
10. New liquid control device
11. Replacement liquid line filter-drier(s)
12. New lubricant, Mobil EAL Arctic 22 CC or ICI EMKARATE RL 32CF (POE)
13. R-134a pressure temperature chart
14. R-134a refrigerant

CHANGEOVER PROCEDURE

NOTE: R-134a is not compatible with the seal material used in Moduload unloading. If your system has Moduload, it MUST be changed. Consult your Copeland wholesaler for the proper part number.

1. The system should be thoroughly leak tested with the R-12 still in the system. All leaks should be repaired before the R-134a refrigerant is added.

2. It is advisable that the system operating conditions be recorded with the R-12 still in the system. This will provide the base data for comparison when the system is put back into operation with the R-134a.

3. It is necessary to thoroughly remove the existing mineral oil lubricant from the system before the refrigerant is changed. No more than 5% residual mineral oil may be left in the system when it is recharged with R-134a for proper compressor operation. 1 to 2% residual mineral oil may be required to assure no loss of heat transfer if enhanced tube heat exchangers are used in the system.

I. Systems with service valves

a. Disconnect electrical power to system.

b. Front seat the service valves to isolate the compressor.

c. Properly remove the R-12 from the compressor.

d. Remove the mineral oil lubricant from the compressor. Hermetic compressors will have to be removed from the system and tipped up to drain the lubricant out through the suction stub.

e. Those systems that have oil separators, oil reservoirs, oil floats and suction line accumulators must have the mineral oil drained from them. Add POE lubricant to the oil separator and to the oil reservoir.

f. Replace the liquid line filter-drier with one that is compatible with R-134a.

g. Fill the compressor with the proper amount of POE lubricant. The oil charge is on the label of Copelaweld® compressors. Copelametic® compressor oil charges can be found in Application Engineering Bulletin 4-1281. If the lubricant charge is unknown, an authorized Copeland wholesaler can provide the technician with the information.

h. Reinstall the compressor in the system. Evacuate it to 250 microns. A vacuum decay test is suggested to assure the system is dry and leak free.

i. Recharge the system with R-12.

j. Operate the compressor in the system for a minimum of 24 hours, longer is better.

k. Repeat steps 3.I.a through j two more times. This will have provided three flushes of the system's lubricant.

l. To date, three complete flushes of the lubricant has shown to lower the mineral oil content down to 5% or less in the system. To be sure of the mineral oil content between flushes and to be sure that the system ultimately has 5% or less mineral oil, Copeland recommends the use of a refractometer.

m. Properly dispose of the lubricant removed from the system after each flush.

II. Systems without service valves

a. Disconnect electrical power to system.

b. Properly remove the R-12 from the system.

c. Remove the mineral oil lubricant from the compressor. Hermetic compressors will have to be removed from the system and tipped up to drain the lubricant out through the suction stub.

d. It may be advisable to add service valves at the compressor suction and discharge connections. The compressor will have to have its lubricant changed generally three times.

e. Those systems that have oil separators, oil reservoirs, oil floats and suction line accumulators must have the mineral oil drained from them. Add POE lubricant to the oil separator and to the oil reservoir.

f. Replace the liquid line filter-drier with one that is compatible with R-134a.

g. Fill the compressor with the proper amount of POE lubricant. The oil charge is on the label of Copelaweld compressors. Copelametic compressor oil charges can be found in Application Engineering Bulletin 4-1281. If the lubricant charge is unknown, an authorized Copeland wholesaler can provide the technician with the information.

h. Reinstall the compressor in the system. Evacuate it to 250 microns. A vacuum decay test is suggested to assure the system is dry and leak free.

i. Recharge the system with R-12.

j. Operate the compressor in the system for a minimum of 24 hours, longer is better.

k. Repeat steps 3.II.a through j two more times. This will have provided three flushes of the system's lubricant.

l. To date, three complete flushes of the lubricant has shown to get the mineral oil content down to 5% or less in the system. To be sure of the mineral oil level between flushes and to be sure that the system has 5% or less mineral oil, Copeland recommends the use of a refracto-meter to determine the residual mineral oil in the system. The refractometer (P/N 998-RMET-00) is available from your Copeland Wholesaler.

4. With the proper amount of polyol ester in the system, the R-12 can now be removed. Measure and note the amount removed.

5. Before the final flush, be sure all leaks are repaired, liquid control devices and any other system components are changed. Install the correct liquid line filter-drier. Driers must be compatible with the refrigerant and lubri-cant.

6. Be advised that POEs are very hygroscopic. They will very quickly absorb moisture from the air once the container is opened. Once the lubricant is added to the compressor, the compressor should be quickly installed. Like an open container, an open compressor with POE will absorb moisture. Add the correct amount of lubricant to the compressor. It is important that the system contain not more than 5% mineral oil. More than 5% may contribute to premature compressor failure and or system capacity problems. Mineral oils are not miscible with R-134a. The lubricant may log in the evaporator resulting in system capacity loss. It is for this reason that the flushing process must be done with the R-12 in the system.

7. Once the compressor is installed and the system is closed, the system must be evacuated to and hold 250 microns or lower.

8. Charge the system with the R-134a. Charge to 90% of the refrigerant removed in item 4.

9. Operate the system. Record the data and compare to the data taken in item 2. Check and adjust the TEV superheat setting if necessary. Make adjustments to other controls as needed. Additional R-134a may have to be added to obtain optimum system performance.

10. Properly label the components. Tag the compressor with the refrigerant used (R-134a) and the lubricant used (Mobil EAL Arctic 22 CC or ICI EMKARATE RL 32CF). The proper color code for R-134a is Light Sky Blue PMS (Paint Matching System) 2975.

11. Clean up and properly dispose of the removed lubri-cant. Check local and state laws regarding the disposal of refrigerant lubricants. Recycle or reclaim the removed refrigerant.

CAUTION: These guidelines are intended for use with R-134a only, not for refrigerants which are similar to R-134a. Other refrigerants may not be compatible with the materials used in our compressors or the lubricants recommended in this bulletin resulting in unacceptable reliability and durability of the compressor.

Note: Retrofit videos are available from your Authorized Copeland Wholesaler. Ask for VT-026.

R-134a Saturated Vapor/Liquid Temperature/Pressure Chart

Temperature °F	Pressure PSIG	Temperature °F	Pressure PSIG
-10	1.8		
-9	2.2	30	25.6
-8	2.6	31	26.4
-7	3.0	32	27.3
-6	3.5	33	28.1
-5	3.9	34	29.0
-4	4.4	35	29.9
-3	4.8	40	34.5
-2	5.3	45	39.5
-1	5.8	50	44.9
		55	50.7
0	6.2		
1	6.7	60	56.9
2	7.2	65	63.5
3	7.8	70	70.7
4	8.3	75	78.3
5	8.8	80	86.4
6	9.3	85	95.0
7	9.9	90	104.2
8	10.5	95	113.9
9	11.0	100	124.3
		105	135.2
10	11.6		
11	12.2	110	146.8
12	12.8	115	159.0
13	13.4	120	171.9
14	14.0	125	185.5
15	14.7	130	199.8
16	15.3	135	214.8
17	16.0		
18	16.7		
19	17.3		
20	18.0		
21	18.7		
22	19.4		
23	20.2		
24	20.9		
25	21.7		
26	22.4		
27	23.2		
28	24.0		
29	24.8		

Form No. 93-04-R4 Revised 11-94
Supersedes Form No. 93-04-R3
© 1994 Copeland Corporation
Printed in U.S.A.
Copeland Corporation
Sidney, OH 45365-0669

Printed on Recycled Paper

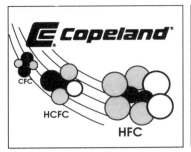

Refrigerant Changeover Guidelines
CFC-12 to R-401A (MP39)

Copeland does not advocate the wholesale changeover of CFC refrigerants to HCFCs or HFCs. If a system is not leaking refrigerant to the atmosphere, and is operating properly, there is no technical need to replace the CFC refrigerant. In fact, changing the refrigerant may void the U.L. listing of the unit. However, once the decision has been made to make the change from CFC-12 (R-12) to the interim R-401A, the following guidelines are recommended.

CONSIDERATIONS

1. Retrofitting systems that employ compressors manufactured prior to 1973 is not recommended. This is due to the different materials used in motor insulation that have not been evaluated for compatibility with the new refrigerants and lubricants. Failure to heed this advice will violate U.L. Standard For Field Conversion/Retrofit Of Alternate Refrigerants In Refrigeration and Air Conditioning Equipment (U.L. 2170-2172).

2. Copeland's lubricant recommendation for use with R-401A is a mixture of 50% mineral oil and 50% alkylbenzene (Zerol 200TD or Soltex AB200A). Polyol ester lubricants, Mobil EAL™ Arctic 22 CC or ICI EMKARATE RL 32CF, can also be used with R-401A if the system is expected to be changed in the near future to an HFC refrigerant such as HFC-134a. This will eliminate the need of having to flush the system again when making the R-134a retrofit.

Refer to item 11 in this section for a list of other approved lubricants for use with R-401A or Application Engineering Bulletin 17-1248 for a complete list of all Copeland approved lubricants.

3. R-401A should be used only in systems where the saturated suction temperature is maintained at -10°F or higher. **It should not be mixed with any other refrigerant!**

4. The expansion valve may need to be changed. The existing R-12 valve when used with R-401A will have approximately 25% more capacity. Oversized expansion valves can result in hunting and refrigerant floodback. Consult with the thermostatic expansion valve manufacturer for the correct valve and size.

5. Filter-driers must be changed at the time of conversion. This is proper air conditioning/refrigeration practice.

   a. Solid core driers such as ALCO ADK, are compatible with either R-12 or R-401A.

   b. Compacted bead driers can use XH6 or XH9 molecular sieve material such as found in the ALCO UltraFlow, EK or EKH series.

   c. If a loose fill type drier is to be used, XH9 molecular sieve is required.

6. R-401A exhibits marginally higher pressures than R-12 at normal condensing temperatures. We do not believe this will require readjustment of safety controls; however, you should verify this with the system manufacturer or component suppliers.

7. Systems that use a low pressure controller to maintain space temperature may have to have the cut-out and cut-in points changed. With R-401A, the pressure setting must reflect an average temperature of the refrigerant in the evaporator. Because of refrigerant glide (see Copeland booklet 92-81, "Guide To Refrigerant Mixtures"), the refrigerant entering the evaporator for a specific suction

WARNING: Use only Copeland approved refrigerants and lubricants in the manner prescribed by Copeland. In some circumstances, other refrigerants and lubricants may be dangerous and could cause fires, explosions or electrical shorting. Contact Copeland Corp., Sidney, Ohio.

pressure will be approximately 8°F colder than the refrigerant vapor at the outlet of the evaporator (not considering superheat). Therefore, the average refrigerant temperature will be at a midpoint pressure/temperature equivalent.

Example: A 35°F refrigerated space usually requires that the refrigerant temperature in the evaporator be approximately 25°F. Using R-401A, the liquid entering the evaporator may be as cold as 21°F, and the vapor temperature before superheat may be 29°F. Taking the saturated vapor pressure at 29°F gives us the exit pressure at the evaporator of 24.8 psig. Considering a 2 psig pressure drop in the suction line, the pressure control cut-out should be set at 22.8 psig.

The cut-in point will be based on the vapor pressure/temperature value. Let's assume that the space temperature can rise to 37°F before the compressor is turned on. 37°F vapor pressure is 31.6 psig. Set the cut-in at 32 psig.

8. Because of glide, pressure regulators such as EPRs may have to be reset. Contact the EPR manufacturer for correct settings.

9. Due to refrigerant glide, it is important that when measuring and/or adjusting TEV superheat, the pressure and SATURATED VAPOR TABLES be used. Example: The pressure measured at the TEV bulb is 30 psig. The Pressure/Temperature (P/T) chart shows that the saturated vapor temperature for 30 psig is 35.2°F. If the actual refrigerant temperature measured is 45.0°F, the superheat is 9.8°F.

To measure sub-cooling at the condenser outlet or at the TEV inlet to verify that a solid column of liquid is present, measure the pressure and the refrigerant temperature at the location that the sub-cooling information is needed. Compare it to the saturated liquid temperature from the SATURATED LIQUID TABLES. Example: A pressure of 140 psig is measured at the condenser coil outlet. From the P/T chart, 140 psig is 100.2°F saturated liquid temperature. If the actual refrigerant temperature is 95°F, the liquid is sub-cooled 5.2°F.

10. Systems using R-401A may have a lower system pressure drop than with R-12. Because of the lower pressure drop, pilot operated solenoid valves and pressure regulators may not operate. Check with the manufacturer of any pressure regulators and pilot operated solenoid valves used in the system to be sure that they will operate properly. These controls may have to be downsized.

11. Mineral oil lubricant only, such as 3GS, cannot be used as the compressor lubricant. Copeland recommends the following lubricant choices:

1. A mixture of 3GS Mineral Oil (MO) and Zerol 200TD or Soltex AB200A Alkyl Benzene (AB) with a minimum of 50% AB

2. Virginia KMP MS2212 or Witco R-195-0 (a 70/30 mixture of AB/MO)

3. A mixture of 3GS Mineral Oil (MO) and Polyol Ester (POE) ie; Mobil EAL Arctic 22 CC or ICI EMKARATE RL 32CF with a minimum of 50% POE

4. 100% Mobil EAL Arctic 22 CC or ICI EMKARATE RL 32CF.

Before starting the changeover, it is suggested that at least the following items be ready:

1. Safety glasses
2. Gloves
3. Refrigerant service gauges
4. Electronic thermometer
5. Vacuum pump capable of pulling 250 microns
6. Thermocouple micron gauge
7. Leak detector
8. Refrigerant recovery unit including refrigerant cylinder
9. Proper container for removed lubricant
10. New liquid control device
11. Replacement liquid line filter-drier(s)
12. New lubricant
    a. Zerol 200TD or Soltex AB200A; or Virginia KMP MS 2212 or Witco R-195-0
    b. Mobil EAL Arctic 22 CC or ICI EMKARATE RL 32CF (POE)
13. R-401A pressure temperature chart
14. R-401A refrigerant

CHANGEOVER PROCEDURE

1. The system should be thoroughly leak tested with the R-12 refrigerant still in the system. All leaks should be repaired before the R-401A refrigerant is added.

2. It is recommended that system operating conditions be recorded with the R-12 still in the system. This will provide the base data for comparison when the system is put back into operation with the R-401A.

3. The system should be electrically shut off and the refrigerant properly removed from the system. Measure the quantity of refrigerant removed. This will provide a guide for recharging the system with R-401A (see item 9 this section).

4. The mineral oil must be removed from the compressor crankcase. Hermetic compressors will have to be removed from the piping and the lubricant drained out through the suction stub. It is advisable to do an acid test on the oil.

5. Measure the amount of lubricant removed. It should be within 4 to 6 ounces of the compressor's factory oil charge. The lubricant charge is indicated on the name plate of Copelaweld® compressors. Copelametic® compressor oil charges can be found in Application Engineering Bulletin 4-1281. If the lubricant charge is unknown, an authorized Copeland wholesaler can provide the technician with the information.

If the amount of lubricant removed is less than 50% of the factory charge, it will be necessary to flush the excess lubricant from the system.

Those systems that have oil separators, oil reservoirs, oil floats and suction line accumulators must have the oil drained from them. If the liquid control device is going to be replaced, it is advisable that the suction line, liquid line and evaporator coil be blown clean using properly regulated dry nitrogen.

NOTE: Properly dispose of the lubricant.

6. Before the new lubricant is installed into the compressor, be sure all leaks are repaired, liquid control devices and any other system components are changed. Install the correct liquid line filter-drier. Driers must be compatible with the refrigerant and lubricant.

7. Be advised that POEs are very hygroscopic. They will very quickly absorb moisture from the air once the container is opened. Once the lubricant is added to the compressor, the compressor should be quickly installed. Like an open container, an open compressor with POE will absorb moisture. Add the correct amount of lubricant to the compressor. It is important that the system contain at least 50% POE. On systems using enhanced surfaces in the heat exchanger, excessive mineral oil can adversely effect the heat transfer due to logging. Therefore, it is desirable to have no more than 20% mineral oil in systems employing these type surfaces.

8. Once the compressor is installed and the system is closed, the system must be evacuated to 500 microns or lower. A vacuum decay test is suggested at this time to assure the system is dry and free of leaks.

9. REFRIGERANT CHARGING WITH "NEAR AZEO-TROPES." Refrigerant R-401A is a near azeotropic mixture (see Copeland booklet 92-81 "Guide to Refrigerant Mixtures"). It is important that during initial charging or "topping" off a system that the refrigerant be removed from the charging cylinder in the liquid phase. Many of the cylinders for the newer refrigerants use a dip tube so that in the upright position liquid is drawn from the cylinder. DO NOT vapor charge out of a cylinder unless the entire cylinder is to be charged into the system. Refer to charging instructions provided by the refrigerant manufacturer.

With the system in a 500 micron or lower vacuum, liquid can be charged into the system "high side." The initial charge should be about 80% of the amount of refrigerant removed from the system.

Put the system into operation and observe its performance. Additional refrigerant may have to be added to the operating system to obtain optimum performance.

When adding refrigerant to an operating system, it may be necessary to add the refrigerant through the compressor suction service valve. Because the refrigerant leaving the refrigerant cylinder must be in the liquid phase, care must be exercised to avoid damage to the compressor. It is suggested that a sight glass be connected between the charging hose and the compressor suction service valve. This will permit your adjusting the cylinder hand valve so that liquid can leave the cylinder while allowing vapor to enter the compressor.

10. Operate the system and record the operating conditions. Compare this data to the base data taken in item 2. Check and adjust the expansion valve superheat setting if necessary. Make adjustment to other controls as needed.

11. Properly label the components. Tag the compressor with the refrigerant used (R-401A) and the lubricant used. The proper color code for R-401A is Coral Red PMS (Paint Matching System) 177.

12. Clean up and properly dispose of removed lubricant. Check local and state laws regarding the disposal of refrigerant lubricants. Recycle or reclaim the removed refrigerant.

CAUTION: These guidelines are intended for use with R-401A only, not for refrigerants which are similar to R-401A. Other refrigerants may not be compatible with the materials used in our compressors or the lubricants recommended in this bulletin resulting in unacceptable reliability and durability of the compressor.

Note: Retrofit Videos are available from your Authorized Copeland Wholesaler. Ask for VT-025.

The information contained herein is based on technical data and tests which we believe to be reliable and is intended for use by persons having technical skill, at their own discretion and risk. Since conditions of use are beyond Copeland's control, we can assume no liability for results obtained or damages incurred through the application of the data presented.

R-401A Saturated Vapor/Liquid
Pressure/Temperature Chart

Press. PSIG	Vapor Temp. °F	Liquid Temp. °F	Press. PSIG	Vapor Temp. °F	Liquid Temp. °F	Press. PSIG	Vapor Temp. °F	Liquid Temp. °F	Press. PSIG	Vapor Temp. °F	Liquid Temp. °F
1	-13.7	-24.3	51	55.8	45.7	101	90.3	80.3	151	113.9	105.1
2	-9.8	-22.0	52	56.7	46.5	102	91.3	80.9	152	114.3	105.5
3	-7.4	-18.5	53	57.5	48.2	103	91.8	81.5	153	114.8	106.0
4	-5.1	-16.2	54	58.3	49.1	104	91.4	82.0	154	115.2	106.4
5	-2.9	-14.0	55	59.1	49.6	105	92.0	82.6	155	115.6	106.8
6	-0.8	-11.9	56	60.0	50.2	106	92.5	83.1	156	116.0	107.2
7	1.2	-9.8	57	60.8	50.7	107	93.0	83.7	157	116.4	107.7
8	3.2	-7.8	58	61.5	51.5	108	93.5	84.2	158	116.8	108.1
9	5.1	-5.9	59	62.3	52.2	109	94.1	84.8	159	117.2	108.6
10	6.9	-4.1	60	63.1	53.1	110	94.7	85.3	160	117.6	108.9
11	8.7	-2.3	61	63.9	53.9	111	95.2	85.9	161	118.0	109.3
12	10.4	-0.5	62	64.6	54.6	112	95.7	86.4	162	118.4	109.7
13	12.1	1.2	63	65.4	55.4	113	96.2	86.9	163	118.8	110.1
14	13.7	2.8	64	66.1	56.2	114	96.7	87.4	164	119.2	110.6
15	15.3	4.4	65	66.9	57.0	115	97.2	88.0	165	119.6	111.0
16	16.8	6.0	66	67.6	57.7	116	97.7	88.5	166	120.0	111.4
17	18.3	7.5	67	68.3	58.4	117	98.2	89.0	167	120.4	111.8
18	19.8	9.0	68	69.1	59.2	118	98.7	89.5	168	120.8	112.2
19	21.2	10.5	69	69.8	59.9	119	99.2	90.0	169	121.2	112.6
20	22.6	11.9	70	70.5	60.6	120	99.7	90.5	170	121.6	113.0
21	24.0	13.3	71	71.2	61.3	121	100.2	91.0	171	122.0	113.4
22	25.4	14.7	72	71.9	62.0	122	100.7	91.5	172	122.4	113.8
23	26.7	16.0	73	72.6	62.8	123	101.2	92.0	173	122.8	114.2
24	28.0	17.3	74	73.2	63.5	124	101.7	92.5	174	123.2	114.6
25	29.2	18.6	75	73.9	64.1	125	102.2	93.0	175	123.6	115.0
26	30.5	19.9	76	74.6	64.8	126	102.6	93.5	176	124.0	115.4
27	31.7	21.1	77	75.3	65.5	127	103.0	94.0	177	124.3	115.8
28	32.9	22.3	78	76.0	66.2	128	103.5	94.5	178	124.7	116.2
29	34.0	23.5	79	76.6	66.9	129	104.1	95.0	179	125.1	116.5
30	35.2	24.7	80	77.2	67.5	130	104.6	95.5	180	125.4	116.9
31	36.3	25.9	81	78.4	68.2	131	105.0	97.2	181	125.8	117.2
32	37.4	27.0	82	78.8	68.8	132	105.5	96.4	182	126.2	117.5
33	38.5	28.1	83	79.1	69.5	133	105.9	96.9	183	126.6	118.1
34	39.6	29.2	84	79.8	70.1	134	106.4	97.4	184	126.9	118.5
35	40.7	30.3	85	80.4	70.8	135	106.9	97.9	185	127.3	118.5
36	41.7	31.3	86	81.0	71.4	136	107.3	98.3	186	127.6	119.2
37	42.7	32.4	87	81.6	72.5	137	107.8	98.8	187	128.0	119.6
38	43.8	33.4	88	82.2	72.6	138	108.2	99.2	188	128.4	120.0
39	44.8	34.4	89	82.8	73.2	139	108.7	99.7	189	128.7	120.4
40	45.8	35.5	90	83.4	73.9	140	109.1	100.2	190	129.1	120.8
41	46.7	36.4	91	84.0	74.5	141	109.5	100.6	191	129.5	121.1
42	47.7	37.4	92	84.6	75.1	142	110.0	101.0	192	129.8	121.5
43	48.6	38.4	93	85.2	75.7	143	110.4	101.5	193	130.2	121.8
44	49.6	39.3	94	85.8	76.3	144	110.9	101.9	194	130.6	122.2
45	50.5	40.3	95	86.4	76.9	145	111.3	102.4	195	130.9	122.6
46	51.4	41.2	96	87.0	77.5	146	111.8	102.9	196	131.3	122.9
47	52.3	42.1	97	87.5	78.0	147	112.2	103.3	197	131.6	123.3
48	53.2	43.0	98	88.1	78.6	148	112.6	102.9	198	132.0	123.7
49	54.1	43.9	99	88.7	79.2	149	113.0	104.2	199	132.3	124.0
50	54.7	44.8	100	89.2	79.8	150	113.4	104.7	200	132.7	124.6

R-401A Saturated Vapor/Liquid
Temperature/Pressure Chart

Temp. °F	Vapor Pressure PSIG	Liquid Pressure PSIG	Temp. °F	Vapor Pressure PSIG	Liquid Pressure PSIG	Temp. °F	Vapor Pressure PSIG	Liquid Pressure PSIG
-10	1.9	6.9	40	34.4	44.7	90	101.4	119.0
-9	2.3	7.4	41	35.3	45.8	91	103.2	121.0
-8	2.7	7.9	42	36.3	46.9	92	105.1	123.0
-7	3.2	8.4	43	37.3	48.0	93	106.9	125.0
-6	3.6	9.0	44	38.2	49.0	94	108.8	127.0
-5	4.0	9.5	45	39.2	50.2	95	110.7	129.0
-4	4.5	10.0	46	40.3	51.4	96	112.6	131.1
-3	4.9	10.6	47	41.3	52.5	97	114.6	133.2
-2	5.4	11.1	48	42.3	53.7	98	116.6	135.4
-1	5.9	11.7	49	43.4	54.9	99	118.6	137.5
0	6.4	12.3	50	44.5	56.1	100	120.6	139.7
1	6.9	12.9	51	45.6	57.4	101	122.6	141.9
2	7.4	13.5	52	46.7	58.6	102	124.7	144.1
3	7.9	14.1	53	47.8	59.9	103	126.8	146.3
4	8.4	14.7	54	48.9	61.1	104	128.9	148.6
5	9.0	15.4	55	50.0	62.4	105	131.0	150.9
6	9.5	16.0	56	51.2	63.7	106	133.1	153.2
7	10.0	16.6	57	52.4	65.1	107	135.3	155.5
8	10.6	17.3	58	53.6	66.4	108	137.5	157.9
9	11.2	18.0	59	54.8	67.8	109	139.8	160.2
10	11.8	18.9	60	56.1	69.1	110	142.0	162.4
11	12.4	19.4	61	57.3	70.5	111	144.3	165.1
12	13.0	20.0	62	58.6	71.9	112	146.6	167.5
13	13.5	20.8	63	59.9	73.4	113	148.9	170.0
14	14.2	21.5	64	61.2	74.8	114	151.3	172.5
15	14.8	22.2	65	62.5	76.3	115	153.6	175.0
16	15.5	23.0	66	63.8	77.7	116	156.0	177.6
17	16.1	23.7	67	65.2	79.2	117	159.5	180.1
18	16.8	24.5	68	66.5	80.8	118	160.9	182.8
19	17.4	25.3	69	67.9	82.3	119	163.4	185.4
20	18.1	26.0	70	69.3	83.9	120	165.9	188.0
21	18.8	26.9	71	70.7	85.4	121	168.4	190.7
22	19.5	27.7	72	72.1	87.0	122	171.0	193.4
23	20.3	28.6	73	73.6	88.6	123	174.6	196.1
24	21.0	29.4	74	75.1	90.2	124	176.2	198.9
25	21.7	30.3	75	76.6	91.9	125	178.8	201.7
26	22.5	31.1	76	78.1	93.6	126	181.5	204.5
27	23.3	32.0	77	79.7	95.2	127	184.2	207.3
28	24.0	32.9	78	81.2	97.0	128	186.9	210.2
29	24.8	33.8	79	82.8	98.7	129	189.7	213.1
30	25.6	34.7	80	84.4	100.4	130	192.5	216.0
31	26.4	35.7	81	86.0	102.2			
32	27.3	36.6	82	87.6	104.0			
33	28.1	37.6	83	89.3	105.8			
34	29.0	38.6	84	91.0	107.6			
35	29.8	39.6	85	92.6	109.4			
36	30.7	40.6	86	94.4	111.3			
37	31.6	41.6	87	96.1	113.2			
38	32.5	42.6	88	97.8	115.1			
39	33.4	43.7	89	99.6	117.0			

Copeland Corporation • 1675 W. Campbell Road, Sidney, Ohio 45365-0669 • Phone (513) 498-3011

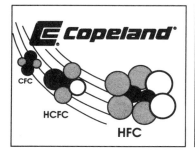

Refrigerant Changeover Guidelines
CFC-12 to R-401B (MP66)

Leading the Industry with Environmentally Responsible Refrigerant Solutions

Copeland does not advocate the wholesale changeover of CFC refrigerants to HCFCs or HFCs. If a system is not leaking refrigerant to the atmosphere, and is operating properly, there is no technical reason to replace the CFC refrigerant. In fact, changing the refrigerant may void the U.L. listing of the unit. However, once the decision has been made to make the change from CFC-12 (R-12) to the interim R-401B, the following guidelines are recommended.

CONSIDERATIONS

1. Retrofitting systems that employ compressors manufactured prior to 1973 is not recommended. This is due to the different materials used in motor insulation that have not been evaluated for compatibility with the new refrigerants and lubricants. Failure to heed this advice will violate the proposed U.L. Standard For Field Conversion/Retrofit Of Alternate Refrigerants In Refrigeration and Air Conditioning Equipment (U.L. 2170-2172).

2. Copeland's lubricant recommendation for use with R-401B is a mixture of 50% mineral oil and 50% alkylbenzene (Zerol 200TD). Polyol ester lubricants, Mobil EAL™ Arctic 22 CC or ICI EMKARATE RL 32S, can also be used with R-401B if the system is expected to be changed in the near future to an HFC refrigerant such as R-401B. This will eliminate the need of having to flush the system again when making the R-134a retrofit.

Refer to item11 in this section for a list of other approved lubricants for use with R-401B or Application Engineering Bulletin 17-1248 for a complete list of all Copeland approved lubricants.

3. R-401B should be used only in systems where the saturated suction temperature is maintained between -40°F and -10°F. **It should not be mixed with any other refrigerant!**

4. The expansion valve may need to be changed. The existing R-12 valve when used with R-401B will have approximately 25% more capacity. Oversized expansion valves can result in hunting and refrigerant floodback. Consult with the thermostatic expansion valve manufacturer for the correct valve and size.

5. Filter-driers must be changed at the time of conversion. This is proper air conditioning/refrigeration practice.

    a. Solid core driers such as ALCO ADK, are compatible with either R-12 or R-401B.

    b. Compacted bead driers can use XH6 or XH9 molecular sieve material such as found in the ALCO UltraFlow or EK or EKH series, or Copeland PureFlow.

    c. If a loose fill type drier is to be used, XH9 molecular sieve is required.

6. R-401B exhibits marginally higher pressures than R-12 at normal condensing temperatures. We do not believe this will require readjustment of safety controls; however, you should verify this with the system manufacturer or component suppliers.

7. Systems that use a low pressure controller to maintain space temperature may have to have the cut-out and cut-in points changed. With R-401B, the pressure setting must reflect an average temperature of the refrigerant in the evaporator. Because of refrigerant glide (see Copeland booklet 92-81, "Guide To Refrigerant Mixtures"), the refrigerant entering the evaporator for a specific suction pressure will be approximately 8°F colder than the refrigerant vapor at the outlet of the evaporator (not considering superheat). Therefore, the average refrigerant

WARNING: Use only Copeland approved refrigerants and lubricants in the manner prescribed by Copeland. In some circumstances, other refrigerants and lubricants may be dangerous and could cause fires, explosions or electrical shorting. Contact Copeland Corp., Sidney, Ohio.

temperature will be at a midpoint pressure/temperature equivalent.

Example: A -5°F refrigerated space usually requires that the refrigerant temperature in the evaporator be approximately -15°F. Using R-402B, the entering liquid temperature may be as cold as -19°F and the vapor temperature before superheat may be -11°F. Taking the saturated vapor pressure at -11° F gives us the exit pressure at the evaporator of 3.3 psig. Considering a 2 psig pressure drop in the suction line, the pressure control cut-out should be set at 1.3 psig.

The cut-in point will be based on the vapor pressure/temperature value. Let's assume that the space temperature can rise to -2°F before the compressor is turned on. -2°F vapor pressure is 7.6 psig. Set the cut-in at 8 psig.

8. Because of glide, pressure regulators such as EPRs may have to be reset. Contact the EPR manufacturer for correct settings.

9. Due to refrigerant glide, it is important that when measuring and/or adjusting TEV superheat, the pressure and SATURATED VAPOR TABLES be used. Example: The pressure measured at the TEV bulb is 7 psig. The Pressure/Temperature (P/T) chart shows that the saturated vapor temperature for 7 psig is -3.2°F. If the actual refrigerant temperature measured is 6°F, the superheat is 6.2°F.

To measure sub-cooling at the condenser outlet or at the TEV inlet to verify that a solid column of liquid is present, measure the pressure and the refrigerant temperature at the location that the sub-cooling information is needed. Compare it to the saturated liquid temperature from the SATURATED LIQUID TABLES. Example: A pressure of 175 psig is measured at the condenser coil outlet. From the P/T chart, 175 psig is 110.2°F saturated liquid temperature. If the actual refrigerant temperature is 105°F, the liquid is sub-cooled 5.2°F.

10. Systems using R-401B may have a lower system pressure drop than with R-12. Because of the lower pressure drop, pilot operated solenoid valves and pressure regulators may not operate. Check with the manufacturer of any pressure regulators and pilot operated solenoid valves used in the system to be sure that they will operate properly. These controls may have to be downsized.

11. Mineral oil lubricant only, such as 3GS, cannot be used as the compressor lubricant. Copeland recommends the following lubricant choices:

1. A mixture of 3GS Mineral Oil (MO) and Zerol 200TD Alkyl Benzene (AB) with a minimum of 50% AB

2. Virginia KMP MS2212 (a 70/30 mixture of AB/MO)

3. A mixture of 3GS Mineral Oil (MO) and Polyol Ester

(POE) ie; Mobil EAL Arctic 22 CC or EMKARATE RL 32S with a minimum of 50% POE

4. 100% Mobil EAL Arctic 22 CC or ICI EMKARATE RL 32S

Before starting the changeover, it is suggested that at a minimum the following items be ready:

1. Safety glasses
2. Gloves
3. Refrigerant service gauges
4. Electronic thermometer
5. Vacuum pump capable of pulling 250 microns
6. Thermcouple micron gauge
7. Leak detector
8. Refrigerant recovery unit including refrigerant cylinder
9. Proper container for removed lubricant
10. New liquid control device
11. Replacement liquid line filter-drier(s)
12. New lubricant
    a. Zerol 200TD (AB) or Virginia KMP MS 2212 (AB)
    b. Mobil EAL Arctic 22 CC or ICI EMKARATE RL 32S (POE)
13. R-401B pressure temperature chart
14. R-401B refrigerant

CHANGEOVER PROCEDURE

1. The system should be thoroughly leak tested with the R-12 refrigerant still in the system. All leaks should be repaired before the R-401B refrigerant is added.

2. It is advisable that the system operating conditions be recorded with the R-12 still in the system. This will provide the base data for comparison when the system is put back into operation with the R-401B.

3. The system should be electrically shut off and the refrigerant properly removed from the system. Measure the quantity of refrigerant removed. This will provide a guide for recharging the system with R-401B (see item 9 this section).

4. The mineral oil lubricant must be removed from the compressor crankcase. Hermetic compressors will have to be removed from the piping and the lubricant drained out through the suction stub. It is advisable to do an acid test on the oil.

5. Measure the amount of lubricant removed. It should be within 4 to 6 ounces of the compressor's factory oil charge. The lubricant charge is indicated on the name plate of Copelaweld® compressors. Copelametic® compressor oil charges can be found in Application Engineering Bulletin 4-1281. If the lubricant charge is unknown, an authorized

Copeland wholesaler can provide the technician with the information.

If the amount of lubricant removed is less than 50% of the factory charge, it will be necessary to flush the excess lubricant from system.

Those systems that have oil separators, oil reservoirs, oil floats and suction line accumulators must have the oil drained from them. If the liquid control device is going to be replaced, it is advisable that the suction line, liquid line and evaporator coil be blown clean using properly regulated dry nitrogen.

NOTE: Properly dispose of the lubricant.

6. Before the new lubricant is installed into the compressor, be sure all leaks are repaired, liquid control devices and any other system components are changed. Install the correct liquid line filter-drier. Driers must be compatible with the refrigerant and lubricant.

7. Be advised that POEs are very hygroscopic. They will very quickly absorb moisture from the air once the container is opened. Once the lubricant is added to the compressor, the compressor should be quickly installed. Like an open container, an open compressor with POE will absorb moisture. Add the correct amount of lubricant to the compressor. It is important that the system contain at least 50% POE. On systems using enhanced surfaces in the heat exchanger, excessive mineral oil can adversely effect the heat transfer due to logging. Therefore, it is desirable to have no more than 20% mineral oil in systems employing these type surfaces.

8. Once the compressor is installed and the system is closed, the system must be evacuated to 250 microns or lower. A vacuum decay test is suggested to assure the system is dry and leak free.

9. REFRIGERANT CHARGING WITH "NEAR AZEO-TROPES." Refrigerant R-401B is a near azeotropic mixture (see Copeland booklet 92-81 "Guide to Refrigerant Mixtures"). It is important that during initial charging or "topping" off a system that the refrigerant be removed from the charging cylinder in the liquid phase. Many of the cylinders for the newer refrigerants use a dip tube so that in the upright position liquid is drawn from the cylinder. DO NOT vapor charge out of a cylinder unless the entire cylinder is to be charged into the system. Refer to charging instructions provided by the refrigerant manufacturer.

With the system in a 250 micron or lower vacuum, liquid can be charged into the system high side. The initial charge should be about 80% of the amount of refrigerant removed from the system.

Put the system into operation and observe its performance. Additional refrigerant may have to be added to the operat-

ing system to obtain optimum performance.

When adding refrigerant to an operating system, it may be necessary to add the refrigerant through the compressor suction service valve. Because the refrigerant leaving the refrigerant cylinder must be in the liquid phase, care must be exercised to avoid damage to the compressor. It is suggested that a sight glass be connected between the charging hose and the compressor suction service valve. This will permit your adjusting the cylinder hand valve so that liquid can leave the cylinder while allowing vapor to enter the compressor.

10. Operate the system and record the operating conditions. Compare this data to the base data taken in item 2 this section. Make adjustments as needed.

11. Properly label the components. Tag the compressor with the refrigerant used (R-401B) and the lubricant used. The proper color code for R-401B is Light Gray Green PMS (Paint Matching System) 413.

12. Clean up and properly dispose of removed lubricant. Check local and state laws regarding the disposal of refrigerant lubricants. Recycle or reclaim the removed refrigerant.

CAUTION: These guidelines are intended for use with R-401B only, not for refrigerants which are similar to R-401B. *Other refrigerants may not be compatible with the materials used in our compressors or the lubricants recommended in this bulletin resulting in unacceptable reliability and durability of the compressor.*

Note: Retrofit videos are avialable from your Authorized Copeland Wholesaler. Ask for VT-025.

R-401B Saturated Vapor/Liquid
Pressure/Temperature Chart

Pressure PSIG	Vapor Temp. °F	Liquid Temp. °F	Pressure PSIG	Vapor Temp. °F	Liquid Temp. °F
(13)	-40.5	-51.1	38	39.0	29.6
(12)	-38.8	-49.1	39	40.0	30.6
(11)	-36.5	-47.1	40	41.0	31.6
(10)	-34.6	-45.2	45	45.7	35.4
(9)	-32.8	-43.4	50	50.1	40.8
(8)	-31.1	-41.6	55	53.4	45.0
(7)	-29.4	-40.0	60	58.1	49.0
(6)	-27.8	-38.4	65	61.9	52.9
(5)	-26.3	-36.8	70	65.1	56.5
(4)	-24.8	-35.3	75	68.8	60.0
(3)	-23.3	-33.8	80	72.1	63.3
(2)	-21.9	-32.4	85	75.2	66.5
(1)	-20.5	-31.0	90	78.2	69.6
0	-19.2	-29.6	95	81.1	72.5
1	-16.6	-26.9	100	83.9	75.4
2	-14.1	-24.4	105	86.6	76.2
3	-11.7	-22.0	110	89.3	80.9
4	-9.4	-19.7	115	91.8	83.5
5	-7.3	-17.5	120	94.3	86.0
6	-5.2	-15.4	125	96.7	88.5
7	-3.2	-13.4	130	99.1	91.4
8	-1.2	-11.4	135	101.3	93.3
9	0.6	-9.5	140	103.6	95.5
10	2.5	-7.6	145	105.7	98.8
11	4.2	-5.8	150	107.9	100.4
12	5.9	-4.1	155	109.9	102.1
13	7.6	-2.4	160	112.0	104.2
14	9.2	-0.8	165	114.0	106.2
15	10.8	0.8	170	115.9	108.2
16	11.7	2.3	175	117.8	110.2
17	13.8	3.9	180	119.7	112.1
18	15.3	5.3	185	121.4	114.0
19	16.7	6.8	190	123.3	115.9
20	18.1	8.2	195	125.1	117.7
21	19.4	9.6	200	126.8	119.8
22	20.8	10.9	205	128.5	121.2
23	22.1	12.3	210	130.0	123.0
24	23.3	13.6	215	131.8	124.7
25	24.6	14.9	220	133.5	126.3
26	25.8	16.1	225	135.0	128.0
27	27.0	17.3	230	138.6	129.5
28	28.2	18.5	235	138.2	131.1
29	29.4	19.7			
30	30.5	20.9	( ) Inches Vacuum		
31	31.6	22.0			
32	32.7	23.2			
33	33.8	24.3			
34	34.9	25.4			
35	35.9	26.4			
36	37.0	27.5			
37	38.0	28.5			

Copeland Corporation • 1675 W. Campbell Road, Sidney, Ohio 45365-0669 • (513) 498-3011

R-401B Saturated Vapor/Liquid
Temperature/Pressure Chart

Temp. °F	Vapor Press. PSIG	Liquid Press. PSIG	Temp. °F	Vapor Press. PSIG	Liquid Press PSIG	Temp. °F	Vapor Press. PSIG	Liquid Press. PSIG	Temp. °F	Vapor Press. PSIG	Liquid Press. PSIG
-40	(12.9)	(7.0)	10	14.5	21.3	60	62.5	75.0	110	155.1	174.5
-39	(12.3)	(6.4)	11	15.1	22.0	61	63.8	76.6	111	157.6	177.0
-38	(11.8)	(5.8)	12	15.8	22.8	62	65.2	78.1	112	160.1	179.7
-37	(11.3)	(5.1)	13	16.5	23.6	63	66.5	79.6	113	162.2	182.3
-36	(10.7)	(4.5)	14	17.1	24.3	64	68.0	81.1	114	165.2	185.0
-35	(10.2)	(3.8)	15	17.8	25.1	65	69.4	82.7	115	167.7	187.7
-34	(9.7)	(3.1)	16	18.5	25.9	66	70.8	84.2	116	170.2	190.4
-33	(9.1)	(2.4)	17	19.2	26.7	67	72.3	85.8	117	172.9	193.1
-32	(8.5)	(1.8)	18	19.9	27.5	68	73.8	87.4	118	175.5	195.9
-31	(7.9)	(1.0)	19	20.7	28.4	69	75.3	89.1	119	178.2	198.7
-30	(7.3)	(0.3)	20	21.4	29.3	70	76.8	90.7	120	180.9	210.5
-29	(6.7)	0.2	21	22.2	30.0	71	78.3	92.4	121	183.5	204.4
-28	(6.1)	0.6	22	22.9	31.0	72	79.9	94.1	122	186.4	207.2
-27	(5.5)	1.0	23	23.7	31.9	73	81.5	95.8	123	189.1	210.3
-26	(4.8)	1.4	24	24.5	32.8	74	83.1	97.5	124	191.9	213.1
-25	(4.2)	1.8	25	25.3	33.7	75	84.7	99.3	125	195.8	216.0
-24	(3.5)	2.2	26	26.1	34.6	76	86.3	101.1	126	197.7	219.0
-23	(2.8)	2.6	27	27.0	35.5	77	88.0	102.8	127	200.6	222.0
-22	(2.1)	3.0	28	27.8	36.5	78	89.6	104.7	128	203.5	225.1
-21	(1.3)	3.4	29	28.9	37.5	79	91.3	106.5	129	206.5	228.2
-20	(0.6)	3.9	30	29.5	38.4	80	93.1	108.3	130	209.5	231.3
-19	0.1	4.3	31	30.4	39.4	81	94.8	110.2			
-18	0.5	4.8	32	31.2	40.6	82	96.6	112.1	( ) Inches Vacuum		
-17	0.8	5.2	33	32.2	41.5	83	98.3	114.0			
-16	1.2	5.7	34	33.2	42.5	84	100.1	116.0			
-15	1.6	6.2	35	34.1	43.6	85	102.0	117.9			
-14	2.0	6.7	36	35.1	44.6	86	103.8	119.4			
-13	2.5	7.2	37	36.0	45.7	87	105.7	121.9			
-12	2.9	7.7	38	37.0	46.8	88	107.6	124.0			
-11	3.3	8.2	39	38.0	47.9	89	109.5	126.0			
-10	3.8	8.7	40	39.0	49.1	90	111.4	128.1			
-9	4.2	9.3	41	40.0	50.2	91	113.4	130.2			
-8	4.7	9.8	42	41.0	51.4	92	115.4	132.3			
-7	5.1	10.4	43	42.1	52.6	93	117.4	134.4			
-6	5.6	10.9	44	43.2	53.7	94	119.4	119.4			
-5	6.1	11.5	45	44.3	55.0	95	121.4	138.8			
-4	6.6	12.1	46	45.4	56.2	96	123.5	123.5			
-3	7.1	12.7	47	46.5	57.4	97	143.2	143.2			
-2	7.6	13.3	48	47.6	58.7	98	127.7	145.5			
-1	8.1	13.9	49	48.8	59.9	99	129.9	147.8			
0	8.7	14.5	50	49.9	61.2	100	132.0	150.1			
1	9.2	15.1	51	51.1	62.5	101	134.2	152.4			
2	9.7	15.8	52	52.3	63.9	102	136.5	154.8			
3	10.3	16.4	53	53.5	65.2	103	138.7	157.1			
4	10.9	17.1	54	54.7	66.5	104	141.0	159.6			
5	11.5	17.8	55	56.0	67.9	105	143.3	162.0			
6	12.0	18.5	56	57.3	69.3	106	145.6	164.4			
7	12.6	19.2	57	58.5	70.7	107	147.9	166.9			
8	13.2	19.9	58	59.8	72.2	108	150.3	169.4			
9	13.9	20.6	59	61.1	73.6	109	152.7	172.0			

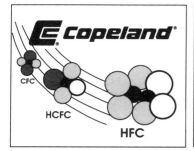

Refrigerant Changeover Guidelines
R-502 to R-402A /R-408A

Leading the Industry with Environmentally Responsible Refrigerant Solutions

Copeland does not advocate the wholesale changeover of CFC refrigerants to HCFCs or HFCs. If a system is not leaking refrigerant to the atmosphere, and is operating properly, there is no technical reason to replace the CFC refrigerant. In fact, changing the refrigerant may void the U.L. listing of the unit. However, once the decision has been made to make the change from R-502 to the interim R-402A (SUVA® HP-80) or R-408A (Forane® FX-10), the following guidelines are recommended.

CONSIDERATIONS

1. Retrofitting systems that employ compressors manufactured prior to 1973 is not recommended. This is due to the different materials used in motor insulation that have not been evaluated for compatibility with the new refrigerants and lubricants. Failure to heed this advice will violate the U.L Standard For Field Conversion/Retrofit Of Alternate Refrigerants In Refrigeration And Air Conditioning Equipment (U.L. 2170).

2. Copeland's lubricant recommendation for use with R-402A/R-408A is a mixture of 50% mineral oil and 50% alkyl benzene (Zerol 200TD or Soltex AB200A). Polyol ester lubricants (Mobil EAL™ Arctic 22 CC or ICI Emkarate RL 32CF) can also be used with R-402A/R-408A if the system is expected to be changed in the near future to an HFC refrigerant such as R-404A. This will eliminate the need of having to flush the system again when making the R-404A retrofit.

Refer to item 12 this section for a list of other approved lubricants for use with R-402A/R-408A or Application Engineering Bulletin 17-1284 for a complete list of all Copeland approved lubricants.

3. R-402A/R-408A should be used only in systems that currently use R-502. It should not be mixed with R-502 or any other refrigerant nor should it be used to replace CFC-12 or HCFC-22.

4. The capacity of the existing R-502 thermal expansion valve (TEV) will be approximately the same when using R-402A/R-408A. However, the superheat setting must be checked and may have to be readjusted after the system is put back into operation.

Consult with the TEV manufacturer for correct sizing and superheat settings.

5. Filter-driers must be changed at the time of conversion. This is proper air conditioning/refrigeration practice.

   a. Solid core driers such as ALCO ADK are compatible with either R-502 or R-402A/R-408A.

   b. Compacted bead type driers can use XH6 or XH9 molecular sieve material such as found in the ALCO EK or EKH series.

   c. If a loose fill type drier is to be used, XH9 molecular sieve is required.

6. Because of glide, pressure regulators such as EPR valves may have to be reset. Contact the EPR manufacturer for the correct settings.

7. R-402A/R-408A exhibits higher pressures than R-502 at normal condensing temperatures. This may require the high pressure safety controls be reset in order to operate as intended.

8. The higher pressure characteristics exhibited by R-402A will in some cases exceed the industry accepted safety

WARNING: Use only Copeland approved refrigerants and lubricants in the manner prescribed by Copeland. In some circumstances, other refrigerants and lubricants may be dangerous and could cause fires. explosions or electrical shorting. Contact Copeland Corp., Sidney, Ohio.

factors on the compressor crankcase (low side). This will require the addition of a pressure relief valve on the compressor crankcase, set at a maximum of 375 psig to adequately protect the compressor from the possibility of excessive pressure. **Semi-Hermetic compressors that require this additional valve are:**

Discus 3D and 4D

All Other Semi-Hermetic (Non-Discus Models)

> **WARNING: IT IS POSSIBLE THAT EXCESS PRESURE BUILD-UP ON MODELS INDICATED COULD RESULT IN THE COMPRESSOR EXPLODING UNLESS THE PRESSURE RELIEF VALVE SPECIFIED HAS BEEN PROPERLY INSTALLED ON THE ORIGINALLY BUILT COPELAND COMPRESSOR.**

Pressure relief valves can be purchased from your Authorized Copeland wholesaler as part number 998-0051-02.

Note: Due to the lower presure of R-408A, pressure relief valves are not required when retrofitting to this refrigerant.

9. Systems that use a low pressure controller to maintain space temperature may need to have the cut in and cut out points changed. With R-402A, the pressure settings must reflect an average temperature of the refrigerant in the evaporator. Because of refrigerant glide (see Copeland booklet 92-81, "Guide To Refrigerant Mixtures"), the refrigerant entering the evaporator for a specific suction pressure will be approximately 2°F colder than the refrigerant vapor at the outlet of the evaporator (not considering superheat). Therefore, the average refrigerant temperature will be at a midpoint pressure/temperature equivalent.

Example: A -10°F refrigerated space usually requires that the refrigerant temperature in the evaporator be approximately -20°F. Using R-402A, the liquid entering the evaporator may be as cold as -21°F and the vapor temperature before superheat may be -19°F. Taking the saturated vapor pressure at -19°F gives us the exit pressure at the evaporator of 19.1 psig. Considering a 2 psig pressure drop in the suction line, the pressure control cut out should be set at 17.1 psig.

10. Due to refrigerant glide, it is important that when measuring and/or adjusting TEV superheat, the pressure and SATURATED VAPOR TABLES be used. Example: The pressure measured at the TEV bulb is 18 psig. The Pressure/Temperature (P/T) chart shows that the saturated vapor temperature of R-402A for 18 psig is -20.7°F. If the actual refrigerant temperature is -15.7°F, the superheat is 5°F.

Note: The glide with R-408A is so small as to be negligible. Use saturated vapor tables to set pressure controls.

To measure sub-cooling at the condenser outlet or at the TEV inlet to verify that a solid column of liquid is present, measure the pressure and the temperature at the location that the sub-cooling information is needed. Compare it to the SATURATED LIQUID TABLES. Example: A pressure of 250 psig is measured at the condenser outlet. From the R-402A P/T chart, 250 psig is 99°F saturated liquid temperature. If the actual refrigerant temperature is 89°F, the liquid is sub-cooled 10°F.

11. Systems using R-402A/R-408A may have a lower system pressure drop than with R-502. Check with the manufacturer of any pressure regulators and pilot operated solenoid valves used in the system to be sure that they will operate properly. These controls may have to be downsized.

12. Mineral oil lubricant only, such as 3GS, <u>cannot</u> be used as the compressor lubricant. Copeland recommends the following lubricant choices :

1. A mixture of 3GS Mineral Oil (MO) and Zerol 200TD or Soltex AB200A Alkyl Benzene (AB) with a minimum of 50% AB

2. Virginia KMP MS2212 or Witco R-195-0 (a 70/30 mixture of AB/MO)

3. A mixture of 3GS Mineral Oil (MO) and Polyol Ester (POE) ie; Mobil EAL Arctic 22 CC or ICI Emkarate RL32CF with a minimum of 50% POE

4. 100% Mobil EAL Arctic 22 CC or ICI Emkarate RL32CF

Before starting the changeover, it is suggested that at least the following items be ready:

1. Safety glasses
2. Gloves
3. Refrigerant service gauges
4. Electronic thermometer
5. Vacuum pump capable of pulling 250 microns
6. Thermocouple micron gauge
7. Leak detector
8. Refrigerant recovery unit including refrigerant cylinder
9. Proper container for removed lubricant
10. New liquid control device
11. Replacement liquid line filter-drier(s)

12.New lubricant
    a. Zerol 200TD or Soltex AB200A (AB) or
    b. Virginia KMP MS 2212 or Witco R-195-0 (AB/MO mixture) or
    c. Mobil EAL Arctic 22CC or ICI Emkarate RL32CF(POE)
13. R-402A/R-408A pressure temperature chart
14. R-402A/R-408A refrigerant

CHANGEOVER PROCEDURE

1. The system should be thoroughly leak tested with the R-502 still in the system. All leaks should be repaired before the R-402A/R-408A refrigerant is added.

2. It is advisable that the system operating conditions be recorded with the R-502 still in the system. This will provide the base data for comparison when the system is put back into operation with the R-402A/R-408A.

3. The system should be electrically shut off and the refrigerant properly removed from the system. Measure the quantity of refrigerant removed. This will provide a guide for recharging the system with R-402A/R-408A (see item 9 this section).

4. The mineral oil must be removed from the compressor crankcase. Hermetic compressors will have to be removed from the piping and the lubricant drained out through the suction stub. It is advisable to do an acid test on the lubricant removed.

5. Measure the amount of lubricant removed. It should be within 4 to 6 ounces of the compressors factory oil charge. The lubricant charge is indicated on the name plate of Copelaweld compressors. Copelametic compressor oil charges can be found in Application Engineering Bulletin 4-1281. If the lubricant charge is unknown, an authorized Copeland wholesaler can provide the technician with the information.

If the amount of lubricant removed is less than 50% of the factory charge, it will be necessary to clean the excess lubricant from the system.

Those systems that have oil separators, oil reservoirs, oil floats and suction line accumulators must have the oil drained from them. If the liquid control device is going to be replaced, it is advisable that the suction line, liquid line and evaporator coil be blown clean using properly regulated dry nitrogen.

NOTE: Properly dispose of the lubricant.

6. Before the new lubricant is installed into the compressor, be sure all leaks are repaired, liquid control device and any other system components are changed. Also assure that a pressure relief valve has been added if required. Install the correct liquid line filter-drier. Driers must be compatible with the refrigerant and lubricant.

7. Be advised that POE's are very hygroscopic. They will very quickly absorb moisture from the air once the container is opened. Once the lubricant is added to the compressor, the compressor should be quickly installed. Like an open container, an open compressor with POE will absorb moisture. Add the correct amount of lubricant to the compressor. It is important that the system contain a minimum of 50% AB or POE. On systems using enhanced surfaces in the heat exchanger, excessive mineral oil can adversely effect the heat transfer due to logging. Therefore, it is desirable to have no more than 20% mineral oil in systems employing these type surfaces.

8. Once the compressor is installed and the system is closed, the system must be evacuated to 250 microns or lower.

9. REFRIGERANT CHARGING WITH "NEAR AZEO-TROPES". R-402A/R-408A is a near azeotropic mixture (see Copeland booklet 92-81, "Guide To Refrigerant Mixtures"). It is important that during initial charging or "topping" off a system that the refrigerant be removed from the charging cylinder in the liquid phase. Many of the cylinders for the newer refrigerants use a dip tube so that in the upright position liquid is drawn from the cylinder. DO NOT vapor charge out of a cylinder unless the entire cylinder is to be charged into the system. Refer to charging instructions provided by the refrigerant manufacturer.

With the system in a 250 micron or lower vacuum, liquid can be charged into the system "high side." The initial charge should be approximately 80% of the amount of refrigerant removed from the system.

10. Start the system and observe its operation. Additional refrigerant may have to be added to obtain optimum performance.

When adding refrigerant to an operating system, it may be necessary to add the refrigerant through the compressor suction service valve. Because the refrigerant leaving the refrigerant cylinder must be in liquid phase, care must be exercised to avoid damage to the compressor. It is suggested that a sight glass be connected between the charging hose and the compressor suction service valve. This will permit you to adjust the cylinder hand valve so that liquid can leave the cylinder while allowing vapor to enter the compressor.

11. Operate the system and record the operating conditions. Compare this data to the data taken in item 2 this section. Check and adjust the TEV superheat setting if necessary. Make adjustments to other controls as needed.

12. Properly label the components, Tag the compressor with the refrigerant used (R-402A/R-408A) and the lubricant used. The proper color code for R-402A is Light Brown PMS (Paint Matching System) 461.The color for R-408A is Medium Purple PMS 248.

13. Clean up and properly dispose of the removed lubricant. Check local and state laws regarding the disposal of refrigerant lubricants. Recycle or reclaim the removed refrigerant.

Note: These guidelines are intended for use with R-402A/ R-408A only, not for refrigerants which are similar to R-402A/R-408A. Other refrigerants may not be compatible with the materials used in our compressors or the lubricants recommended in this bulletin resulting in unacceptable reliability and durability of the compressor.

The information contained herein is based on technical data and tests which we believe to be reliable and is intended for use by persons having technical skill, at their own discretion and risk. Since conditions of use are beyond Copeland's control, we can assume no liability for results obtained or damages incurred through the application of the data presented.

R-402A Saturated Vapor/Liquid
Temperature/Pressure Chart

Temp. °F	Vapor Press. PSIG	Liquid Press. PSIG	Temp. °F	Vapor Press. PSIG	Liquid Press PSIG	Temp. °F	Vapor Press. PSIG	Liquid Press. PSIG	Temp. °F	Vapor Press. PSIG	Liquid Press. PSIG
-40	5.8	7.6	10	47.1	50.2	60	132.7	137.3	110	285.7	290.9
-39	6.4	8.1	11	48.3	51.5	61	135.0	139.6	111	289.6	294.9
-38	6.9	8.7	12	49.5	52.7	62	137.4	142.0	112	293.6	298.8
-37	7.4	9.2	13	50.8	54.0	63	139.7	144.4	113	297.6	302.8
-36	8.0	9.8	14	52.0	55.3	64	142.1	146.8	114	301.6	306.9
-35	8.5	10.4	15	53.3	56.6	65	144.6	149.2	115	305.7	311.0
-34	9.1	11.0	16	54.6	57.9	66	147.0	151.7	116	309.8	315.1
-33	9.7	11.6	17	55.9	59.3	67	149.5	154.2	117	314.0	319.2
-32	10.3	12.2	18	57.2	60.6	68	152.0	156.7	118	318.2	323.4
-31	10.9	12.9	19	58.6	62.0	69	154.6	159.3	119	322.4	327.6
-30	11.5	13.5	20	60.0	63.4	70	157.1	161.9	120	326.7	331.9
-29	12.1	14.2	21	61.4	64.8	71	159.7	164.5	121	331.0	336.2
-28	12.8	14.8	22	62.8	66.3	72	162.2	167.1	122	335.3	340.5
-27	13.4	15.5	23	64.2	67.8	73	165.0	169.8	123	339.8	344.9
-26	14.1	16.2	24	65.7	69.2	74	167.7	172.5	124	344.2	349.3
-25	14.8	16.9	25	67.2	70.7	75	170.4	175.3	125	348.7	353.8
-24	15.4	17.6	26	68.6	72.3	76	173.1	178.0	126	353.2	358.3
-23	16.2	18.3	27	70.2	73.8	77	175.9	181.0	127	357.8	365.9
-22	16.9	19.1	28	71.7	75.4	78	178.7	183.6	128	362.4	367.4
-21	17.6	19.8	29	73.2	77.0	79	181.5	186.6	129	367.1	372.1
-20	18.3	20.6	30	74.8	78.6	80	184.4	189.6	130	371.7	376.7
-19	19.1	21.4	31	76.4	80.2	81	187.3	192.4			
-18	19.8	22.1	32	78.0	81.8	82	190.2	195.2			
-17	20.6	23.0	33	79.7	83.5	83	193.2	198.2			
-16	21.4	23.8	34	81.3	85.2	84	196.2	201.2			
-15	22.2	24.6	35	83.0	86.9	85	199.2	204.2			
-14	23.0	25.4	36	84.7	88.6	86	202.2	207.3			
-13	23.8	26.3	37	86.5	90.4	87	205.3	210.4			
-12	24.7	27.2	38	88.2	92.2	88	208.4	213.5			
-11	25.6	28.1	39	90.0	94.0	89	211.6	216.7			
-10	24.4	29.0	40	91.8	95.8	90	214.7	219.9			
-9	27.3	29.9	41	93.6	97.6	91	218.0	223.1			
-8	28.2	30.8	42	95.4	99.5	92	221.2	226.4			
-7	29.1	31.8	43	97.3	101.4	93	224.5	229.7			
-6	30.0	32.7	44	99.2	103.3	94	227.8	233.0			
-5	31.0	33.7	45	101.1	105.2	95	231.2	236.4			
-4	32.0	34.7	46	103.0	107.2	96	234.6	239.8			
-3	33.0	35.7	47	105.0	109.2	97	238.0	243.2			
-2	33.9	36.7	48	107.0	111.2	98	241.4	246.6			
-1	34.9	37.8	49	109.0	113.2	99	244.9	250.2			
0	36.0	38.8	50	111.0	115.3	100	248.4	253.7			
1	37.0	39.9	51	113.1	117.4	101	252.0	257.3			
2	38.0	41.0	52	115.1	119.5	102	255.6	260.8			
3	39.1	42.0	53	117.3	121.6	103	259.2	264.5			
4	40.2	43.2	54	119.4	123.8	104	262.9	268.1			
5	41.3	44.3	55	121.5	126.0	105	266.6	271.9			
6	42.4	45.5	56	123.7	128.2	106	270.4	275.6			
7	43.6	46.6	57	125.9	130.4	107	274.2	279.4			
8	44.7	47.8	58	128.2	132.7	108	277.9	283.2			
9	45.9	49.0	59	130.4	135.0	109	281.8	287.1			

Copeland Corporation • 1675 W. Campbell Road, Sidney, Ohio 45365-0669 • Phone (513) 498-3011

R-402A Saturated Vapor/Liquid
Pressure/Temperature Chart

Press. PSIG	Vapor Temp. °F	Liquid Temp. °F	Press. PSIG	Vapor Temp. °F	Liquid Temp. °F	Press. PSIG	Vapor Temp. °F	Liquid Temp. °F	Press. PSIG	Vapor Temp. °F	Liquid Temp. °F
5	-41.6	-45.0	55	16.2	13.6	105	47.0	44.9	335	121.9	120.7
6	-39.8	-42.9	56	16.9	14.3	106	47.5	45.4	340	123.0	121.9
7	-37.9	-41.1	57	17.6	15.2	107	48.0	45.9	345	124.2	123.0
8	-36.0	-39.6	58	18.3	16.0	108	48.5	46.2	350	125.3	124.1
9	-34.6	-37.7	59	19.1	16.7	109	49.0	46.9	355	126.4	125.3
10	-33.2	-35.8	60	20.0	17.4	110	49.5	47.4	360	127.5	126.3
11	-31.4	-34.0	61	20.7	18.1	115	51.9	49.9	365	128.6	127.5
12	-29.8	-32.7	62	21.4	18.9	129	54.1	52.1	370	129.6	128.5
13	-27.8	-31.4	63	22.0	19.5	125	56.6	54.6	375	130.7	129.6
14	-26.6	-29.6	64	22.8	20.2	130	58.8	56.8	380	131.7	130.7
15	-25.3	-27.9	65	23.4	20.9	135	61.0	59.0	385	132.8	131.7
16	-23.6	-26.6	66	24.1	21.6	140	63.1	61.2	390	133.8	132.8
17	-21.9	-25.4	67	24.8	22.2	145	65.2	63.3	395	134.8	133.8
18	-20.7	-23.7	68	25.5	22.9	150	67.2	65.3	400	135.8	134.9
19	-19.6	-22.6	69	26.1	23.6	155	69.2	67.3			
20	-17.9	-21.4	70	26.8	24.3	160	71.1	69.3			
21	-16.8	-19.7	71	27.4	24.9	165	73.0	71.2			
22	-15.6	-18.6	72	28.1	25.6	170	74.9	73.1			
23	-14.5	-17.5	73	28.8	26.2	175	76.7	74.9			
24	-13.4	-15.9	74	29.4	26.9	180	78.5	76.7			
25	-11.8	-14.8	75	30.0	27.5	185	80.2	78.5			
26	-10.8	-13.7	76	30.7	28.2	190	81.9	80.1			
27	-9.7	-12.6	77	31.4	28.8	195	83.6	81.9			
28	-8.6	-11.5	78	32.0	29.5	200	85.3	83.6			
29	-7.6	-10.0	79	32.3	30.1	205	86.9	85.2			
30	-6.5	-8.9	80	33.2	30.8	210	88.5	86.9			
31	-5.5	-7.9	81	33.8	31.4	215	90.0	88.5			
32	-4.0	-6.9	82	34.2	32.1	220	91.6	90.0			
33	-3.0	-5.9	83	35.9	32.7	225	93.1	91.6			
34	-2.0	-4.6	84	35.6	33.3	230	94.6	93.1			
35	-1.0	-3.8	85	36.1	33.9	235	96.1	94.6			
36	0.0	-2.8	86	36.7	34.2	240	97.6	96.1			
37	0.8	-1.9	87	37.3	35.1	245	99.0	97.6			
38	1.5	-0.9	88	37.9	35.6	250	100.4	99.0			
39	2.2	0.1	89	38.2	36.1	255	101.8	100.4			
40	3.2	1.1	90	39.0	36.8	260	103.2	101.8			
41	4.2	2.0	91	39.6	37.3	265	104.6	103.1			
42	5.2	2.8	92	40.1	37.9	270	105.9	104.5			
43	6.2	3.5	93	40.7	38.5	275	107.2	105.8			
44	7.1	4.4	94	41.2	39.0	280	108.5	107.2			
45	8.1	5.3	95	41.8	39.6	285	109.8	108.5			
46	8.9	6.2	96	42.2	40.1	290	111.1	109.8			
47	9.6	7.2	97	42.8	40.7	295	112.4	111.0			
48	10.4	8.1	98	43.4	41.2	300	113.6	112.3			
49	10.4	8.1	98	43.4	41.2	300	113.6	112.3			
49	11.3	8.8	99	43.9	41.7	305	114.8	113.5			
50	12.2	9.6	100	44.2	42.1	310	116.0	114.8			
51	12.9	10.3	101	45.0	42.8	315	117.2	116.0			
52	13.6	11.2	102	45.7	43.3	320	118.4	117.2			
53	14.4	12.1	103	46.0	43.8	325	119.6	118.4			
54	15.3	12.8	104	46.3	44.2	330	120.8	119.6			

Copeland Corporation • 1675 W. Campbell Road, Sidney, Ohio 45365-0669 • Phone (513) 498-3011

R-408A Saturated Vapor/Liquid
Temperature/Pressure Chart

Temp. °F	Vapor Press. PSIG	Liquid Press. PSIG	Temp. °F	Vapor Press. PSIG	Liquid Press PSIG	Temp. °F	Vapor Press. PSIG	Liquid Press. PSIG	Temp. °F	Vapor Press. PSIG	Liquid Press. PSIG
-40	2.7	2.3	10	39.1	38.2	60	116.2	114.7	110	255.7	253.6
-39	3.1	2.7	11	40.2	39.3	61	118.4	116.9	111	259.3	257.3
-38	3.6	3.2	12	41.3	40.4	62	120.5	119.0	112	262.9	260.9
-37	4.1	3.6	13	42.4	41.5	63	122.7	121.2	113	266.6	264.6
-36	4.5	4.1	14	43.5	42.5	64	124.9	123.4	114	270.2	268.3
-35	5.0	4.5	15	44.6	43.6	65	127.1	125.5	115	273.8	272.0
-34	5.5	5.1	16	45.8	44.8	66	129.3	127.8	116	277.7	275.9
-33	6.1	5.6	17	47.0	46.0	67	131.6	130.0	117	281.7	279.7
-32	6.6	6.1	18	48.2	47.2	68	133.9	132.3	118	285.6	283.6
-31	7.2	6.6	19	49.4	48.4	69	136.1	134.5	119	289.5	287.5
-30	7.7	7.2	20	50.6	49.6	70	138.4	136.7	120	293.4	291.4
-29	8.2	7.7	21	51.9	50.8	71	140.8	139.1	121	297.5	295.5
-28	8.8	8.2	22	53.1	52.1	72	143.3	141.6	122	301.6	299.5
-27	9.3	8.8	23	54.4	53.3	73	145.7	144.0	123	305.6	303.6
-26	9.9	9.3	24	57.0	55.8	74	148.1	146.4	124	309.7	307.7
-25	10.4	9.9	25	58.4	57.2	75	150.6	148.8	125	313.8	311.7
-24	11.1	10.4	26	59.8	58.6	76	153.1	151.4	126	317.8	315.8
-23	11.7	11.0	27	61.1	60.0	77	155.6	153.9	127	321.9	319.9
-22	12.3	11.5	28	62.5	61.4	78	158.2	156.5	128	326.0	323.9
-21	13.0	12.1	29	63.9	62.7	79	160.7	159.1	129	330.0	328.0
-20	13.6	12.7	30	63.9	62.7	80	163.3	161.6	130	334.1	332.1
-19	14.3	13.3	31	65.3	64.2	81	165.9	164.3			
-18	15.0	14.0	32	66.8	65.6	82	168.6	167.0			
-17	15.6	14.6	33	68.3	67.1	83	171.3	169.7			
-16	16.3	15.3	34	69.8	68.6	84	174.0	172.4			
-15	17.0	15.9	35	71.2	70.0	85	176.7	175.0			
-14	17.7	16.6	36	72.8	71.6	86	179.6	177.9			
-13	18.5	17.3	37	74.4	73.2	87	182.4	180.7			
-12	19.2	18.0	38	76.0	74.7	88	185.3	183.5			
-11	20.0	18.7	39	77.6	76.3	89	188.2	186.3			
-10	20.7	19.4	40	79.2	77.9	90	191.0	189.1			
-9	21.5	20.2	41	80.9	79.5	91	194.0	192.1			
-8	22.3	20.9	42	82.6	81.2	92	197.0	195.1			
-7	23.1	21.7	43	84.2	82.9	93	200.0	198.1			
-6	23.9	22.5	44	85.9	84.5	94	202.9	201.1			
-5	24.7	23.2	45	87.6	86.2	95	205.9	204.1			
-4	25.6	24.3	46	89.4	88.0	96	209.1	207.2			
-3	26.5	25.3	47	91.2	89.8	97	212.3	210.4			
-2	27.4	26.3	48	93.0	91.6	98	215.4	213.5			
-1	28.3	27.3	49	94.8	93.3	99	218.6	216.7			
0	29.1	28.3	50	96.6	95.1	100	221.8	219.8			
1	30.1	29.3	51	98.5	97.0	101	225.1	223.1			
2	31.1	30.2	52	100.4	99.0	102	228.4	226.4			
3	32.0	31.2	53	102.3	100.9	103	231.7	229.7			
4	33.0	32.1	54	104.2	102.8	104	235.0	233.0			
5	33.9	33.0	55	106.1	104.7	105	238.3	236.3			
6	35.0	34.1	56	108.1	106.7	106	241.8	239.8			
7	36.0	35.1	57	110.1	108.7	107	245.2	243.2			
8	37.0	36.1	58	112.1	110.7	108	248.7	246.7			
9	38.0	37.2	59	114.2	112.7	109	252.2	250.1			

Copeland Corporation • 1675 W. Campbell Road, Sidney, Ohio 45365-0669 • Phone (513) 498-3011

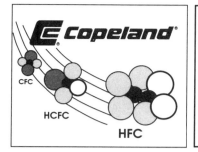

Refrigerant Changeover Guidelines
CFC R-502 to HFC R-404A/R-507

Leading the Industry with Environmentally Responsible Refrigerant Solutions

Copeland does not advocate the wholesale changeover of CFC refrigerants to HCFCs or HFCs. If a system is not leaking refrigerant to the atmosphere, and is operating properly, there is no technical reason to replace the CFC refrigerant. In fact, changing the refrigerant may void the U.L. listing of the system. However, once the decision has been made to make the change from CFC R-502 to HFC R-404A or R-507, the following guidelines are recommended.

CONSIDERATIONS

1. Retrofitting systems that employ compressors manufactured prior to 1973 is not recommended. This is due to the different materials used in motor insulation systems that have not been evaluated for compatibility with the new refrigerants and lubricants. Failure to heed this advice will violate the proposed U.L. Standard For Field Conversion/Retrofit Of Alternate Refrigerants In Refrigeration and Air Conditioning Equipment (U.L.2154).

2. Copeland's lubricant recommendation for use with R-404A/R-507 is a Polyol Ester (POE), either Mobil EAL™ Arctic 22 CC or ICI EMKARATE™ RL 32CF. These are the only POE lubricants approved for use in Copeland compressors and are available from all authorized Copeland wholesalers. The use of any other POE lubricant may void the compressor warranty.

It is important that the system contain not more than 5% residual mineral oil. More than 5% may contribute to premature compressor failure and or system capacity short-fall. Because mineral oils are not miscible with R-404A/R-507, they may log in the evaporator resulting in system capacity loss. It is for this reason that the flushing process must be done with the R-502 in the system.

3. R-404A/R-507 can be used in either low or medium temperature systems. **R-404A/R-507 should not be mixed with any other refrigerant!**

4. The expansion valve will not need to be changed. The existing R-502 valve when used with R-404A/R-507 will have virtually the same capacity; however, it may be necessary to adjust the superheat.

5. Filter-driers must be changed at the time of conversion. This is proper air conditioning/refrigeration practice.

   a. Solid core driers such as ALCO ADK are compatible with either R-502 or R-404A/R-507.

   b. Compacted bead type driers can use XH6 or XH9 molecular sieve material such as found in the ALCO EK or EKH series.

   c. If a loose fill type drier is to be used, XH9 molecular sieve material is required.

6. Pressure regulators such as EPR valves may have to be reset. Contact the EPR manufacturer for the correct settings.

7. R-404A/R-507 exhibits higher pressures than R-502 at normal condensing temperatures. This may require that the high pressure safety controls be reset in order to operate as intended.

8. The higher pressure characteristics exhibited by R-404A/R-507 will in some cases exceed the industry accepted safety factors on the compressor crankcase (low side). This will require the addition of a pressure relief valve on the compressor crankcase, set at a maximum of 375 psig to adequately protect the compressor from the possibility of excessive pressure. Pressure relief valves can be purchased from your

WARNING: Use only Copeland approved refrigerants and lubricants in the manner prescribed by Copeland. In some circumstances, other refrigerants and lubricants may be dangerous and could cause fires, explosions or electrical shorting. Contact Copeland Corp., Sidney, Ohio for more information.

Authorized Copeland wholesaler as part number 998-0051-02. Copeland semi-hermetic compressors that require this additional valve are:

* Discus® 3D, 4D, 9D and MD

* All Other Semi-Hermetic (Non-Discus Models)

WARNING: IT IS POSSIBLE THAT EXCESS PRESSURE BUILD-UP ON MODELS INDICATED COULD RESULT IN THE COMPRESSOR EXPLODING UNLESS THE PRESSURE RELIEF VALVE SPECIFIED HAS BEEN PROPERLY INSTALLED ON THE ORIGINALLY BUILT COPELAND COMPRESSOR.

9. Systems that use a low pressure controller to maintain space temperature may need to have the cut in and cut out points changed. Although R-404A/R-507 does exhibit "glide", the average evaporator or condenser temperature is within 0.5°F of the saturated vapor temperature; therefore no correction is required.

10. Systems using R-404A/R-507 should have approximately the same system pressure drop as with R-502. Check with the manufacturer of any pressure regulators and pilot operated solenoid valves used in the system to be sure that they will operate properly.

11. Mineral oil lubricants, such as 3GS, must not be used as the compressor lubricant with R-404A/R-507. Polyol Ester (POE) lubricant, Mobil EAL Arctic 22 CC or ICI Emkarate™ RL 32CF, are the only lubricants that can be used in a Copeland compressor when using R-404A/R-507.

Before starting the changeover, it is suggested that at least the following items be ready:

1. Safety glasses
2. Gloves
3. Refrigerant service gauges
4. Electronic thermometer
5. Vacuum pump capable of pulling 250 microns
6. Thermocouple micron gauge
7. Leak detector
8. Refrigerant recovery unit including refrigerant cylinder
9. Proper container for removed lubricant
10. New liquid control device
11. Replacement liquid line filter-drier(s)
12. New POE lubricant, Mobil EAL Arctic 22 CC/ICI EMKARATE RL 32CF
13. R-404A/R-507 pressure temperature chart
14. R-404A/R-507 refrigerant

CHANGEOVER PROCEDURE

NOTE: R-404A/R-507 is not compatible with the seal material used in the R-502/R-22 Moduload Unloading System. If your system has Moduload, it MUST be changed. Consult your Copeland wholesaler for the proper part number.

1. The system should be thoroughly leak tested with the R-502 refrigerant still in the system. All leaks should be repaired before the R-404A/R-507 refrigerant is added.

2. It is advisable that the system operating conditions be recorded with the R-502 still in the system. This will provide the base data for comparison when the system is put back into operation with the R-404A/R-507.

3. It is necessary to thoroughly remove the existing mineral oil lubricant from the system before the refrigerant is changed. No more than 5% residual mineral oil may be left in the system when it is recharged with R-404A/R-507 for proper compressor operation. No more than 1 to 2% residual mineral oil may be required to assure no loss of heat transfer if enhanced tube heat exchangers are used in the system.

I. Systems with service valves

  a. Disconnect electrical power to system.

  b. Front seat the service valves to isolate the compressor.

  c. Properly remove the R-502 from the compressor.

  d. Remove the mineral oil lubricant from the compressor. Hermetic compressors will have to be removed from the system and tipped up to drain the lubricant out through the suction/process fitting.

  e. Those systems that have oil separators, oil reservoirs, oil floats and suction line accumulators must have the mineral oil drained from them. Add POE lubricant to the oil separator and to the oil reservoir.

  f. Replace the liquid line filter-drier with one that is compatible with R-404A/R-507.

  g. Fill the compressor with the proper amount of POE lubricant. The oil charge is on the label of Copelaweld® compressors. Copelametic® compressor oil charges can be found in Application Engineering Bulletin 4-1281. If the lubricant charge is unknown, an authorized Copeland wholesaler can provide the technician with the information.

  h. Reinstall the compressor in the system. Evacuate it to 250 microns. A vacuum decay test is suggested to assure the system is dry and leak free.

  i. Recharge the system with R-502.

j. Operate the compressor in the system for a minimum of 24 hours. On large systems with long piping runs, experience indicates operating for several days to allow for thorough mixing of the remaining mineral oil and POE will minimize the number of flushes.

k. Repeat steps 3.I.a through j until the residual mineral oil is less than 5%. This may require three flushes of the system's lubricant.

l. In most cases, three complete flushes of the lubricant lowers the mineral oil content down to 5% or less in the system. To be sure of the mineral oil content between flushes and to be sure that the system ultimately has 5% or less mineral oil, test kits are available from Virginia KMP or Nu Calgon. A Refractometer may also be used to determine the residual mineral oil in the system. The Refractometer (p/n 998-RMET-00) is available from your Copeland Wholesaler.

m. Properly dispose of the lubricant removed from the system after each flush.

II. Systems without service valves

a. Disconnect electrical power to system.

b. Properly remove the R-502 from the system.

c. Remove the mineral oil lubricant from the compressor. Hermetic compressors will have to be removed from the system and tipped up to drain the lubricant out through the suction/process fitting.

d. It may be advisable to add service valves at the compressor suction and discharge connections. The compressor will have to have its lubricant changed generally three times.

e. Those systems that have oil separators, oil reservoirs, oil floats and suction line accumulators must have the mineral oil drained from them. Add POE lubricant to the oil separator and to the oil reservoir.

f. Replace the liquid line filter-drier with one that is compatible with R-404A/R-507.

g. Fill the compressor with the proper amount of POE lubricant. The oil charge is on the label of Copelaweld compressors. Copelametic compressor oil charges can be found in Application Engineering Bulletin 4-1281. If the lubricant charge is unknown, an authorized Copeland wholesaler can provide the technician with the information.

h. Reinstall the compressor in the system. Evacuate it to 250 microns. A vacuum decay test is suggested to assure the system is dry and leak free.

i. Recharge the system with R-502.

j. Operate the compressor in the system for a minimum of 24 hours. On large systems with long piping runs, experience indicates operating for several days to allow for thorough mixing of the remaining mineral oil and POE will minimize the number of flushes.

k. Repeat steps 3.II.a through j until the residual mineral oil is less than 5%. This may require three flushes of the system's lubricant.

l. To date, three complete flushes of the lubricant has shown to lower the mineral oil content down to 5% or less in the system. To be sure of the mineral oil content between flushes and to be sure that the system ultimately has 5% or less mineral oil, test kits are available from Virginia KMP or Nu Calgon. A Refractometer may also be used to determine the residual mineral oil in the system. The Refractometer (p/n 998-RMET-00) is available from the Copeland Wholesaler.

m. Properly dispose of the lubricant after each flush.

4. With the proper amount of polyol ester in the system, the R-502 can now be removed. Measure and note the amount removed.

5. Before the final flush, be sure all leaks are repaired, liquid control devices and any other system components are changed. Install the correct liquid line filter-drier. Driers must be compatible with the refrigerant and lubricant.

6. Be advised that POEs are very hygroscopic. They will very quickly absorb moisture from the air once the container is opened. Once the lubricant is added to the compressor, the compressor should be quickly installed. Like an open container, an open compressor with POE will absorb moisture. Add the correct amount of lubricant to the compressor. It is important that the system contain not more than 5% mineral oil. More than 5% may contribute to premature compressor failure and or system capacity problems. Mineral oils are not miscible with R-404A/R-507. The lubricant may log in the evaporator resulting in system capacity loss. It is for this reason that the flushing process must be done with the R-502 in the system.

7. Once the compressor is installed and the system is closed, the system must be evacuated to and hold 250 microns or lower.

8. Charge the system with the R-404A/R-507. Charge to 90% of the refrigerant removed in item 4. R-404A/R-507 must leave the charging cylinder in the liquid phase. It is suggestd that a sight glass be connected between the charging hose and compressor suction service valve. This will permit adjustment of the cylinder valve to assure the refrigerant enters the compressor in the vapor state.

9. Operate the system. Record the data and compare to the data taken in item 2. Check and adjust the TEV superheat setting if necessary. Make adjustments to other controls as needed. Additional R-404A/R-507 may have to be added to obtain optimum system performance.

10. Properly label the components. Tag the compressor with the refrigerant used (R-404A/R-507) and the lubricant used (Mobil EAL Arctic 22 CC or ICI EMKARATE RL 32CF). The proper color code for R-404A is Pantone Orange PMS (Paint Matching System) 021C; for R-507, Teal, PMS 326.

11. Clean up and properly dispose of the removed lubricant. Check local and state laws regarding the disposal of refrigerant lubricants. Recycle or reclaim the removed refrigerant.

CAUTION: These guidelines are intended for use with R-404A and/or R-507 only. Other refrigerants may not be compatible with the materials used in our compressors or the lubricants recommended in this bulletin resulting in unacceptable reliability and durability of the compressor.

Copeland Corporation • 1675 W. Campbell Road, Sidney, Ohio 45365-0669 • Phone (513) 498-3011

R-404A/507 Pressure/Temperature Chart

Vapor Pressure PSIG	Vapor Temp.°F R-404A	Vapor Temp °F R-507	Vapor Pressure PSIG	Vapor Temp.°F R-404A	Vapor Temp °F R-507	Vapor Pressure PSIG	Vapor Temp.°F R-404A	Vapor Temp °F R-507
0	-50	-52.1	41	8	5.5	82	38.2	35.4
1	-48	-49.7	42	9	6.4	83	38.8	36
2	-46	-47.3	43	10	7.3	84	39.4	36.6
3	-43.5	-46	44	11	8.2	85	40	37.2
4	-41	-43	45	11.5	9	86	40.6	37.8
5	-39	-41	46	12	9.9	87	41.2	38.4
6	-37	-39	47	13	10.7	88	41.8	39
7	-35	-37	48	14	11.6	89	42.4	39.5
8	-33	-35.3	49	15	12.3	90	43	40.1
9	-31.5	-33.5	50	16	13.2	91	43.4	40.7
10	-30	-31.8	51	16.5	14	92	43.8	41.2
11	-28	-30.2	52	17	14.8	93	44.2	41.8
12	-26	-28.6	53	18	15.6	94	44.6	42.4
13	-24.5	-27	54	19	16.4	95	45	42.9
14	-23	-25.5	55	19.5	17.2	96	45.6	43.4
15	-21.5	-24	56	20	17.9	97	46.2	44
16	-20	-22.6	57	21	18.7	98	46.8	44.5
17	-19	-21.2	58	22	19.4	99	47.4	45.1
18	-18	-19.8	59	22.5	20.2	100	48	45.6
19	-16.5	-18.4	60	23	20.9	101	48.6	46.1
20	-15	-17.1	61	24	21.6	102	49.2	46.7
21	-13.5	-15.8	62	25	22.3	103	49.8	47.2
22	-12	-14.6	63	25.5	23.1	104	50.4	47.7
23	-11	-13.4	64	26	23.8	105	51	48.2
24	-10	-12.2	65	27	24.5	106	51.4	48.7
25	-8.5	-11	66	28	25.2	107	51.8	49.2
26	-7	-9.8	67	28.5	25.8	108	52.2	49.7
27	-6	-8.7	68	29	26.5	109	52.6	50.2
28	-5	-7.5	69	29.5	27.2	110	53	50.7
29	-4	-6.4	70	30	27.9	115	56	53.2
30	-3	-5.4	71	31	28.5	120	58	55.6
31	-2	-4.3	72	32	29.2	125	60	57.9
32	-1	-3.3	73	32.5	29.8	130	63	60.1
33	0	-2.2	74	33	30.5	135	65	62.3
34	1	-1.2	75	33.5	31.1	140	67	64.4
35	2	-0.2	76	34	31.7	145	69	66.5
36	3	0.8	77	34.5	32.4	150	71	68.6
37	4	1.7	78	35	33	155	73	70.6
38	5	2.7	79	36	33.6	160	75	72.5
39	6	3.6	80	37	34.2	165	77	74.4
40	7	4.6	81	37.6	34.8	*Chart continued next page*		

R-404A/507 Pressure/Temperature Chart

Vapor Pressure PSIG	Vapor Temp.°F R-404A	Vapor Temp °F R-507	Vapor Pressure PSIG	Vapor Temp.°F R-404A	Vapor Temp °F R-507
Continued from previous page			285	114	111.1
170	79	76.3	290	115	112.4
175	81	78.1	295	116.5	113.6
180	82	79.9	300	118	114.8
185	84	81.7	305	119	116
190	86	83.4	310	120	117.2
195	88	85.1	315	121	118.4
200	89	86.7	320	122	119.6
205	90.5	88.3	325	123.5	120.7
210	92	89.9	330	125	121.8
215	94	91.5	335	126	123
220	96	93.1	340	127	124.1
225	97.5	94.6	345	128	125.2
230	99	96.1	350	129	126.3
235	100.5	97.5	355	130.5	127.3
240	102	99	360	132	128.4
245	103	100.4	365	133	129.4
250	104	101.8	370	134	130.5
255	105.5	103.2	375	135	131.5
260	107	104.6	380	136	132.5
265	108.5	105.9	385	137	133.5
270	110	107.2	390	138	134.5
275	111.5	108.6	395	139	135.5
280	113	109.8	400	140	136.4

Note: Saturated Vapor Temperatures are shown. The average evaporator or condenser temperature is within .5°F of the saturated vapor temperature; therefore, no correction is required.

Copeland Corporation • 1675 W. Campbell Road, Sidney, Ohio 45365-0669 • Phone (513) 498-3011

Figures A-1, A-2, and A-3 illustrate some of the procedures discussed in the previous pages.

Figure A-1. Draining oil from a hermetic compressor (Photo by Bill Bitzinger, Office of University Communication Services, Ferris State University)

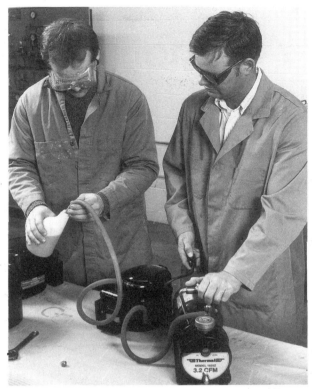

Figure A-3. Filling a hermetic compressor with POE lubricant (Photo by Bill Bitzinger, Office of University Communication Services, Ferris State University)

Figure A-2. Silver soldering on a new filter drier (Photo by Bill Bitzinger, Office of University Communication Services, Ferris State University)

Alternative Refrigerant Update

R-407C

R-407C is a near-azeotropic, ternary, refrigerant blend consisting of HFC-32, HFC-125, and HFC-134a (23%/25%/52%) respectively. Its chemical name is Difluoromethane, Pentafluoroethane, Tetrafluoroethane. It cylinder color is Brown. R-407C is an HFC-based refrigerant blend with a high temperature glide (10 degrees) that does not cause any operational related problems in typical systems. R-407C does have the ability to fractionate. This may cause service related leakage issues regarding fractionation potential of the blend. It is a long-term replacement for HCFC-22 in commercial and residential air conditioning applications. R-407C is being used by original equipment manufacturers in new equipment and can also be used as a retrofit refrigerant blend. It can be used in the same equipment R-22 is used in with only minor equipment modifications. It has a very close capacity to R-22, but has lower efficiencies. R-407C systems use synthetic polyol ester (POE) lubricants in their crankcases.

R-407C		
Temperature, F	Bubble Pressure, psig	Dew Pressure, psig
−60	9.1 ("Hg Vac)	15.9 ("Hg Vac)
−55	5.9 ("Hg Vac)	13.5 ("Hg Vac)
−50	2.4 ("Hg Vac)	10.9 ("Hg Vac)
−45	0.7	7.9 ("Hg Vac)
−40	2.9	4.5 ("Hg Vac)
−35	5.2	0.7 ("Hg Vac)
−30	7.9	1.7
−25	10.7	4.0
−20	13.9	6.5
−15	17.3	9.3
−10	21.1	12.4
−5	25.2	15.8
0	29.6	19.5
5	34.4	23.6
10	39.6	28.0
15	45.2	32.7
20	51.3	37.9
25	57.8	43.6
30	64.7	49.6
35	72.2	56.2
40	80.2	63.2
45	88.7	70.7
50	97.8	78.8
55	107.5	87.5
60	117.7	96.8
65	128.7	106.7
70	140.2	117.2
75	152.5	128.4
80	165.5	140.4
85	179.2	153.1
90	193.6	166.5
95	208.8	180.8
100	224.9	195.8
105	241.8	211.8
110	259.6	228.7
115	278.2	246.5
120	297.8	265.3
125	318.3	285.2
130	339.9	306.1
135	362.4	328.2
140	386.0	351.4
145	410.7	375.9
150	436.5	401.7

HCFC-22 (R-22) Replacement Refrigerants

For years, chemical companies have been researching refrigerants and refrigerant blends in order to find a permanent substitute for HCFC-22. HCFC-22 is scheduled for a total phaseout in the year 2030 under the Montreal Protocol, with a gradual phaseout leading up to that date.

R-410A

R-410A is considered an azeotropic, binary refrigerant blend consisting of HFC-32 and HFC-125 (50%/50%) respectively. Its chemical name is Difluoromethane, Pentafluoroethane. Its cylinder color is Rose. This binary blend has a very small temperature glide (0.2 degrees) in normal commercial and residential air conditioning applications and temperature ranges. It also has negligible fractionate abilities in these same ranges. R-410A is an HFC-based refrigerant blend, which is being used by original equipment manufacturers on new equipment in place of HCFC-22. It is a long-term replacement for HCFC-22 in commercial and residential air conditioning applications.

R-410A systems operate at about 60% higher pressures than standard HCFC-22 air conditioning systems. In fact, R-22 service equipment (hoses, manifold gage sets, and recovery equipment) cannot be used on R-410A systems because of these higher operating pressures. Service equipment used for R-410A has to be rated to handle higher operating pressures. Safety glasses and gloves should always be worn when working with R-410A. Systems designed for R-410A will have smaller components (heat exchangers, compressors) when compared to R-22 systems to perform the same cooling job. R-410A systems use synthetic polyol ester (POE) lubricants in their crankcases and have higher efficiencies than standard R-22 systems. Listed below is a pressure/temperature comparison of R-22 and R-410A.

R-410A

Temperature, F	Pressure, psig	Temperature, F	Pressure, psig
−60	0.4	50	142.7
−55	2.6	55	156.0
−50	5.1	60	170.1
−45	7.8	65	185.1
−40	10.9	70	201.0
−35	14.2	75	217.8
−30	17.9	80	235.6
−25	22.0	85	254.5
−20	26.4	90	274.3
−15	31.3	95	295.3
−10	36.5	100	317.4
−5	42.4	105	340.6
0	48.4	110	365.1
5	55.1	115	390.9
10	62.4	120	418.0
15	70.2	125	446.5
20	78.5	130	476.5
25	87.5	135	508.0
30	97.2	140	541.2
35	107.5	145	576.0
40	118.5	150	612.8
45	130.2		